Computers in Health Care

Kathryn J. Hannah   Marion J. Ball
Series Editors

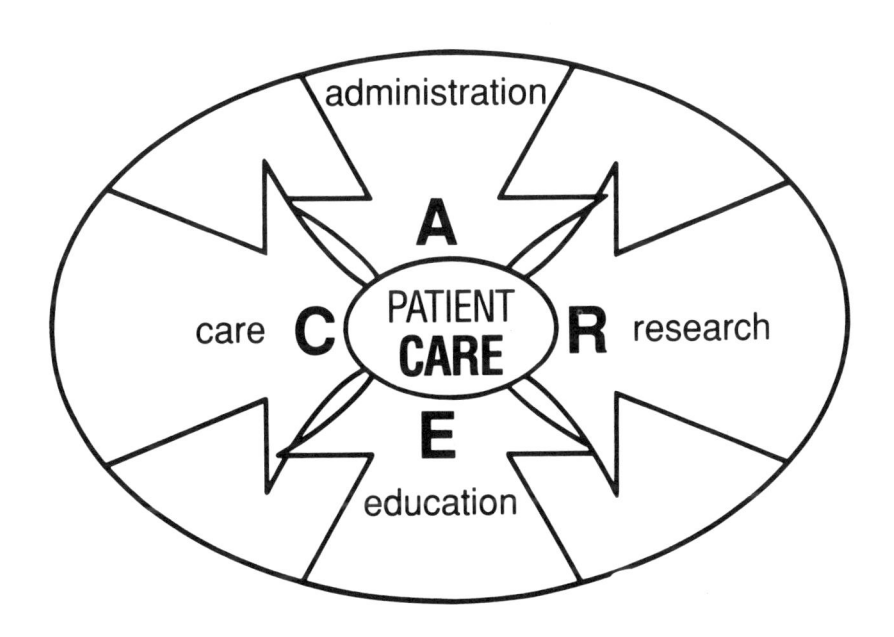

Springer
New York
Berlin
Heidelberg
Barcelona
Budapest
Hong Kong
London
Milan
Paris
Santa Clara
Singapore
Tokyo

Computers in Health Care

Series Editors:
Kathryn J. Hannah   Marion J. Ball

Dental Informatics
Integrating Technology into the Dental Environment
L.M. Abbey and J. Zimmerman

Aspects of the Computer-based Patient Record
M.J. Ball and M.F. Collen

Nursing Informatics
Where Caring and Technology Meet, Second Edition
M.J. Ball, K.J. Hannah, S.K. Newbold, and J.V. Douglas

Healthcare Information Management Systems
A Practical Guide, Second Edition
M.J. Ball, D.W. Simborg, J.W. Albright, and J.V. Douglas

Strategy and Architecture of Health Care Information Systems
M.K. Bourke

Information Networks for Community Health
P.F. Brennan, S.J. Schneider, and E. Tornquist

Introduction to Medical Informatics
P. Degoulet and M. Fieschi

Patient Care Information Systems
Successful Design and Implementation
E.L. Drazen, J.B. Metzger, J.L. Ritter, and M.K. Schneider

Introduction to Nursing Informatics
K.J. Hannah, M.J. Ball, and M.J.A. Edwards

Computerizing Large Integrated Health Networks
The VA Success
R.M. Kolodner

Organizational Aspects of Health Informatics
Managing Technological Change
N.M. Lorenzi and R.T. Riley

Transforming Health Care Through Information
Case Studies
N.M. Lorenzi, R.T. Riley, M.J. Ball, and J.V. Douglas

Knowledge Coupling
New Premises and New Tools for Medical Care and Education
L.L. Weed

Robert M. Kolodner
Editor

Computerizing Large Integrated Health Networks

The VA Success

With the Assistance of Judith V. Douglas

With 82 Illustrations

Springer

Robert M. Kolodner, MD
Associate Chief Information Officer
Veterans Health Administration
Department of Veterans Affairs
810 Vermont Avenue, NW
Washington, DC 20420, USA

Series Editors:

Kathryn J. Hannah, PhD, RN
Leader, Health Informatics Group
Sierra Systems Consultants, Inc.
and
Professor, Department of Community
 Health Science
Faculty of Medicine
The University of Calgary
Calgary, Alberta, Canada

Marion J. Ball, EdD
Professor, Department of Epidemiology
University of Maryland School of
 Medicine
and
Vice President
First Consulting Group
Baltimore, MD, USA

Library of Congress Cataloging-in-Publication Data
Kolodner, Robert M.
　Computerizing Large Integrated Health Networks: the VA success/Robert M. Kolodner.
　　p. cm.—(Computers and health care)
　Includes bibliographical references and index.
　ISBN 0-387-94837-6 (hardcover: alk. paper)
　1. United States. Veterans Health Administration—Data
processing.　2. Health services administration—United States—Data
processing.　I. Title.　II. Series.
UB369.K65　1997
362.1′1′0285—dc20　　　　　　　　　　　　　　　　　　　　　　　　　　96-25988

Printed on acid-free paper.

© 1997 Springer-Verlag New York, Inc.
In recognition of the authors' work undertaken as part of their official duties as U.S. Government employees, reproduction of this work in whole or in part for any purpose of the U.S. Government is permitted.
All rights reserved. This work may not be translated or copied in whole or in part without the written permission of the publisher (Springer-Verlag New York, Inc., 175 Fifth Avenue, New York, NY 10010, USA), except for brief excerpts in connection with reviews or scholarly analysis. Use in connection with any form of information storage and retrieval, electronic adaptation, computer software, or by similar or dissimilar methodology now known or hereafter developed is forbidden.
The use of general descriptive names, trade names, trademarks, etc., in this publication, even if the former are not especially identified, is not to be taken as a sign that such names, as understood by the Trade Marks and Merchandise Marks Act, may accordingly be used freely by anyone.
While the advice and information in this book are believed to be true and accurate at the date of going to press, neither the authors nor the editors nor the publisher can accept any legal responsibility for any errors or omissions that may be made. The publisher makes no warranty, express or implied, with respect to the material contained herein.

Production coordinated by Chernow Editorial Services, Inc., and managed by Francine McNeill; manufacturing supervised by Joe Quatela.
Typeset by Best-set Typesetter Ltd., Hong Kong.
Printed and bound by Maple-Vail Book Manufacturing Group, York, PA.
Printed in the United States of America.

9 8 7 6 5 4 3 2 1

ISBN 0-387-94837-6 Springer-Verlag New York Berlin Heidelberg　　SPIN 10543555

To Ted O'Neill and Marty Johnson, whose vision, efforts, and careers laid the foundation for the accomplishments described in this book, and to the thousands of programmers and users since 1977, both inside and outside VA, whose dedication and contributions brought VA to the forefront of medical computing.

Foreword

This book has been a long time in the making. The computerization activities described in these pages began in 1977 at the Department of Veterans Affairs (VA), but we devoted most of our focus and efforts to building and then implementing the extensive hospital information system known as the Decentralized Hospital Computer System (DHCP) throughout VA. Delivering the product has been our primary goal. We spent relatively little time documenting or describing our experiences or lessons learned. Except for some presentations at national meetings and a relatively few publications, almost none of which were in the standard trade journals read by Chief Information Officers (CIOs) and equivalent top managers in the private and nonprofit sectors, VA's accomplishments remained a well-kept secret. In 1988, Helly Orthner encouraged VA staff to consider writing a book, but the press of day-to-day activities always seemed to take precedence, and the book languished on the back burner.

Due to changes since that time in both VA and the rest of the healthcare system, VA's experience in creating, maintaining, and, most importantly, evolving an extensive, clinically oriented healthcare information system has become even more relevant to other large integrated healthcare organizations. Today, the information needs of VA and other healthcare institutions are much closer to each other than they were previously. An important difference between them is the direction from which their information systems grew. VA's advantage is that its information system is built on one that is rich with clinical information and that has been expanding and improving for the past 15 years, rather than being based on systems that started out being primarily financial in nature. It is our firm belief that a clinical system can be extended to generate management and financial information far more easily and effectively than a financial system can be extended into becoming a clinical one.

In 1988, a billing system was not an essential function in the initial implementation of DHCP, because the vast majority of VA's budget is based on its allocation from Congress. This contrasted with most other hospital information systems, with a few notable exceptions, which concentrated on billing and financial functions. Instead, VA addressed challenges in the implementation of clinical and management systems in a large healthcare network. We concentrated on the needs of the individual healthcare facility for patient identification and tracking (registration, admissions/discharge/transfer) and for efficiently managing high-cost operations, such as laboratory and pharmacy, which are often the profit centers in other institutions. For these latter services, the emphasis was on managing and recording the workload and on decreasing the costs, including the distribution of results to the front-line clinicians, rather than generating a bill. As new modules were added, a similar focus was maintained. Moreover, front-line VA users, including clinicians, were involved in the design of the software in order to improve the volume and accuracy of the information they needed to deliver care to veterans. The result is a combined clinical and management information system that is firmly based on gathering information during the course of clinical care and deriving management data as a byproduct of these interactions.

As VA established the authority and mechanisms for collecting reimbursement from the subset of veterans who were not entitled to free care, billing became a more important consideration. As VA reorganizes into its multi-facility networks (see Chapter 1 for more details), accurate management data across facilities become not just desirable, but vital to its future. Emphasis in the networks has been shifting from inpatient care to ambulatory care settings and to the formation of both integrated and virtual healthcare networks. Today, VA is moving into a managed care operation, charged with meeting the needs of its enrolled population. Now that VA no longer simply takes and reacts to whoever comes in the door for care, a managed care information system capability is becoming a necessity. Thus, VA's information needs are moving closer to those of other healthcare institutions, even though its budget currently remains primarily dependent on congressional appropriations.

At the same time that VA has been changing, other healthcare organizations have been shifting their emphasis from pure financial information systems to those that are more clinically oriented. As healthcare becomes more competitive and forces increase to consolidate resources and decrease costs, more and more healthcare institutions are recognizing the critical importance of their information systems and are increasing their investments accordingly. The most effective information systems will be those that contribute not only to cutting costs, but also to increasing both access to and quality of care. These multiple goals require that clinical information systems be an integral part of healthcare information systems.

The VA experience, detailed in this book, offers a model based on a long track record of developing and implementing a low-cost, highly integrated healthcare information system that has been designed to evolve and grow gracefully—changing hardware platforms, operating systems, and computer languages, and integrating a variety of commercial technologies—

without requiring the gut-wrenching and expensive conversion from one computer system to another. Moreover, the information system is remarkably robust and locally extensible, meeting the needs of an impressive range of healthcare facilities. These facilities range from outpatient clinics and nursing homes to large, multi-divisional tertiary care medical centers. A variety of foreign countries have even adapted portions of the system to use in their native languages, such as German, Spanish, Arabic, and Chinese.

Even the problems encountered as the system grew and changed over these 15 years provide opportunities for CIOs and their counterparts in the private sector to understand the potential pitfalls and successful strategies for overcoming them. Our attempt has been to include these "warts" in our presentations, so that others might learn from them.

These past 15 years, and the 5 years that preceded them before DHCP was launched, have been an exciting journey for those of us who have participated. We sometimes tried to do "impossible" tasks, such as implementing healthcare information systems in 169 facilities in a 3-year period where no infrastructure or expertise existed before. And we often succeeded. Our information system was originally designed for individual facilities, yet it has been able to grow and evolve to support multi-facility networks. Critics accused VA of using out-of-date technology or being a closed system, often because they viewed DHCP at a point in time, rather than as a large, dynamic system that was undergoing a metamorphosis over time. Key components of the system were chosen because of their utility as robust solutions, but our commitment is to the goal, not to the specific solution. So, as better technologies arrived, such as client–server and graphical user interfaces, we have moved on to incorporate these in VA's healthcare information system. The flexibility derives, in large part, from the creativity and dedication of the technical staff within VA, both at the national and the local level, and at collaborative institutions such as the Indian Health Service, who generate innovative ideas to overcome potential barriers in the evolutionary process. Without them, DHCP would have become just another short-term information system that would have had to be discarded and replaced by yet another system with a typical life cycle of development, maturity, and obsolescence over an 8 to 10 year period. Instead, VA's system remains dynamic and vital today, even if every component has been or will be swapped out over the course of its evolution.

This book was constructed to provide the reader with an idea of the breadth of the automation activities that have been underway at VA to support the day-to-day operations at the local medical facility level. We have not tried to describe either the myriad of national level databases or the commercial decision support system that aggregates cost and workload data so that management can determine the value of the care VA delivers. Since 1982, DHCP has been central to facilitating this daily activity. Although DHCP represented the total automation activity at most VA medical centers in 1985, it became clear that by 1996, DHCP—the national system developed by VA staff—was only one part of the overall information resource at the local facility level. Office automation, local software development, and commercial-off-the-shelf healthcare information systems such as those used to support intensive care units, anesthesia services, and telemedicine activities have been added to the mix of information resources available today in VA medical facilities. Thus, by the time this book is published, VA will be using a new term to encompass this broader information resource environment that is installed throughout VA. This new term—the Veterans Health Information Systems and Technology Architecture (*V*IST*A*)—will be used to describe VA's healthcare information system instead of DHCP. As such, *V*IST*A* incorporates all of the benefits of DHCP as well as including the rich array of other information resources that are vital to the day-to-day operations at VA medical facilities. The switch to *V*IST*A* does not represent the discarding of DHCP. Instead, it represents the culmination of DHCP's evolution and metamorphosis into a new, open system, client–server-based environment, that takes full advantage of commercial solutions including those provided by Internet technologies. The underlying principles, goals, and vision from DHCP form the heart of *V*IST*A* as well.

Robert M. Kolodner

Series Preface

This series is intended for the rapidly increasing number of health care professionals who have rudimentary knowledge and experience in health care computing and are seeking opportunities to expand their horizons. It does not attempt to compete with the primers already on the market for novices. Eminent international experts will edit, author, or contribute to each volume in order to provide comprehensive and current accounts of innovations and future trends in this quickly evolving field. Each book will be practical, easy to use, and well referenced.

Our aim is for the series to encompass all of the health professions by focusing on specific professions, such as nursing, in individual volumes. However, integrated computing systems are only one tool for improving communication among members of the health care team. Therefore, it is our hope that the series will stimulate professionals to explore additional means of fostering interdisciplinary exchange.

This series springs from a professional collaboration that has grown over the years into a highly valued personal friendship. Our joint values put people first. If the Computers in Health Care series lets us share those values by helping health care professionals to communicate their ideas for the benefit of patients, then our efforts will have succeeded.

Kathryn J. Hannah
Marion J. Ball

Preface

The Federal High Performance Computing and Communications (HPCC) initiative was begun as a bipartisan effort in 1989 and passed into legislation in 1991. A confederation of Federal agencies with major scientific and research commitments, the HPCC initiative was remarkable in several respects, one being the absence of the medical sciences in the initial planning. This has been corrected. The National Institutes of Health, National Library of Medicine, the Agency for Health Care Policy and Research, and the Veterans Health Administration are now supporting a number of advanced applications of computer and communications technologies to biomedical research and healthcare delivery. The Department of Veterans Affairs (VA) has been an excellent participant and has proved to be a perfect testbed for applying many of the new technologies in healthcare settings from coast to coast.

The VA's Veterans Health Administration has a long history of efforts to bring computer-based information systems to help in healthcare. Quietly, without much fanfare, VA has steadily added to early computerized methods, and these systems have penetrated deeply into the organization of this vast medical undertaking. Stating this truism so flatly ignores, of course, the many dedicated VA scientists, healthcare professionals, administrators, congressional supporters, and public advocates who have "kept the faith" in these efforts over the decades. All deserve our praise and thanks.

Currently, the Veterans Health Administration is a beehive of activity. To adapt to the rapidly changing healthcare management and financing scene in the United States necessitates a serious focus on information systems. Not only the HPCC initiative, but also the National Information Infrastructure (NII) now present timely opportunities for VA to enhance and expand its services. I personally believe that telemedicine is now the strongest strategy by which biomedical applications can make good use of the NII. My picture of telemedicine includes three elements: providing information to support medical decision-making; signal processing, including physiological data and images; and arrangements to practice medicine at a distance. This book presents examples of such work in the VA system.

In many other settings, the arrangements to practice medicine at a distance have been the most daunting of the three elements. Within the VA system, however, the arrangements are potentially somewhat simpler, given the absence of interstate medical licensure and malpractice problems, and the presence of relatively consistent administrative patterns throughout the nationwide VA system.

Two last obstacles to optimal, widespread use of telemedicine are fully addressed here, but by no means overcome. Even within the VA system, we need to learn how to evaluate the medical and economic worth of telemedicine (and other technology-assisted medical practices). And, regrettably, our ability to satisfy the public as to medical data privacy (or at least confidentiality) also largely remains beyond the range of the excellent medical and technical "fixes" described in this book. Americans have yet to arrive at a practical consensus concerning the balance between privacy of data, accessibility of medical records for proper purposes, and the costs of both, so that these excellent VA systems can be exploited fully for the better healthcare that, when more widely introduced, they will surely give us.

Donald A.B. Lindberg

Preface

In March 1988, the Department of Defense awarded a $1.1 billion contract to Science Applications International Corporation (SAIC) for the Composite Health Care System (CHCS) for its military hospitals (Collen, 1995; General Accounting Office, 1988). This new system, the CHCS, was to be based on the Decentralized Hospital Computer Program (DHCP) software developed by the Department of Veterans Affairs (VA). Clearly, the "VA DHCP Story" had to be told.

I approached Dr. Ingeborg M. Kuhn, then the Director of the VA Information Systems Center in San Francisco, who enthusiastically agreed. In January 1989, after some work on the contents of the book, Dr. Kuhn signed a publishing contract with Springer-Verlag. Dr. Kuhn's tragic and sudden death shelved this publishing effort.

To my delight, Dr. Robert M. Kolodner has now finished the "VA DHCP Story." Many thanks also to Mrs. Judith V. Douglas and Dr. Marion J. Ball for making it possible to read this story finally in print. There are valuable lessons to be learned on how the VA DHCP System started, matured, and evolved. The evolution continues, not only with functional enhancements (e.g., images and clinical decision support), but also in their technical implementation. In my judgment, the VA DHCP System is among the functionally richest and most extensive clinical information systems in use today. Because it is in the public domain, the system also offers a cost-effective model that is eminently affordable.

My association with the VA DHCP began in the fall of 1978 when I was finalizing the program for the second Symposium on Computer Applications in Medical Care (SCAMC). I invited Joseph Ted O'Neill and Martin E. Johnson of the Department of Medicine and Surgery (DM&S) in the VA Central Office in Washington, D.C., to organize a panel on their innovative approach to creating clinical applications for their hospitals. This discussion, recorded by Medisett, Inc.,[1] was the first publicly advertised meeting of what later became the VA DHCP (Figure I.1).

---
[1] The Audio Recording of this panel, detailed in Figure I.1, was made by Medisette, Inc., Dallas, TX. A transcript is available.

Second Annual Sumposium on
Computer Applications in Medical Care

SHERATON INN -- WASHINGTON NORTHWEST
8727 Colesville Road, Silver Spring, Maryland 20910

November 5 - 9, 1978
Program Chair: Helmuth F. Orthner, Ph.D.

```
Monday, November 6, 1978

MONDAY EVENING PROGRAM

8:00 p.m. - 10:00 p.m.   PANEL DISCUSSION GROUPS

II.  THE VETERANS ADMINISTRATION: AUTOMATED HEALTH
     CARE APPLICATIONS (Fiesta East Room)

Moderator:  Martin Johnson, VA Central Hospital in
            Washington DC

Participants:

   George Timson, San Francisco VA Hospital,
      California
   Arden Forrey, Ph.D., Seattle VA Hospital,
      Washington
   Robert Wickizer, Columbia VA Hospital, Missouri
```

FIGURE I.1. Program Listing from the Second Annual Symposium on Computer Applications in Medical Care (SCAMC)

Martin Johnson articulated the principles underlying DHCP: the use of interactive, low-cost dumb terminals connected to minicomputers; the use of the high-level American National Standard (ANS) programming language and integrated database then known as MUMPS; table-driven application modules that can be tailored and reused in other VA hospitals; and a decentralized development approach involving clinicians and healthcare professionals. Widely accepted today, these principles were revolutionary and went against the "glass-house" mentality of the traditional hospital information systems industry prevalent in the late 1970s.

George Timson reported on the Medical Administrative System (MAS), which included patient registration, admission, discharge, transfer, census, and patient scheduling. He stressed simplicity, using the computer only for basic storage, retrieval, and reporting functions. "Present the user with information as quickly as possible but let the user make the complex decision he/she would normally make," he said. This package became the basis for the admission, discharge, and transfer (ADT) and scheduling package of DHCP.

Although George never mentioned FileMan, the core database management system of the VA DHCP, for which he is known best, he did discuss the mental health and the psycho-diagnostic testing modules. These modules were refined by Gordon Moreshad's group in Salt Lake City, Utah, and by Dr. Robert E. Lushene's group in Bay Pines, Florida, respectively. Later they became the Mental Health Package of DHCP.

Robert Wickizer reported on a radiology package that was based on the Missouri Automated Radiology System (MARS). The package consisted of several modules and included a radiology reporting system with a large lexicon of anatomic and pathology terms, a film file tracking system, a quality assurance system, and a film retrieval system for education.

Arden Forrey described a dietetic and food management package to assist clinical nutritionists and professional dietetics personnel in managing the nutritional intake of elderly veterans in the hospital. The package included modules for nutrition analysis, recipe analysis, menu planning, and more. This package was later refined by Dr. Lushene in the Bay Pines VA Hospital in Florida.

The rapid progress of these clinical systems and their acceptance by healthcare providers in the VA hospitals surprised many. Attendees at a coordinating meeting in Oklahoma City in December 1978 agreed on basic programming and data dictionary standards (Munnecke and Kuhn, 1989) that promised to accelerate the development process. However, proponents of the traditional mainframe-based approach had serious misgivings. The Office of Data Management and Telecommunications (ODMT) in the Central Office, which had the official mandate for hospital automation, declared a stop to DHCP development. VA employees participating in the DHCP revolution, including Ted O'Neill, who started it, were dismissed. Employment contracts were not renewed. The DM&S capital and operating budgets for the acquisition and maintenance of computers and software were slashed, and computers actively used in hospitals and clinics were removed and placed into storage to be used as communication controllers in the future.

Despite the circumstances, progress in the core technologies of the VA DHCP continued. The era of the "Underground Railroad" had arrived. Developers avoided the Central Office-based ODMT and found shelter among sympathetic regional and local medical directors who recognized the cost effectiveness of their approach. Developers formed a network, using telephone and e-mail and building on personal contacts at national meetings like SCAMC and MUMPS User Group conferences. Politically powerful allies, with important contacts in the Congress, were found in the National Association of VA Physicians (NAVAP).

In 1981, the DHCP developers created a working prototype of an integrated hospital information system based on an enhanced VA Kernel. This enhanced VA Kernel provided independence from computer hardware and operating systems. At its technical core was FileMan, an advanced database management system developed by George Timson; MailMan, an advanced e-mail system developed by Tom Munnecke; a sign-on/security system; a device handler; a task management system; and a menu manager. One of the earliest e-mail systems completely integrated with a database management system, MailMan was key to many advanced features of the VA DHCP.

In 1981, the prototype hospital information system included the following applications: patient registration, admission/discharge/transfer, outpatient scheduling, outpatient and inpatient pharmacies, laboratory, mental health, radiology, dietetics, and some others. All applications shared a common data dictionary, used a common integrated database, and were based on a common simple software environment. There was no need for complex gateways bridging incompatible computing environments.

In February 1982, the administrator of the VA, Robert Nimmo, issued an executive order creating the Decentralized Hospital Computer Program (DHCP). This cleared the way for installation of the DHCP in all VA hospitals. The "Underground Railroad" was emerging from the dark tunnel. By 1984, the DHCP was installed in nearly all 170 VA hospitals. Major technical and functional upgrades followed quickly.

The VA DHCP started out as a classical legacy architecture and evolved across a variety of hardware platforms—clusters of Digital PDP II minicomputers, clusters of Intel-based PCs, very large Digital VAX clusters, or UNIX computers in several flavors. The migration to a client–server architecture was enabled by advances in the M technology and the development of the "Silent FileMan." The latter permitted the coexistence of dumb terminals and PC-based clients on the same database. This was essential since it allowed an orderly migration from a dumb terminal-based to a client–server-based architecture. In fact, the next generation of clinical workstations used by the VA DHCP is based on PC clients with Visual Basic or Delphi user interfaces. From a technical perspective, the M-based software modules may be viewed as "stored procedures" of the M database.

My personal experience with the VA Kernel goes back to 1980–1981. I found this computing environment to be sufficiently robust and secure to implement the educational testing services for the George Washington University School of Medicine, while also supporting operations for the Continuing Medical Education Office. It also formed the base of the SCAMC Conference Management System for 11 years, from the fourth

SCAMC in 1980 through the fourteenth SCAMC in 1990, when the American Medical Informatics Association (AMIA) assumed responsibility for the meeting.

Over the years, the VA DHCP has proved itself to be a functionally rich health information system. Its efficient character-based database can be used with multi-tier client–server architectures. In the future, the VA DHCP will be ready to move from a client–server architectures to multi-tiered web-based architecture.

References

Collen MF. 1995. *A history of medical informatics in the United States 1950 to 1990.* Washington, DC: American Medical Informatics Association, pp. 446–457.

General Accounting Office. 1988 (March 4). Medical ADP systems: Composite health care system acquisition—fair, reasonable, supported. GAO/IMTEC-88-26. Gaithersburg, MD: Author.

Munnecke TH and Kuhn IM. 1989. Large-scale portability of hospital information system software within the Veterans Administration. In HF Orthner and BI Blum (eds): Implementing health care information systems. New York: Springer-Verlag, pp. 133–148.

Helmuth F. Orthner

Acknowledgments

First and foremost, I want to thank Marion Ball for making this book possible. It is only because of her guidance, encouragement, and support that the book turned into a reality. She has been an inspiration for me since I first met her at the beginning of my medical informatics activities, and I have been honored to have had the privilege to work with her more closely in recent years.

Helly Orthner, too, has been instrumental, first by initiating the idea for a book, and then by keeping the idea of this book alive. He has doggedly pursued me and my VA colleagues at national meetings and on e-mail for the past several years to put our experience down in one place. Although we did not ultimately publish the book in the series Helly edits for the same publisher, Springer-Verlag, I am delighted that he has written a preface for this book, and I value his ideas and his friendship.

Inspiration and encouragement are important, but the assistance that Judy Douglas provided is what ultimately brought this book to press. The coordination of our authors, many of whom had not published works on a national level, was a monumental task. Judy was more than a match for the challenges, of which I am sure I was among the biggest. I now have a more complete understanding of just how valuable she has been to Marion Ball and to the Computers in Health Care series. I appreciate the assistance that David Johnson provided to allow Judy to work on this book.

An edited book such as this one owes a great deal to the authors of the individual chapters, and in usual fashion, I called on those who were already very busy to fit yet another task into their long list of things to do. My thanks to all of them for their wonderful contributions.

I am indebted to Joe Radford for his generosity in designing and producing the artwork for the front cover and to Mary Ann Tatman for coordinating this activity.

Catherine Pfeil and the late Ingeborg Kuhn are owed a special debt of gratitude for beginning the dialogue to create a book capturing our experiences and accomplishments.

In addition to Ingeborg, we lost other friends too early along the journey who had been part of the team of people who helped to create the systems described in this book. These friends included Susan Richie, Bruce Beebe, C.J. Deese, Michael Distaso, Jesse Floyd, Michael Harris, Paul Keltz, Harry Lamb, Len Lawson, Ted Lindsey, Tom Lynch, Dennis May, Linda Phipps, Susan Richie, Georgia Sehon, Rudy Rumbaoa, Dave Van Hooser, and LaVonne Williams.

Writing a book takes time away from other duties and responsibilities, and I am grateful to Dave Albinson, Nancy Wilck, and others at the new VA Headquarters for letting me pursue this goal while they covered the crisis du jour.

The people who are most affected by the demands created by a book are the families closest to the authors and editors. My thanks go to the families of all those who contributed to this book, including Judy Douglas's husband, Paul, and sons, Matt and Justin. Most especially, I would like to thank my wife, Jane Porter Kolodner, and sons, Ben Kolodner and Alex Kolodner, for their understanding, support, and love through this process.

Robert M. Kolodner

Contents

Foreword   vii
Series Preface   xi
Preface by Donald A.B. Lindberg   xiii
Preface by Helmuth F. Orthner   xv
Acknowledgments   xxi

Section 1—Networking for Healthcare Delivery

CHAPTER 1—Vision for Change: An Integrated Service Network
    Kenneth W. Kizer and Thomas L. Garthwaite   3
CHAPTER 2—The Role of Systems at the Facility and Network Level
    Robert H. Roswell   14
CHAPTER 3—Teleinformatics: The Future of Healthcare
    Quality Management
    John W. Williamson   19

Section 2—System Design

CHAPTER 4—Creating a Robust Multi-Facility Healthcare
    Information System
    Robert M. Kolodner   39
CHAPTER 5—Infrastructure and Tools for Flexibility
    Catherine N. Pfeil   57
CHAPTER 6—Structuring Databases for Data Sharing
    Cameron Schlehuber   81
CHAPTER 7—Collaborating for Integration
    Steven A. Wagner   87

CHAPTER 8—Supporting Systems After Installation
    Roy H. Swatzell, Jr., and Virginia S. Price 98
CHAPTER 9—Privacy and Security Protections for
    Healthcare Information
    Gail Belles 116

Section 3—Clinical Information Technology

CHAPTER 10—Integrating a Clinical System
    Rusty W. Andrus 137
CHAPTER 11—Capturing Data in Ambulatory-Care Settings
    Susan H. Fenton 148
CHAPTER 12—Moving to Clinical Workstations
    Sharon Carmen Chávez Mobley 155
CHAPTER 13—Controlled Representation in Patient Records and
    Healthcare Delivery Systems
    Kenric W. Hammond 164
CHAPTER 14—Clinical Decision Support in a
    Distributed Environment
    Curtis L. Anderson 183
CHAPTER 15—The CARE Decision Support System
    Douglas K. Martin 203

Section 4—Clinical and Support Applications

CHAPTER 16—Meeting Clinical Needs in Ambulatory Care
    John G. Demakis 231
CHAPTER 17—Surgical Systems
    Shukri F. Khuri 240
CHAPTER 18—Nursing Use of Systems
    *Bobbie D. Vance, Joan Gilleran-Strom,
    Margaret Ross Kraft, Barbara Lang,
    and Mary E. Mead* 253
CHAPTER 19—Developing Clinical Computer Systems:
    Applications in Cardiology
    Ross D. Fletcher and Christopher McManus 275
CHAPTER 20—Anesthesiology Systems
    *Franklin L. Scamman, Holly M. Forcier,
    and Matthew Manilow* 293

CHAPTER 21—The Library Network: Contributions to the
VA's Integrated Information System
Wendy N. Carter and Christiane J. Jones 308
CHAPTER 22—Using Data for Quality Assessment and Improvement
Galen L. Barbour 330
CHAPTER 23—Developing and Implementing the Problem List
Michael J. Lincoln 349

Section 5—System Evolution

CHAPTER 24—Conceptual Integrity and Information Systems:
VA and DHCP
Tom Munnecke 385
CHAPTER 25—Hybrid Open Systems Technology
Virginia S. Price 395
CHAPTER 26—Using DHCP Technology in Another
Public Environment
A. Clayton Curtis 405
CHAPTER 27—International Installations
Marion J. Ball and Judith V. Douglas 426

Section 6—Telemedicine and Telehealth

CHAPTER 28—Telemedicine: Taking a Leadership Role
John C. Scott and Neal I. Neuberger 435
CHAPTER 29—Telemedicine at the Veterans Health
Administration
Roger H. Shannon and Daniel L. Maloney 447
CHAPTER 30—Telehealth for the Consumer
Peter Groen 466
CHAPTER 31—Digital Imaging Within and Among
Medical Facilities
Ruth E. Dayhoff and Eliot L. Siegel 473

Index 491
Contributors 503

Section 1
Networking for Healthcare Delivery

Chapter 1
Vision for Change: An Integrated Service Network
 Kenneth W. Kizer and Thomas L. Garthwaite   3

Chapter 2
The Role of Systems at the Facility and Network Level
 Robert H. Roswell   14

Chapter 3
Teleinformatics: The Future of Healthcare Quality Management
 John W. Williamson   19

1
Vision for Change: An Integrated Service Network

KENNETH W. KIZER AND THOMAS L. GARTHWAITE

Introduction

Across the United States, technological advances, economic factors, demographic changes, and the rise of managed health care, among other things, are causing a dramatic shift away from inpatient care and a corresponding increase in ambulatory care. For the Veterans Health Administration (VHA)—the largest integrated healthcare provider in the nation—these forces are compounded by demands for downsizing its work force, cutting its costs, and shutting down facilities. At the same time, VHA, commonly known simply as "VA," is being asked to be more responsive to those it serves and to provide better quality care.

Today, VA is making fundamental changes to the way that veterans health care is provided. These include increasing ambulatory care access points, emphasizing primary care, decentralizing decision making, and integrating the delivery assets to provide an interdependent, interlocking system of care. The structural vehicle for these changes is known as the Veterans Integrated Service Network (VISN). As shown in Table 1.1, this new VISN structure is intended to optimize the organization's ability to function as an integrated and a virtual healthcare organization, and, at the same time, to encourage efficiency, quality, and improved access. In addition to reviewing its structure and those of other federal agencies providing health care, VA also considered the organizational structure of large private healthcare entities such as Kaiser Permanente and others.

The concept of an integrated healthcare organization is based on the success of various manufacturing and retailing firms that use horizontal and vertical integration to improve their market position. The fundamental business concept is that if an entity controls and coordinates (i.e., integrates) supply, production, distribution, marketing and all other facets of the enterprise, then it will be able to incorporate all the profits of the otherwise necessary "middlemen" into the parent organization, and thereby accrue cost and service advantages over less integrated competitors. Said in a more healthcare relevant manner, the basic concept of an

TABLE 1.1. Values guiding the establishment of veterans integrated service networks (VISNs).

Category 1: Patient Care
1. Enhance timely access to medical care and other VA services.
2. Maximize resource allocation to direct patient-care services.
3. Facilitate health promotion, disease prevention, and early diagnosis of disease.
4. Enhance appropriate patient referral and service utilization.
5. Keep patient-care decision making as close as possible to the patient.
6. Promote horizontal, patient-focused processes.
7. Provide a community-based focus.
8. Minimize disruption of the system during implementation.

Category 2: Quality
1. Facilitate the development of integrated systems of care.
2. Ensure systemwide consistency in quality and coverage.
3. Minimize fragmentation of functions.
4. Enhance capacity for continuous improvement.
5. Facilitate systemwide data acquisition and performance measurement.

Category 3: Flexibility
1. Facilitate sharing and collaborative agreements.
2. Accommodate state and local healthcare reform initiatives.
3. Facilitate local flexibility and decision making.

Category 3: Efficiency
1. Promote innovation and creativity.
2. Provide clear lines of authority and responsibility and enhance managerial accountability.
3. Minimize organizational redundancies and maximize administrative efficiencies.
4. Maximize information flow and the timeliness of information flow to appropriate decision makers and internal and external stakeholders.
5. Enable decision making at all levels of the organization.
6. Maximize field organization control over support functions.
7. Ensure that each organizational layer or higher level oversight provides "added value."

Category 4: Responsiveness
1. Maximize responsiveness to individual patient needs.
2. Maximize responsiveness to external stakeholders (e.g., Veterans Service Organizations, Congress).
3. Provide for a manageable span of control at all levels of the organization.
4. Enhance VA competitiveness with private and other government sponsored healthcare providers.

These values, 27 in all, have been categorized according to their intended main result, although many of them overlap with other categories.

integrated healthcare organization is that it is one that will be accountable for providing a coordinated range of physician, hospital, and other medical care services for a defined population, and generally for a fixed amount. The assumption is that it will be easier and more efficient to provide for all the needs of the population if all the pieces of the healthcare system needed to provide the care are integrated into and under the control of a single entity.

In an integrated healthcare system, physicians, hospitals, and all other components share the risks and rewards and support one another. In doing so they blend their talents and pool their resources; they focus on delivering "best value" care. To be successful, the integrated healthcare system requires management of total costs; a focus on populations rather than individuals; and a data-driven, process-focused customer orientation. Private healthcare providers started to emulate these business concepts in the 1970s (with some notable earlier efforts), and they are now being pursued in the private sector at a frenzied pace.

Another organizational model that arose in the 1980s, based largely on the experience of the biotechnology industry, is the "virtual healthcare organization." Under this model, integration is achieved by a wide array of discrete corporate arrangements to develop and market specific products. These arrangements are "tailor made" to individual products, markets or corporate competencies of strategic allies.

A number of private healthcare companies have formed virtual organizations and experienced great success as reflected in market share and profitability. What holds these virtual organizations together are: (1) the operating framework (i.e., the aggregate of agreements and protocols that governs how patients are cared for and the information systems that monitor patient flow) and (2) the framework of incentives that governs how physicians and hospitals are paid. Both frameworks are "learning systems," which evolve and change as more is learned about how to improve the provision of care, conserve resources, and manage the system. Virtual healthcare systems invest substantial resources in developing and maintaining their provider networks, focusing on community-based networks of physicians participating in the plan.

The VHA has the relatively unique advantage of being able to function as an integrated and a virtual healthcare organization, although it has not been organizationally aligned and managed as such in the past.

The new structure is based on the concept of coordinating and integrating VA's healthcare delivery assets and the creation of 22 VISNs, replacing what were 4 regions, 33 networks, 173 VA hospitals, 132 nursing homes, and over 400 outpatient clinics. The VISN structure is derived from a model of organizational management that emphasizes quality patient care, customer satisfaction, innovation, personal initiative, and accountability.

Under the VISN strategy, the basic budgetary and planning unit of healthcare delivery shifts from individual medical centers to integrated service networks providing for populations of veteran beneficiaries in defined geographic areas. Decision-making authority is shifted closer to those affected by the decision. These network service areas and their veteran populations are defined on the basis of VA's natural patient referral patterns; aggregate numbers of beneficiaries and facilities needed to support and provide primary, secondary, and tertiary care; and, to a lesser extent, political jurisdictional boundaries such as state borders.

Under the new organizational paradigm, services are provided via better integrated VA resources and through strategic alliances among neighboring VA medical centers, sharing agreements with other government providers, direct purchase of services from the private sector, and other such relationships. The model promotes the benefits of an integrated and a virtual healthcare system.

The new VISN structure places a premium on improved patient services, rigorous cost management, process improvement, outcomes, and "best value" care. As an integrated system of care, the VISN model promotes a pooling of resources and an expansion of community-based access points for primary care. In this scheme, the hospital will remain an important, albeit less central, component of a larger, more coordinated community-based network of care. In such a system, emphasis is placed on the integration of ambulatory care and acute and extended inpatient services so as to provide a coordinated continuum of care.

Each VISN is led by a director who reports to the Chief Network Officer in the Office of the Under Secretary for Health. Veterans Administration medical center and other independent facility managers within a VISN report to the VISN Director. Veterans Integrated Service Network Directors do not serve concurrently as facility directors. Their attention must remain focused on the network. The location of the VISN office within a network is determined by several factors, including ease of access, pre-existing staffing, and cost considerations.

Veterans Integrated Service Networks focus on (1) integrating ambulatory services with acute and long-term inpatient services, and (2) achieving the greatest possible healthcare value for the allocated resources provided. Specifically, each VISN Director has the authority and is held accountable for:

- Ensuring that a full range of services is provided, including specialized services and programs for disabled veterans;
- Developing and implementing the VISN budgets;
- Areawide (population-based) planning;
- Consolidating and/or realigning institutional functions;
- Maximizing effectiveness of the human resources available to the VISN;
- Moving patients within and outside the VISN to ensure receipt of appropriate and timely care;
- Contracting with non-VA providers for medical and nonmedical services, as needed;
- Maintaining cooperative relationships with other VA field entities.

An important component of the VISN model is the requirement that each VISN establish a formalized structure to assure input from VA's internal and external stakeholders. The recommended way to address this need is to establish a management assistance council. This council is comprised of representative facility directors, chiefs of staff, nurse executives, union representatives, and others from within the VISN. External stakeholders, such as veterans service organizations, state and local government officials, members of academic affiliates, and private sector healthcare entities act as consultants to the council. They regularly participate in meetings and provide input into the operation of and planning for the VISN. Each council, working in concert with its external consultants, formulates plans and recommendations to the VISN Director.

While field facilities remain the sites at which VA medical care is provided, the role and function of the medical center director changed as a result of the VISN structure. Decentralization of a broad range of authorities from headquarters to the field increased the director's ability to effect changes within the facility. However, because the basic planning and budgetary unit is the VISN rather than the individual facility, the role of the facility director in decisions affecting the delivery of patient care services shifts from one of independent action to one of collaboration within the network. Each VISN Director works closely and in a collegial fashion with representatives of all the facilities in the VISN to ensure that the views and concerns of facility managers are fully considered as decisions are made relative to the fulfillment of the goals and objectives of the VISN as a whole.

For VHA to survive and perform effectively in state and local markets, it will need an organizational structure that fosters patient-centered service delivery and allows for rapid decision making by giving authority to local management. The Veterans Health Administration must become more "user friendly" and more efficient. It must promote a customer-centered culture that emphasizes continuous improvement of quality, consistency of quality, and the provision of the most cost-effective care possible. The plan will provide that needed structure.

The VISN plan embodies a fundamentally new way of thinking about providing VA healthcare services. The VISN structure encourages the pooling of resources and places a premium on process improvement, outcomes, cost management, and value engineering. It recognizes that the hospital, while still an important component, is no longer the center of the healthcare delivery system, and it provides incentives for expanding community-based access points and primary care. The VISN model also places flexibility, authority, and accountability at the true operating level.

The overarching goal of this reorganization is to improve VA's ability to fulfill its patient care mission. In determining that the VISN model was best suited for achieving this goal, care was given to assure no disruption or diminution of VHA's ability to support its other missions namely, research and backup to the Department of Defense.

The main effects of the field reorganization, combined with the restructuring of the central office, will be less bureaucracy, more timely decision making, easier access to care, and greater consistency in the quality of care systemwide. The VISN Director will have the authority and responsibility to manage the distribution of the network's resources to maximize the advantages to veterans within the VISN service area. This allocation will be achieved by VISN management working in collaboration with the directors of the component VISN facilities and the input of its Management Assistance Council and other appropriate entities. In addition, there will be greater systemwide direction in strategic planning, quality improvement, clinical protocols, and medical management. Also, because the VISN Director will be able to structure the delivery of patient-care services around the needs of the beneficiaries, the result should be better integration of and access to acute and long-term inpatient and ambulatory services.

In terms of the effects on benefits and services not directly affected by the reorganization, restructuring is intended to facilitate more cooperative, mutually beneficial relationships between VA's healthcare system and its other administrations and staff offices. While the transition to a more corporate management role for headquarters and greater authority for the networks will take dedication and time, the end result will align VHA with the best practices of outstanding healthcare corporations.

Lower recurring costs for VISNs (compared to the current regions) should generate annual savings of over $9 million, allowing these monies to be redirected for other high priority needs within VHA. Initially, there will be significant nonrecurring costs in implementing the VISNs, especially those costs related to the relocation and displacement of current VA region office staff. While staffing is the largest recurring cost associated with the reorganization, there will be other significant recurring costs, particularly leased space and employee relocation expenses. Important to note, though, is the fact that the recurring costs of the VISN management structure are substantially less than those associated with the current regional management structure. Also, while the VISN management will be separate and distinct from any medical center or other facility management, VA will colocate VISN management on the grounds of existing facilities or in currently leased sites, wherever possible, to minimize leasing and other support costs.

A guiding principle in reassigning region responsibilities is that most operational functions will be performed at the field level; for example, healthcare facility or VISN office. Veterans Health Administration headquarters staff will provide advice, national policy direction, and technical expertise to support the field.

The VISN Directors play a direct and significant part in decision making concerning the disposition of all of the current region functions. Many of these functions have implications for the operation and ultimate success of the VISN structure; therefore, the direct involvement of VISN management is viewed as crucial.

As VA moves into the future, the VISN concept represents a profound paradigm shift in the provision of healthcare services to veterans. Central to this shift is the creation of a powerful interdependence that has not previously existed between field operations and the central office. Now is the time to fully integrate all of VHA's critical functions and refocus each on the most important customer—the veteran patient.

This reorganization is not a simple realignment of 173 hospitals, 33 networks, and 4 regions into 22 VISNs. Nor is it a reshuffling of bureaucratic boxes on a central office organizational chart. Rather, it is a fundamental change in the way responsibility is spread across many decision points to imbue the organization with a common sense of purpose. The Veterans Health Administration will become less like a mega-corporation and more like a system of federated networks that are bound together by a determination to provide quality patient care. If roles are properly defined and executed, and if power, authority, and accountability are balanced and dispersed throughout the organization, then the result will be an interdependent and interlocking system whose whole is greater than the sum of its parts.

"Think global, act local" is a fashionable slogan that embodies how VA intends to function in the future. Strategic planning will become integrated with quality improvement to assure that the changing demands of the national healthcare environment are reflected in the consistent delivery of local services. Patient-care services, and most importantly VHA's recognized special programs, will benefit from a new way of thinking in which multifunctional teams will collaborate and offer expert consultative services for clinicians at the point of service delivery. Healthcare delivery will be shifted away from institutional inpatient modalities to network-based ambulatory solutions. And a renewed emphasis on data capture and information management will provide the vehicle for meaningful performance measurement and resultant accountability.

Decentralization of authority will stimulate innovation, but it may also lead to undesired variability in quality, services, or access. To increase the likelihood that desirable standards of uniformity in these areas are being met or exceeded, performance measures are being developed in multiple product lines. These measures are especially important in our special program areas where our patients are often unable to advocate for their own care or to find similar care in the private sector. The ultimate goals of performance monitoring is to focus the entire organization on improving the outcome for patients. Automated systems to collect performance data will be critical because we have multiple programs, patients, providers, and outcomes and because the data must be recent to be relevant.

Information is the "glue" that holds the integrated and virtual networks together. As patients receive their care in multiple settings from various providers, timely and accessible clinical information is critical to providing quality care. Outcome and cost information allow efficient system design and guide buy–build decisions. Ultimately, such data will help define value in health care provided by VHA and will allow us to compare our value with that of non-VA systems. Sophisticated information systems will shorten the route from symptom to diagnosis, detect drug interactions before drugs are used, and allow distant experts to participate in the care of many patients via telemedicine. In the future, access to care in many cases will be more dependent on the access of the decision-maker to accurate patient information than on the need to get the decision-maker and the patient in the same room with the patient's chart.

For these reasons and others, information technology is being used or will be used to facilitate VA's transition from a hospital-based system to a network of VISNs—providers and facilities that assume responsibility for the health of a population of eligible veterans. A fundamental assumption is that the collection and use of information are a natural consequence of a clearly articulated plan, well-defined business practices, and a valid sense of priorities. In the absence of these parameters, more information has little chance of helping the organization succeed.

The development of clear goals and objectives for the VHA had been traditionally a process that involved VA headquarters in Washington often with the input of a committee that had modest VA field representation. During the past year, advances in information technology have facilitated a wider dissemination of draft policy. Such drafts have been circulated by the traditional routes (e.g., copies mailed to each facility), and they also have been made available on VA's bulletin board system (BBS) and Web page (*http://www.va.gov*). Written feedback has been accepted on paper or via e-mail. For the Vision of Change document alone, we received over 1200 pages of comment. The improved distribution to various parts of the organization has resulted in widespread consideration and debate. As an added benefit, participation in policy development has created broad knowledge of the principles and has greatly facilitated their implementation.

One of the first tasks facing the leadership of a network (VISN) has been the need to develop a business plan. The plan must begin with an understanding of the healthcare needs of the population of veterans within the network. Information for this analysis currently resides in various national databases. Based on historical use data, veteran demographics, trends, facilities, existing programs, and national policies, networks are developing plans to enhance care and reduce costs in each network. Many of the networks are basing the decisions to keep or close programs and facilities on the data (i.e., information) available. In other cases, networks are seeking new data to guide the planning decisions—whether to buy or build, how to evaluate programs for efficiency and efficacy, how to benchmark technologies and programs.

Good national and network plans derive from good information. Such plans and resultant systems form the foundation for the delivery of health care. While information about patients is critical to the delivery of health care at each facility, the delivery of care across several facilities and involving multiple providers requires significant national and network standardization of data. Conventions must be established for the definition, collection, storage, and transmission of data. Decisions must be made regarding where the data are stored and how often they are updated. Ultimately, the complete database on each patient might be used to aid clinicians in decision making, to allow analysis of the case by expert systems, to allow analysis of the cost and outcomes of care (i.e., to measure value), and to communicate aspects of the case to various other systems. The Veterans Health Administration, like other health systems, is looking to the development and use of systemwide clinical protocols and practice guidelines, which are clearly information-based.

In addition to the delivery of care to individual patients, information must be generated and analyzed regarding the performance of the system of care. Examples of data that might be generated include outcome measures (mortality rates, postoperative complication rates, immunization rate, medication errors, etc.), and cost data (cost by product, cost per user, and cost per diagnosis).

An exciting aspect of information in a healthcare system is the concept of system learning. In a large system, many problems are identified and most of them are fixed with various amounts of effort. Few systems are able to apply multiple creative individuals to the solution of the problem. Additionally, few systems are able to rapidly disseminate a creative solution to all facilities. An ideal system would easily identify problems and would present the problem to a broad but targeted subset of employees. Identified solutions would not only be implemented locally, but would be communicated widely and implemented quickly. Other information (e.g., a new standard treatment from a high respected medical journal) also would be disseminated quickly.

At the VISN level, the bricks and mortar of individual institutions will no longer be the central point of patient services. While there will be additional flexibility and autonomy for local managers, independent decision making and parochial interests must be subsumed for the greater good of the geographic network. Community involvement and resource sharing will become the vehicles for outreach and expanded services. Program and resource decisions will be built on the shared vision of improved customer satisfaction, quality care, access, and cost-effectiveness.

At the national headquarters level, the focus will move from a centralized organization that exercises a traditionally hierarchical mode of operational control toward a headquarters that supports the field, through governance and leadership, in its critical role of serving patients. Patient-care decision making will be exercised as close to the patient as possible, allowing headquarters to concentrate on leading the system through the dynamic and turbulent changes ahead. Ultimately, the goal is that the field will seek advice and counsel from headquarters because headquarters has expertise to offer and adds value to field decision making, not because it is holding operational decisions hostage.

While this VA reorganization is dramatic in scope, the implementation of the new structure will be relatively simple when compared with the cultural change needed to make the operation truly effective. The inherent value and strength lie in the individuals who comprise the organization, and this new enterprise will demand new behaviors and attitudes. The organization's responsibilities for two-way communication, job reengineering, education, and training are tremendous, and its leadership is committed to providing the tools necessary to assure a smooth, orderly, and compassionate transition. Additionally, the entire organization needs to demonstrate greater sensitivity to its various stakeholders, including veterans service organizations, employee unions, affiliated medical schools, and state and local healthcare entities.

Change of this magnitude does not come quickly or easily, and changing an organization's culture is not a task for the fainthearted. The key to managing the change process and facilitating the acquisition of new skills is education and training. As part of the implementation plan, comprehensive education and training needs were identified and options to meet those needs developed. For example, the VISN Directors received an orientation on the scope of responsibilities for this new position, stressing the need to acquire new skill sets, such as business planning, performance measurement, and systems monitoring.

Central to the success of this plan is the future development of financial and information management systems that support integrated networks. The Resource Planning and Management (RPM) is being modified to refine funding distribution at the VISN level, to accurately measure the financial needs of new programs and access points, and to ensure that the model meets the requirements of a financial monitoring system. Further, the leadership of the new structure has tremendous demands for accurate and realtime information to aid decision making and monitor performance. These modifications and resources for the development of new models have high priority and have now begun to be made.

An extremely important aspect of this plan is its flexibility. For example, while the VISN boundaries are largely defined by patient referral patterns and natural planning groups, they can be altered over time, if warranted. State-level healthcare reform is continuing in many parts of the country, and, if state legislation dictates that a network be organized around state borders, the change can be implemented with little disruption to the system as a whole. Or if, after implementation, it becomes clear that a VISN is unwieldy because of its size, complexity, or other factors, then alignment can be adjusted. The provision of healthcare services is a dynamic field, and the integrated networks must retain the agility necessary to thrive. Also VA must recognize that improvement comes from knowledge, knowledge from experience, experience from action, and action from planning and evaluation. Failure and paralysis are predictable when the status quo is blindly defended; success and growth are realized when initiative is valued and honest mistakes are tolerated.

The transition period poses special challenges because change of this magnitude is often accompanied by organizational anxiety and disruption of existing systems. The VHA is committed, however, to minimizing these untoward effects through effective communication and education and by building on the enthusiasm and momentum this new vision creates. An important goal of the restructuring is that VA become an employer of choice, an organization that thrives on the growth and development of its most important asset—its people.

The current quest for healthcare "reform" is a quest for value in health care. Recognizing this need to demonstrate value, the VHA has embarked on a fundamental restructuring of its healthcare system: from a collection of largely independent and often competing hospitals to 22 networks of facilities and providers charged with the responsibility to care for the health of a geographically defined population of patients. Other basic tenets of the restructuring are decentralization, flexibility to use either integrated or virtual organizational structures as appropriate, patient outcome based performance measurement, and enhanced information handling. Clearly, these are ambitious goals—goals that the VHA has pledged to meet as it moves into the future.

Acknowledgments. The authors gratefully acknowledge the document, *Vision for Change: A Plan to Restructure the Veterans Health Administration* (1995), for the ideas presented here. Members of the team who helped craft that document include the following: Nora Egan, Barry Bell, Kathleen Bishop, Holly Harrington, John Charles Herzberg, Maureen Humphrys, Pamela McBride, Robert McQueeney, John Molnar-Suhajda, Norman Morse, Gregory Neuner, and Charles Smith. To them the authors express their thanks.

2
The Role of Systems at the Facility and Network Level

ROBERT H. ROSWELL

Introduction

The concept of networking individual medical centers into an integrated healthcare delivery system is certainly not new, but it represents a significant change for the Department of Veterans Affairs (VA). Many, if not most, of the potential benefits of such an integrated healthcare delivery systems with decentralized authority for planning and budgeting are dependent upon effective information management systems and the ability to move information across the system of medical facilities.

An effective integrated healthcare delivery system should be able to deliver a full range of clinical services from routine preventive health screening to the most complex surgical procedures. Yet despite this wide range of services provided at multiple sites across the network of medical facilities, the patient should perceive all services as being available, based on need, in a "seamless" system of healthcare delivery that does not require reapplication for care or reentry into a new delivery system.

Just as healthcare services must be seamless to the consumer or patient, healthcare information must be seamless to the provider or clinician. Similarly, fiscal data and utilization management information must be readily available to healthcare administrators from all sites across the delivery network. Only when these challenges have been fully met, can the benefits of an integrated healthcare delivery system be realized. Quite simply, healthcare facilities can not be integrated into delivery systems or networks unless the information systems that support the delivery of care are first fully integrated.

Information Resource Advisory Council

Several years before the VA began contemplating the reorganization of healthcare facilities into integrated service delivery networks, the Medical Information Resource Management Office (MIRMO) in the VA's Washington offices was busy trying to evaluate the future direction of its Decentralized Hospital Computer Program (DHCP). This effort took the form of a field task force charged with developing a Strategic Information Systems Plan (SISP). After a great deal of thought and effort—and a fair amount of argument—the first SISP was completed in 1989.

The plan called for the creation of a field advisory board to provide guidance and recommendations in future systems development efforts. It also identified support of patient care as the highest development priority, and the support of management decision making as the second highest priority. The SISP further called for a restructured development process that would allow enhanced integration of individual software modules, while improving field personnel input into the development process.

In response to the first SISP, an Information Resources Advisory Council (IRAC) was created in early 1990. Soon after its formation, this group set about determining what type of organizational structure would best serve its role as stated in the SISP. The result was the creation of three broad applications requirement groups with diverse representation from a wide range of end users. Not surprisingly, the three application requirement groups (ARGs) mirrored the priorities established in the strategic information plan.

The first group was named the Clinical Applications Requirement Group (CARG) and was charged with prioritization and development of clinical applications. The second group was designated the Management Applications Requirement Group (MARG), which was similarly charged with prioritization of management applications. The final group, known as the Integration and Technology Applications Requirements Group (ITARG), was intended to ensure that various application packages be compatible and not redundant when they were integrated into the DHCP system. The ITARG was also charged with staying abreast of emerging technologies that might impact or influence VA's future information systems.

The identification of specific applications requirements was charged to ad hoc expert panels drawn from an extensive registry of field personnel with expertise and interest in various systems applications. Resulting specifications were reviewed by the parent ARG and prioritized prior to actual development by VA's programmers located in seven field-based Information Service Centers.

Fortuitously, this rather complex field advisory system helped evolve an informatics philosophy that squarely addressed the information needs brought about by effort of the VA to transition to a system of 22 Veterans Integrated Service Networks (VISNs). Specifically, the Clinical Applications Requirement Group identified an automated patient record system as the highest clinical development priority, while the Management Applications Requirement Group felt an effective decision support system that would facilitate patient care delivery was the highest management development priority. As discussed in more detail below, these are the two critical pieces necessary to effect the transition from a system of rather autonomous hospitals into an integrated system or network of healthcare services.

Clinical Needs

The major goal of networking small numbers of VA Medical Centers (VAMCs) with a decentralized management structure should be the enhancement of services to the veteran patients served by the system. These enhancements may take the form of improved access to a variety of healthcare services, improved quality of services or increased satisfaction with the delivery of these services, or better healthcare value for the money spent on providing care. By placing administrative oversight of strategic planning, budget execution, and operational management closer to the patient–clinician interface, the VA hopes to realize many of these potential benefits. Ultimately, however, the success of these efforts to decentralize control and integrate facilities and services will be judged through the eyes of the veteran patients the system serves.

With this in mind, it becomes imperative to look beyond potential operational efficiencies that are readily apparent in a networked system of healthcare delivery, and to examine carefully the patient's perspective of the care provided by such a system. Obvious values quickly come to mind:

- Is care readily available?
- Is the care compassionate?
- Is the care well coordinated?
- Is the need for, and process of, delivering care easily understood?
- Is the patient personally involved in his/her own care?
- Is the care comprehensive?

The ability to achieve each of these attributes of patient-oriented comprehensive care delivery is, in large measure, dependent upon the realization of the goal set by the IRAC. Specifically, this goal is the ability to access immediately comprehensive clinical information related to all aspects of care at any point throughout the networked system of healthcare facilities. In other words, this goal is an automated patient record.

Although operational and economic incentives (such as less space required for storage and fewer personnel needed to retrieve, transport, and file records) would clearly support a transition to automated medical records, the real value of a computerized patient record lies in its benefits to the patient. As the complexity of health care increases, the simultaneous need for patient information at multiple locations rises dramatically. This need is nowhere greater than in an integrated network of healthcare facilities where each institution has historically maintained all of its own patient records.

A successful, multi-facility healthcare delivery system must be able to bring all needed clinical information from all parts of the system immediately and directly to the patient–clinician interface. Only then will the patient begin to appreciate the comprehensive services of such a system, and only then will the network resources be harnessed for the enhancement of patient outcomes. Specific components of the clinical record that are particularly important to the successful implementation of a multi-facility integrated delivery network include access to scheduling, test and procedure ordering and reporting, specialty consultation reports, and virtually any other information or service that is not readily available at all locations of the integrated delivery system.

Several other intriguing applications of an automated record within a networked healthcare delivery system lie beyond the immediate goal of efficient medical record support for patient-care activities. Among these are the incorporation of practice guidelines or patterns, online access to expert systems, and realtime quality improvement and utilization management capabilities. When these kinds of additional practice aids can be incorporated directly into the healthcare delivery process through inclusion in an online automated patient record that is readily available precisely at the point of care, the opportunity to influence practice patterns and improve patient outcomes is dramatically enhanced. In addition, retrospective utilization and quality management activities that unfortunately can only identify past deviations from best practice become a thing of the past.

Management Needs

The IRAC identified an effective decision support system that would facilitate efficient patient-care activities as the highest priority management application for VA's DHCP system. The need for such support became even more paramount as VA began the process of organizing multiple medical facilities into 22 integrated service networks, particularly at a time when appropriated federal dollars to support VA care began to shrink.

A number of key objectives must be met by each of the 22 VISNs to be successful in this climate of declining appropriated dollars. Foremost among these is the ability to determine accurately the true cost of services provided to patients. This information is essential in deciding which programs will be eliminated when redundant services exist within the delivery network or, alternatively, in competitively pricing these services so that the excess capacity can be sold to other healthcare providers such as the Department of Defense or affiliated academic institutions. Similar information is also needed in determining whether it is more cost-effective to add new programs or to contract for services when the network delivery capacity is insufficient to meet the current veteran patient demand.

Another key function required at the network level is the ability to compare utilization of services at each facility within the network so that distribution of available resources can be achieved in the most equitable manner. This entails quickly identifying opportunities for management improvements when overutilization occurs.

A final critical dimension of the system support of VA's new VISNs is the ability to move management information instantaneously across the network of medical facilities. While this application may seem obvious, one must remember that the VA DHCP system was developed as a decentralized system intended to support a single facility, with subsequent data roll-ups accomplished by periodically extracting information from each individual site. Thus, new functionality is needed to support a networked delivery system. Current strategies used to meet this demand include wide area networking as well as Internet and Intranet applications, in addition to continuing to utilize data periodically extracted from the individual healthcare facilities.

Summary

In large measure due to thoughtful long-range planning and extensive field input, the VA's DHCP system is evolving in ways that will meet the information needs brought on by its transition to a decentralized system of 22 integrated healthcare service delivery networks. Notable among these requirements is the capability to access vital information at any point throughout the networked system of healthcare facilities. Specific applications, identified several years ahead of the reorganization effort by the VA IRAC, which will help meet this need, are an automated patient record and a robust decision support management information system. When coupled with an effective realtime communication and management information system linking each facility within the healthcare network, these developments will help ensure the success of the new VA VISNs.

3
Teleinformatics: The Future of Healthcare Quality Management

JOHN W. WILLIAMSON

Quality Management: Past, Present, and Future

History of QM

The history of recorded healthcare quality management (QM) dates back 5000 years. Doctors have long been accountable for healthcare outcomes. As early as 3000 BC, in the legal statutes of the Egyptians, heads of state generously rewarded physicians for exceptional medical outcomes. For example, the accomplishments of Imhoptep (2686–2613 BC), who was the physician for King Zoser, won him such acclaim that he was elevated to demigod about 100 years after his death, and to a full deity around 525 BC. (Was this the beginning of the physician "God Complex"?) A similar process took place later in the Homeric Period of Greek history, about 900 BC, with Aesculapius, to whom the caduceus, as the symbol of the medical profession, was first ascribed. Like Imhoptep, he was later made a full god, to be worshipped in the Temple of Aesculapius, along with his daughters Hygeia and Panacea (Bettmann, 1972, pp. 16–17).

One of the earliest examples of what was roughly analogous to modern health outcomes management is found in the Mesopotamian Codes of Hammurabi, dating from the 18th century BC (Harper, 1904). One code states "If a physician should operate on a man for a severe wound with a bronze lancet and cause the man's death, or open an abscess [in the eye] ... and destroy the eye, they shall cut off his [the physician's] fingers." Hammurabi also addressed "economic outcomes" by establishing specific fees for medical care (Garrison, 1914). In this early period of history, it is difficult to find a documented example of what we today call "satisfaction outcomes" measurement. Perhaps the tried and true method of consumers, "voting with their feet" and not returning to the same care provider with whom they are dissatisfied, has been the universal means of consumer management of "satisfaction outcomes."

Thus, throughout the span of recorded history, examples are evident of activity roughly analogous to current quality improvement functions. Such activity can be traced back in history through five streams of healthcare stakeholders: government, academia, health professional organizations, financing groups, and consumer groups. In the past 40 to 50 years, quality management has been systematized and, in many parts of the world, developed on an institutional basis, utilizing formal, if not scientific, tools and methods. Currently, healthcare professionals are slowly learning many of the basic principles of quality management—some of which date back nearly 5000 years.

Generic Categories of Today's QM Systems

Today there are as many different quality management systems as there are healthcare organizations. Each group has its own concept of quality and its own philosophy as to how it should be organized and implemented. However, there are enough common characteristics of these systems that it is possible to group them into three generic families, that is, traditional, industrial, and innovative.

Traditional systems encompass those that place primary emphasis on "structure" (licensing, certification, and accreditation) and, to some extent, "process" such as the use of chart review monitoring systems. *Industrial systems* incorporate the principles popularized by W. Edwards Deming (1986). He won so much fame in revolutionizing Japanese industrial quality. The major focus of these systems is total quality management (TQM) theory that involves organizational cultural change to facilitate continuous quality improvement (CQI). *Innovative systems* include those programs built around computerized medical informatics infrastructure. They include a growing family of expert systems to assess and improve clinical problem-solving in a wide variety of contexts, from direct patient care to administration and support services. The concept of teleinformatics is in this third category and will be the emphasis of this chapter.

Teleinformatics: The Future of QM

In brief, teleinformatics includes workstations for electronic patient records, digital imaging and audio capability, and linkages to local area networks (LANs) and wide area networks (WANs), together with interactive telecommunications hardware and software.

Teleinformatics promises outcomes improvement in diagnosis and treatment through computer-driven QM systems. In a very real sense, the future of quality management will be a function of the growth of teleinformatics.

At present, quality management systems, with one or two exceptions, do not have cost-effective methods for monitoring, assessing, and improving quality of care across the range of healthcare outcomes and healthcare facilities.

Today QM systems are constrained by the dependence on paper chart review, the lack of integration of inpatient and outpatient records, and the time required to create databases for healthcare variables. These QM systems cannot support widespread, cost-effective quality management. Fortunately, new developments in the field of teleinformatics can overcome many of these shortcomings.

Teleinformatics and Outcomes Management

Teleinformatics will provide the essential infrastructure for improving healthcare quality management. Teleinformatics will provide the electronic means of retrieving and integrating data at all levels and sites of health care. In the future, successful healthcare systems will depend upon total integration of services. This integration, in turn, will depend upon pervasive data systems building upon the teleinformatics infrastructure.

Certain concepts and principles clarify the relationship of teleinformatics to healthcare outcomes management. Teleinformatics networks will provide the most immediate and effective means of supporting such quality improvement activity. The most fundamental principle of quality management is the classic improvement cycle, depicted in Figure 3.1. Care is assessed to identify whether improvement is required. If it is, then plans are made to effect needed change. Next these plans are implemented, and finally care is reevaluated to determine whether improvement actions were successful.

As shown in Figure 3.2, the most powerful strategy for implementing the improvement cycle involves first identifying and prioritizing unacceptable healthcare outcomes, and then deducing causal factors related to these outcomes to determine which processes or structures need to be changed. By induction, this change should lead to the outcome improvement previously judged possible.

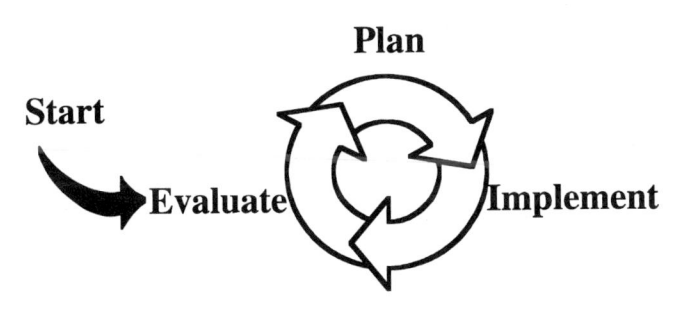

FIGURE 3.1. The basic QI cycle underlying all quality management systems, however detailed they each may be elaborated (Williamson, 1991).

22   Networking for Healthcare Delivery

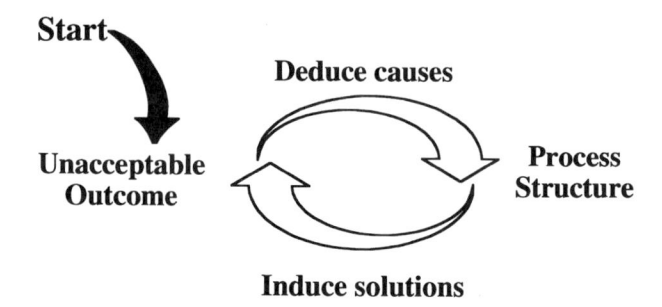

```
Start↘                Deduce causes
        ↓           ⟳
    Unacceptable         Process
      Outcome            Structure
                      
                Induce solutions
```

FIGURE 3.2. The most effective QI strategy: identify an unacceptable high priority outcome, deduce correctable causal factors to be changed, and then deduce the likely outcome improvement to be achieved by this change (Williamson, 1991).

Figure 3.3 shows those generic outcome families that must be considered in determining where to invest scarce QI time and funds. The rows represent final outcomes of care, with the row marked *Other*, including such outcomes as ethical, medical legal, or even religious results of health care. The two columns represent both consumers and providers of care, including clinicians, administrators, purchasers, and financiers (e.g., health insurance

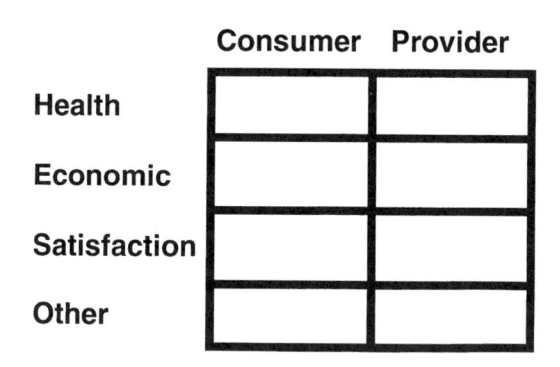

	Consumer	Provider
Health		
Economic		
Satisfaction		
Other		

FIGURE 3.3. Depicted are eight generic families of healthcare outcomes. Each cell encompasses unique population denominators, quality measures, standards, and biases to be managed in their assessment and improvement. Each can be more specifically defined in smaller population clusters, ranging from international, national, regional, local, and even to one-on-one levels. "Consumer" encompasses the total population within the jurisdiction of any healthcare system. "Provider" includes clinicians, administrators, third-party payers, healthcare plans, care purchases, and regulators. In any QI strategy, all of these outcome groups must be addressed to some extent. To emphasise health improvement to the neglect of economic or satisfaction outcomes is folly. To stress patient satisfaction to the neglect of provider job satisfaction is equally counterproductive. "Other" outcomes indicate important, though often lesser priority, results of care such as ethical, legal, educational, or religious.

3. Teleinformatics: The Future of Healthcare Quality Management   23

groups), as well as suppliers and regulators. The major principle is that effort must be balanced across all of these cells, as opposed to focusing all activity in only one or two of these outcome families.

The numbers represented in Figure 3.4 indicate the balance of QI effort established by a heterogeneous group of healthcare providers in Belgium. This group included clinicians, administrators, and members of their national health ministry. While consumer health outcome improvement was given highest priority, consumer satisfaction was judged second. Provider satisfaction (e.g., personnel turnover and burnout measures) ranked third. This figure indicates that QI resource investment must be balanced among the eight generic outcome clusters.

Figure 3.5 depicts a measure of patient health status over time. Hippocrites pointed out several thousand years ago that such an outcome measure alone can *not* indicate quality of care provided. Hippocrites also pointed out that to measure quality, two constructs are essential, namely what would have happened if no care were given, and what would have resulted if optimum care had been provided. Figure 3.6 depicts the area between these two curves as achievable benefit or efficacy of aggregate health care. In Figure 3.7, Area A shows benefit that was actually accomplished, that is, a measure of the effectiveness of care given. Area B depicts the most critical focus for quality management activity, namely Achievable Benefits Not Achieved (ABNA). This model can be applied to any outcome content, such as economic, satisfaction, or ethical-legal. Based on a review of current scientific medical literature, Figure 3.8 depicts ABNA in caring for patients having diabetes mellitus. Evidence of similar areas for effecting improved outcomes of care can be derived for most other health problems.

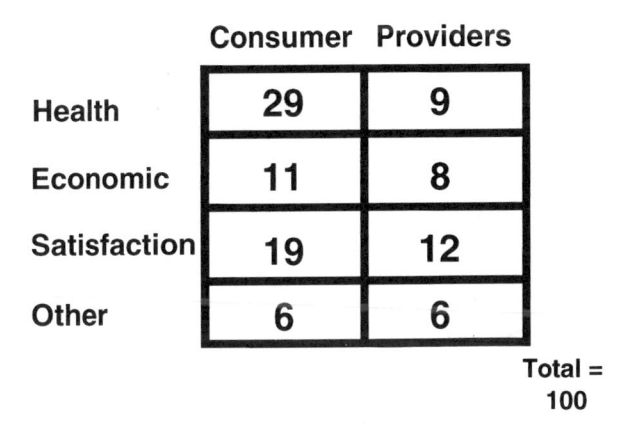

	Consumer	Providers
Health	29	9
Economic	11	8
Satisfaction	19	12
Other	6	6

Total = 100

FIGURE 3.4. Consensus priority weight given each outcome group by a multidisciplinary team. Each point could indicate a percentage of time and dollars to be expended for assessing and improving each outcome group.

24   Networking for Healthcare Delivery

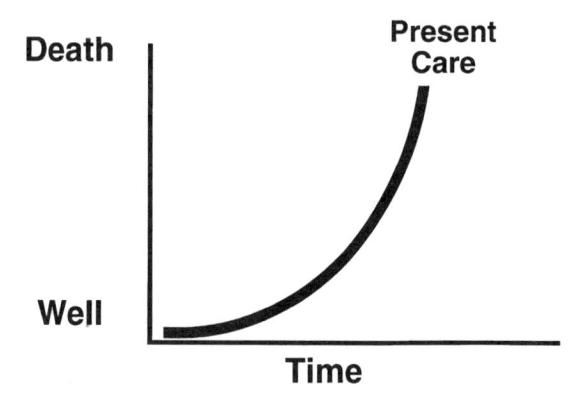

FIGURE 3.5. This construct indicates that any healthcare outcomes can be measured on a value scale over time. This example depicts the health status of a middle-aged man experiencing an acute myocardial infarction, which leads to his death while receiving professional care (Williamson, 1971).

As shown in Figure 3.9, another construct is paramount in identifying where quality management activity might effect the most improvement of healthcare outcomes. Diagnostic outcomes are the aggregate of missed diagnoses (false negatives) and misdiagnoses (false positives). Therapeutic outcomes involve the results of care provided in areas such as health, economics, or satisfaction. Unless diagnoses are accurate, however, it is extremely difficult to improve quality of therapeutic care. The five functional stages of outcomes improvement shown in Figure 3.10 address both categories of outcomes. Data from these studies document the efficacy of applying quality improvement principles to healthcare outcomes.

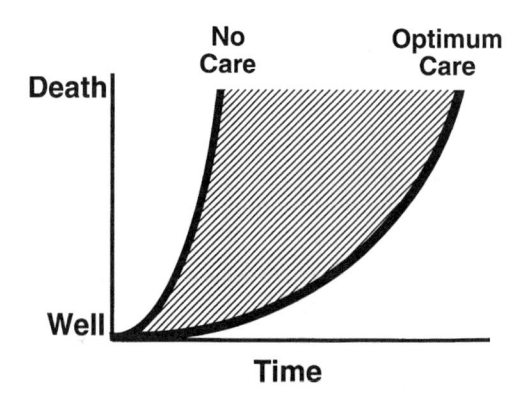

FIGURE 3.6. Hippocrates pointed out that a patient's health status, as shown in Figure 3.5, cannot be quality assessed without knowing two constructs that must usually be approximated, namely, what might have happened if no professional care were rendered, and what might have happened if optimum care were provided. The area between these curves represents efficacy of aggregate care (Williamson, 1971).

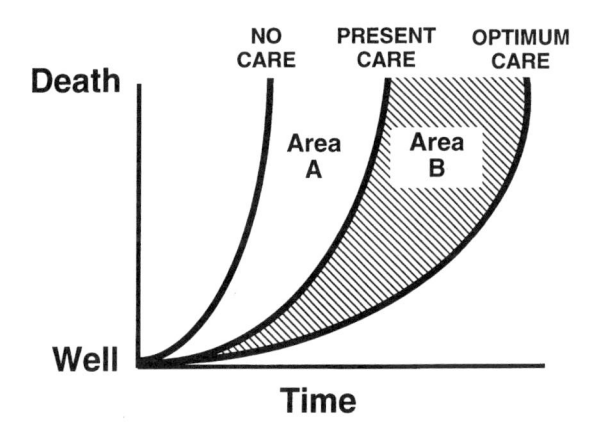

FIGURE 3.7. Comparing the results of present care with the potential efficacy of aggregate care yields two areas: Area A represents Achievable Benefits Achieved (i.e., effectiveness); and Area B represents Achievable Benefits *Not* Achieved (ABNA). Achievable Benefits Not Achieved represent the potential healthcare improvement that might be reasonably accomplished by QI effort. This concept should be at the core of any approach to establish priorities for QI topics (Williamson, 1971).

Action	Result
Appropriate foot care	50% reduction in amputations
Appropriate eye care	60% reduction in severe vision loss
Detection / control of B.P.	85% reduction in stroke
Improved self-management and education	70% reduction in DKA

FIGURE 3.8. Based on literature review, this figure illustrates ABNA for diabetic QI activity (Williamson, 1978).

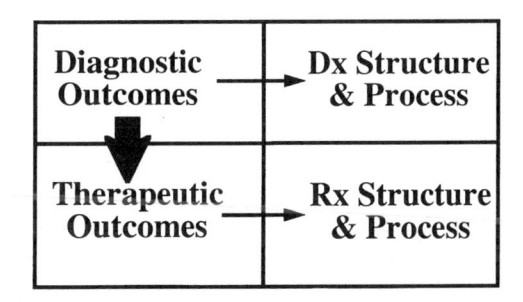

FIGURE 3.9. This construct suggests that it is usually essential to assess diagnostic outcomes first, then move to assess therapeutic outcomes. If either of the outcomes is unacceptable, then evaluate the process or structure which, if changed, might lead to improvement of the key outcomes (Brook, Kamberg, Mayer-Oakes, et al., 1989).

26   Networking for Healthcare Delivery

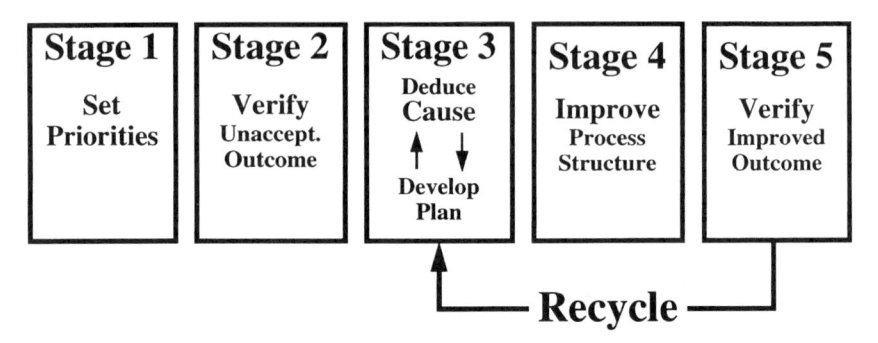

Stage 1	Stage 2	Stage 3	Stage 4	Stage 5
Set Priorities	Verify Unaccept. Outcome	Deduce Cause ↑ ↓ Develop Plan	Improve Process Structure	Verify Improved Outcome

└──── Recycle ────┘

FIGURE 3.10. The five essential stages of an "outcomes-based" QI system. If outcomes improvement goals are met, then monitoring mechanisms are set in place. If improvement goals are not met, it is necessary to recycle back to identify other processes to be changed and to try again (Williamson, 1978).

Teleinformatics and Diagnostic Outcomes Improvement

Diagnostic errors are the most serious healthcare quality problem we face today. Computerized expert systems and guidelines offer the greatest hope for substantially improving diagnostic accuracy. Incorporating such systems into the teleinformatics infrastructure will make these benefits available in small and geographically isolated sites as well as in large urban centers.

The literature is rich in studies documenting the seriousness of diagnostic errors in health care. Health services research, as shown in Figure 3.11, offers evidence of the serious magnitude of the problem, as measured for three different patient populations, in three separate studies. Figure 3.12 summarizes Howard Foculty's and Leon Goldman's autopsy meta-analysis as reported by Hill and Anderson (1988), showing little or no overall improvement since 1959 in either Class I diagnostic errors (directly causing

	% Patients
Geriatric Patients	26-64%
Ambulatory Patients	50-58%
Housestaff Patients	33%

FIGURE 3.11. Evidence from health services research studies confirming the substantial extent of diagnostic errors in specific areas of our healthcare system. Diagnostic errors for (1) geriatric patients (Brook, Kamberg, Mayer-Oakes, et al., 1989); (2) ambulatory patients (Williamson, Fehlauer, Gaiennie, et al., 1991); (3) patients cared for by medical house staff (Wu, Folkman, McPhee, et al., 1991).

3. Teleinformatics: The Future of Healthcare Quality Management   27

Discrepancy Level	1959-1960	1969-1970	1979-1980
Class I	8%	12%	11%
Class II	14%	11%	10%
Total	22%	23%	22%

FIGURE 3.12. Based on three decades of autopsy studies, there is little evidence that diagnostic accuracy has improved. Class I diagnostic discrepancies are direct causes of mortality; Class II errors are contributory causes (Hill & Anderson, 1988).

the death of the patient), or Class II (not causing, but directly contributing to death). Figure 3.13 shows that autopsy diagnoses in five key areas were associated with serious diagnostic errors. Note that roughly half of patients dying from pulmonary embolus did not have a premorbid diagnosis of this problem—a proportion that has not changed since around 1900 (Zarling, Sexton, & Milnor, 1983). Figure 3.14 indicates that the percentage of people dying as a result Class I and II diagnostic errors directly correlates with the patient's age at death. For those dying after the age of 75, up to one in three might be victim of such an error. Equally ominous, for those dying between 45 and 64, one in seven is a victim of such an error.

Based on autopsy evidence, diagnostic competence is often domain specific, that is, relates to the area of training and experience of the physician. As shown in Figure 3.15, cardiologists and general internists have the highest accuracy in diagnosing acute myocardial infarction, while subspecialists practicing outside their domain, especially surgeons, have shockingly poor diagnostic outcomes. Other nonautopsy health services research studies seem to confirm this finding (Rhee, 1975). Improving diagnostic accuracy is not easy. Still, as shown in Figure 3.16, there are many notable examples of effecting such improvement. What is most important to note is that most of

Autopsy Diagnosis	% Missed
Pulmonary Embolus	46.8
Peritonitis	45.1
Post-Op Hemorrhage/Infection	37.9
Intestinal Vascular Insufficiency	37.2
Lung Abscess	34.1

FIGURE 3.13. Five leading diagnostic errors based on autopsy studies (Hill & Anderson, 1990). The accuracy of diagnosing puimonary embolism has not changed since the early 1900s (Cabot, 1912).

Networking for Healthcare Delivery

%
40
30
20
10
0
1-14   15-44   45-64   67-74   ≥75

Age at Death

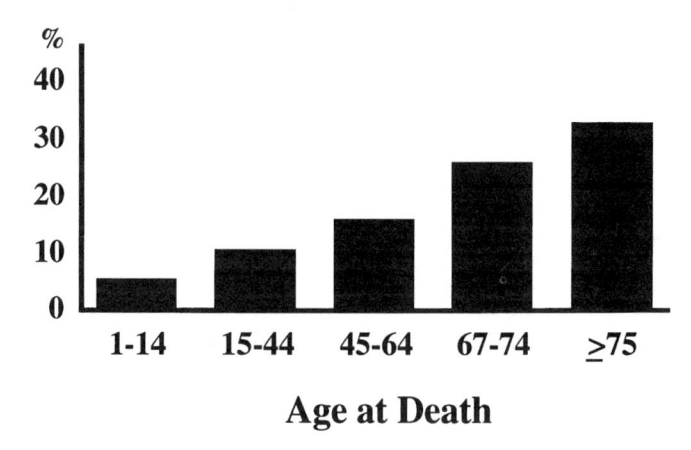

FIGURE 3.14. This figure suggests that diagnostic discrepancies become more frequent as patients age (Zarling, Sexton, & Milnor, 1983).

the successful studies involved use of computer guidelines and expert systems. In one study of diagnosis of abdominal pain, 40% experienced a missed diagnosis, often leading to a ruptured appendix. This rate was reduced to 5% by use of a computer guideline for diagnosing abdominal pain. In Lau and Warner's study (1992) of QI performance by a government professional review organization, nurses and their physician review teams, using traditional chart audit methods, missed 71% of problems having serious but avoidable poor outcomes. When provided with ILIAD, an expert system, these same nurses and physicians improved their QI diagnostic accuracy by 45%.

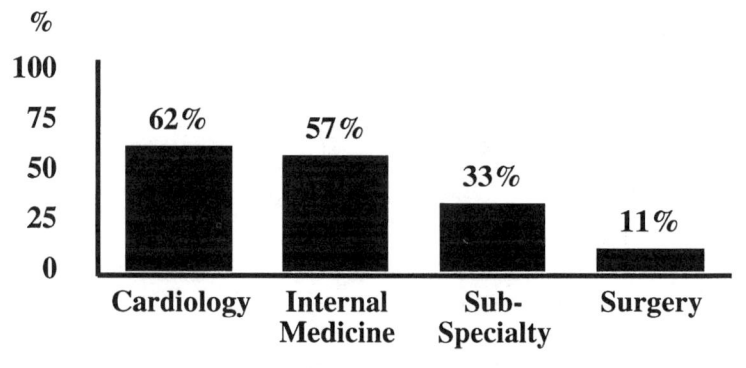

%
100
75   62%   57%
50
25   33%
                              11%
0
Cardiology Internal   Sub-   Surgery
            Medicine   Specialty

FIGURE 3.15. Autopsy evidence that diagnostic accuracy is domain specific. Cardiologists do best, followed closely by general internists. Subspecialists practicing outside the domain of their specialty are worse and surgeons worst of all (Rhee, 1975).

Decreased Dx Error Rate:

- Med students 32 to 14%
- Physicians   40 to 5%
- QI Staff   71 to 39%

FIGURE 3.16. Evidence of diagnostic improvement by use of computerized protocols and expert systems. Medical student diagnosis of general medical patients (Lincoln, Turner, Hau, Warner, Williamson, et al., 1991); physician diagnosis of abdominal pain (DeDombal, Leaper, Horrocks, Staniland, & McCamm, 1974); and Professional Review Organization (PRO) staff screening of patient-care errors (Lau & Warner, 1992).

Teleinformatics and Therapeutic Outcomes Improvement

A landmark study documenting therapeutic outcomes improvement through informatics comes from the LDS Hospital in Salt Lake City, Utah. This institution has an integrated system incorporating guidelines, protocols, and elements of expert systems. In 1989, the LDS Hospital added a guideline to prevent postsurgical deep-wound infections, a problem with disastrous health and economic impact. Elements of the simple guideline are shown in Figure 3.17, and the dramatic results of its use in Figure 3.18. The postsurgical deep-wound infections rate dropped to nearly one-fourth of the national norm; patients were discharged earlier, healthier, and more satisfied. At that one hospital alone, annual savings amounted to a third to a half of a million US dollars per year.

As shown in Figure 3.19, another study documents the results over time of physician compliance with computer alerts warning of potentially serious pharmaceutical errors, such as wrong dosing, using the wrong drug for the illness, or overlooking patient allergies to the drug prescribed (Gardner, Hulse, & Larsen, 1990). When computer alerts were first implemented, clinicians rebelled and became notably less compliant. Gradually, physi-

- **Start antibiotics 1-2 hours before surgery**
- **Stop antibiotics 24 hours after surgery**

FIGURE 3.17. Computerized guidelines for administration of prophylactic antibiotics to prevent postsurgical deep-wound infections in the LDS Hospital in Salt Lake City, Utah (Classen, Evans, Pestotnik, et al., 1992).

30 Networking for Healthcare Delivery

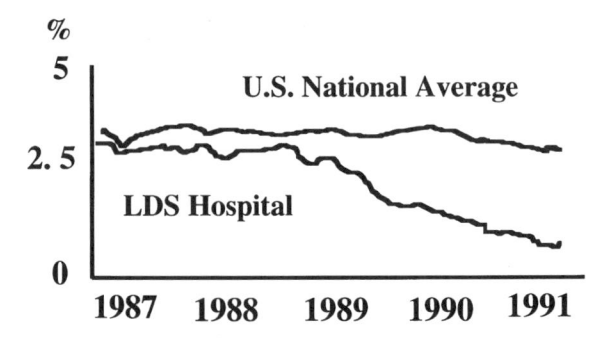

FIGURE 3.18. Measured impact of the computerized guideline for prophylactic antibiotics in surgery at the LDS Hospital (Classen, Evans, Pestotnik, et al., 1992).

cians learned that patient outcomes were improved substantially by adhering to these computer alerts. Today their adherence is close to 98%.

Figure 3.20 illustrates the improvement achieved in an Indian Health Service hospital in Arizona. A computer expert system was used to identify diabetic patients at high risk of lower extremity amputation. Home health care was provided, together with collaboration with a shoe manufacturer to provide custom shoes when needed to prevent foot lesions. The amputation rate was lowered by 90% of the baseline.

Figure 3.21 depicts one of the more dramatic examples of the use of computer protocols to improve therapeutic outcomes. A team of pulmonologists at the LDS Hospital in Salt Lake City evolved a protocol to improve survival rates among patients with Adult Respiratory Distress Syndrome (ARDS) (Henderson, East, Morris et al., 1989). The expected survival rate was doubled, compared to a US and world standard of 32%, reaching 62%.

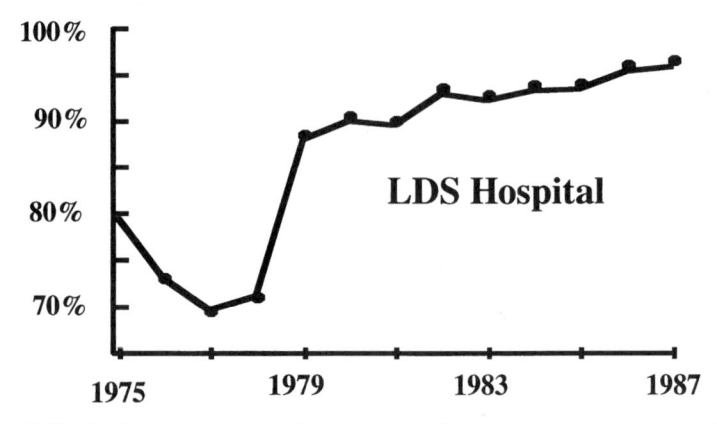

FIGURE 3.19. A 12-year saga to improve physician response to computerized pharmaceutic alerts (Gardner, Hulse, & Larsen, 1990).

3. Teleinformatics: The Future of Healthcare Quality Management   31

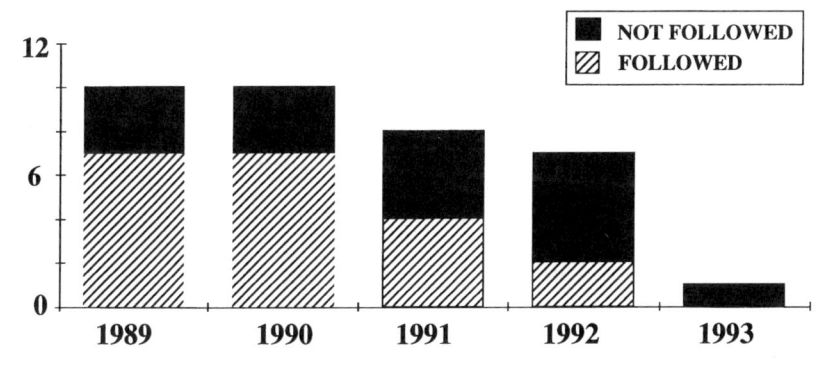

FIGURE 3.20. Impact of an electronic medical record system to identify and follow diabetic patients at risk of lower limb amputations in an Indian Health Service Medical Center (Yamada, 1993).

Future of Teleinformatics and QI

Planning is now underway for an international teleinformatics networking program, a partnership of Interwest Quality of Care, Inc., Pantheon Health Equity Corporation, and Lockheed-Martin Western Development Laboratory.

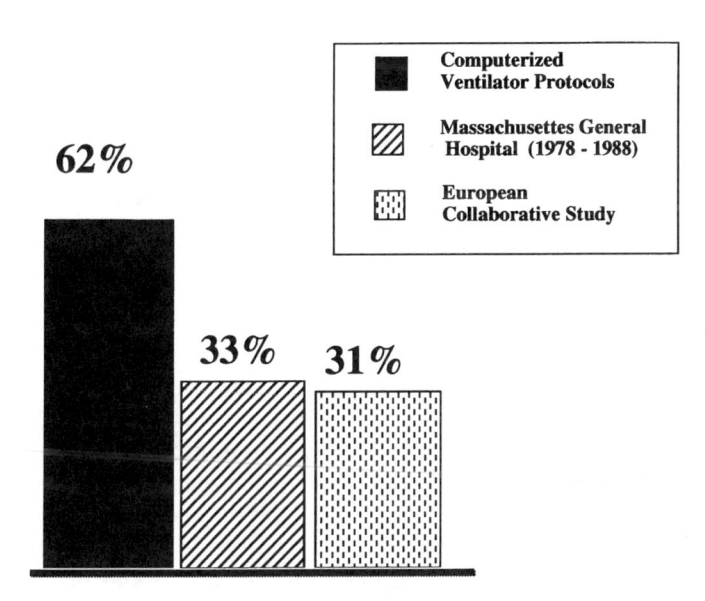

FIGURE 3.21. Impact of a computerized protocol on survival of Adult Respiratory Distress syndrome (ARDS) patients at the LDS Hospital in Salt Lake City, Utah (Henderson, East, Morris, et al., 1989).

Pantheon, a group of investment bankers, has played a key role in developing the overall vision for the program and will provide investment capital for the system. Lockheed-Martin Western Development Laboratory (formerly Loral), was the developer of the international communications network for the United States Department of Defense (DoD) and will develop the macrotelecommunications linkages, including needed hardware and software. The Lockheed group will also coordinate the overall management of the program, and be the guarantor of products and services. Interwest Quality of Care, the developer, will provide a quality improvement framework, a development laboratory, and technical support services. Interwest will also establish a health science information service for keeping up with the recent medical advances essential to the development and validation of guidelines, protocols, practice parameters, and other decision support technologies integral to the system.

The system will be funded by private capital and public funds when feasible. Pantheon has organized a consortium of investment bankers to help fund the effort and, with Lockheed-Martin Western Development Laboratory, will assume much of the management and technological responsibility. Plans call for an effort to utilize equity management financing to minimize the amount of upfront capitalization required by individual facilities or networks. Financing for the initial investment in hardware, software, and physical linkages will be repaid from a share of savings accrued based on cost avoidance and performance enhancement. Eventually, use of the system will be an ongoing cost of doing business, much like current expenditures for telephones or electric power utilities.

The vision of this enterprise is to develop a new public–private utility, to which healthcare providers and consumers can subscribe. The utility will provide a comprehensive system encompassing healthcare elements, quality improvement, education and training, and health services research. Using an "open architecture" base along with a vocabulary server interface technology, this network utility will adapt to most informatics legacy systems.

Program plans include *healthcare elements* for clinical, administrative, supply, and financial decision support. *Quality improvement* will include internal and external functions for ongoing systematic assessment and improvement-of-care performance. External QI will include meeting requirements of current healthcare regulations, monitoring, preapprovals, licensing, certification, and accreditation. *Education and training* will offer multiple activities at various levels (e.g., preprofessional, professional, graduate, and postgraduate) and will include embedded performance support services. *Health services research* will encompass the full range of applied research studies. Of these, the most immediately applicable are outcomes research and quality improvement research, the "applied of the applied" health sciences. This program will involve cooperation between

3. Teleinformatics: The Future of Healthcare Quality Management

public and private segments of participating healthcare systems. Perhaps the most important use of these research findings will be to facilitate healthcare outcomes management throughout the system.

The teleinformatics program will network the full range of healthcare stakeholders, including:

- Consumers and providers;
- Urban and rural facilities;
- Primary through quaternary sites;
- Providers, purchasers, and payers;
- Clinicians, educators, and researchers.

The core of the system will be a standard workstation consisting of a desktop computer, with a high-resolution TV monitor and TV camera, satellite uplink and downlink, and integrated software. Most software needed has been developed to at least an alpha or beta test level in the United States, and includes the following elements:

- The Decentralized Hospital Computer Program (DHCP) developed by the Department of Veterans Affairs;
- The Indian Health Service developed Patient Care Component (PCC);
- Embedded workstation user support (e.g., just-in-time program help or training modules);
- Integrated clinical, administrative, and financial data systems, including decision support modules;
- Telecommunications, telemedicine, and distance learning linkages.

The basic core software will be a variant of the DHCP, developed by the Department of Veterans Affairs; this software provides a standards-based architecture, facilitating linkages, and adaptation to the range of hardware and software at participating facilities. ILIAD and its consumer version Housecall are illustrative of expert systems to be incorporated.

Few commercial systems have a 16-year track record for development as does the DHCP system. Very few offer a standards-based software system integrating multiple care sites and multiple payment sources as does DHCP. Most importantly, DHCP is in the public domain. Hence, this comprehensive healthcare information system is available at a relatively small cost. For sites using DHCP, interface linkages can be developed to our proposed teleinformatics utility, which will include training, service, and maintenance.

Development will encompass users ranging from local demonstration sites to full installation sites in this country and abroad. The teleinformatics system will be used to integrate QI activities into a concurrent or "realtime" mode, providing healthcare clinicians, administrators, and consumers with the most recent "meta-validated" scientific information for immediate decision making. The informatics framework for local offices, institutions, or healthcare plans will include automated data collection to drive these decision support modules. A physician who is about to prescribe a potentially fatal drug will be instantly alerted. A QI team will be able to program an ongoing application for monitoring and managing referral patterns and costs.

With the new capabilities provided by the teleinformatics system, product-line cost accounting will yield data for cost-effectiveness and cost-benefit analyses. "Instant" information will support the development of more efficient practice management patterns. Newly validated guideline modules can instantly be made universally available to care providers in the United States and around the world. Regional servers can make customized training packages available for downloading when needed. Busy clinicians will have instant access to information for any given patient. For physicians, who spend over 30% of their practice time coping with paper, even a small reduction of this need will result in remarkable savings in time and dollars, reduced frustration levels, and improved job satisfaction.

This teleinformatics infrastructure will make QI activities almost totally unobtrusive. It will document these activities and link them directly to healthcare outcome measures. The infrastructure will be the future core of healthcare improvement in all its dimensions—the means for effecting healthcare reform based on documented facts of the costs and benefits of system enhancements as a whole.

The opportunity is now. Widespread integrated effort, supported by available funding, is now within reach. We are on the brink of harnessing existing teleinformatics technologies and laying the foundation for managed collaboration in the future. Only then will we experience the power and the benefits of healthcare quality improvement that visionaries have long predicted.

References

Bettmann OL. 1972. *A pictorial history of medicine.* Springfield, IL: Charles Thomas Publisher.

Brook RH, Kamberg CJ, Mayer-Oakes A, et al. 1989. *Appropriateness of acute medical care for the elderly: An analysis of the literature.* Santa Monica, CA: Rand Corporation.

Cabot RC. 1912. Diagnostic pitfalls identified during a study of 3000 autopsies. *Journal of the American Medical Association, 59,* 2295–2298.

Classen DC, Evans RS, Pestotnik SL, et al. 1992. The timing of prophylactic administration of antibiotics and the risk of surgical-wound infection. *New England Journal of Medicine, 326,* 281–286.

DeDombal FT, Leaper DJ, Horrocks JC, Staniland JR, & McCamm AP. 1974. Human and computer-aided diagnosis of abdominal pain: Further report with emphasis on performance of clinicians. *British Medical Journal, 1,* 376–380.

3. Teleinformatics: The Future of Healthcare Quality Management 35

Derning WE. 1986. *Out of the crisis.* Cambridge, MA: MIT-CAES.
Gardner RM, Hulse RK, & Larsen KG. 1990. Assessing the effectiveness of a computerized pharmacy systems. *Proceedings of the 14th Annual Symposium on Computer Applications in Medical Care,* pp. 668–672.
Garrison FH. 1914. *An introduction to the history of medicine with medical chronology; bibliographic data, and test questions.* Philadelphia, PA: WB Saunders and Company.
Harper RF. 1904. *The code of Hammurabi.* Chicago, IL: University of Chicago Press.
Henderson S, East TD, Morris AH, et al. 1989. Performance evaluation of computerized clinical protocols for management of arterial hypoxemia in ADDS patients. *Proceedings of the 13th Annual Symposium on Computer Applications in Medical Care,* pp. 583–587.
Hill RB, & Anderson RE. 1988. A model for the autopsy-based quality assessment of medical diagnostics. *Human Pathology, 21,* 174–181.
Lau LM, & Warner HR. 1992. Performance of a diagnostic expert system (Iliad) as a tool for quality assurance. *Proceedings of the 16th Annual Symposium on Computer Applications in a Medical Care,* pp. 1005–1010.
Lincoln MJ, Turner CW, Haug PJ, Warner HR, Williamson JW, et al. 1991. Iliad training enhances medical students' diagnostic skills. *Journal of Medical Systems, 15(1),* 93–109.
Rhee S-O. 1975. *Relative influence of specialty status, organization of office care and organization of hospital care on the quality of medical care; a multivariate analysis.* PhD dissertation, University of Michigan, Ann Arbor.
Williamson JW. 1971. Evaluating quality of patient care: a strategy relating outcome and process assessment. *Journal of the American Medical Association, 218,* 564–569.
Williamson JW. 1978. *Assessing and improving health care outcomes: The health accounting approach to quality assurance.* Cambridge, MA: Ballinger Publishing Co.
Williamson JW. 1991. Medical quality management systems in perspective (chap. 2). In JB Couch (ed): *Health Care Quality Management for the 21st Century.* Tampa, FL: Hillsboro Printing Co., pp. 23–72.
Williamson JW, Fehlauer CS, Gaiennie MA, et al. 1991. Assessing quality of ambulatory care. *Quality Assurance Utilization Review, 6,* 8–15.
Wu AW, Folkman CJ, McPhee SJ, et al. 1991. Do house officers learn from their mistakes? *Journal of the American Medical Association, 265,* 2089–2094.
Yamada W. 1993. *Reducing lower extremity amputations in diabetic patients.* PD HuHuKam Indian Health Service Hospital, AZ.
Zarling E, Sexton H, & Milnor P. 1983. Failure to diagnose acute myocardial infarction. *Journal of the American Medical Association, 250,* 1177–1181.

Section 2
System Design

Chapter 4
Creating a Robust Multi-Facility Healthcare Information System
Robert M. Kolodner 39

Chapter 5
Infrastructure and Tools for Flexibility
Catherine N. Pfeil 57

Chapter 6
Structuring Databases for Data Sharing
Cameron Schlehuber 81

Chapter 7
Collaborating for Integration
Steven A. Wagner 87

Chapter 8
Supporting Systems After Installation
Roy H. Swatzell, Jr. and Virginia S. Price 98

Chapter 9
Privacy and Security Protections for Healthcare Information
Gail Belles 116

4
Creating a Robust Multi-Facility Healthcare Information System

ROBERT M. KOLODNER

Evolution of a System

Introduction

July 4th holiday weekend, 1994. Of course! Events like this always occur on holidays or weekends. This crisis combined them both. A mechanical problem at the Department of Veterans Affairs (VA) Medical Center in Syracuse, New York, compromised the ability to pump water into the facility. Hospitals can operate without the electric company—they use emergency generators as backup—but they cannot operate without water. Thus, the medical center had to transfer all of its inpatients to other hospitals.

The VA healthcare information system was used for electronic transfer of administrative and clinical data in the Syracuse system to the receiving VA medical centers. On the paper orders that accompanied some of the patients, two patients were identified as having an antibiotic-resistant bacterial infection that required them to be placed in isolation. When the receiving staff reviewed the electronic records of the patients transferred, they were able to identify four additional patients who were also infected with the same antibiotic-resistant bacteria. Without the information obtained from these electronic records, cross-contamination could have occurred, with the infection being spread to other patients.

Healthcare information systems can and do make a difference. In this case, the ability of the VA information system to transmit reports among facilities was vital to preventing further threats to patient safety. Staff were able to provide patient care at the other VA medical centers until water could be restored at the Syracuse facility.

The Veterans Health Administration (VHA, often shortened simply to VA) is the portion of the Department of Veterans Affairs responsible for providing health care to eligible military veterans. In addition to delivering health care in its medical facilities, VHA provides training for a large number of healthcare professionals, conducts extensive healthcare research, and provides emergency healthcare backup for the Department of

Defense (DoD) in case of military action (Kizer, 1996). In 1982, VA established the Decentralized Hospital Computer Program (DHCP), operational throughout the VA healthcare system since 1985. This healthcare information system is one of the most extensive in the world, deployed throughout the Veterans Health Administration, which is the largest healthcare system under one management in the Western world. Moreover, DHCP forms the basis for information systems in almost all federal healthcare delivery systems (VA, DoD, and Indian Health Service [IHS]) and is used at numerous healthcare facilities in the United States and around the world.

The experiences gained, successful and unsuccessful, during the conceptualization, design, implementation, and evolution of the VA system, provide a rich source of information. This can benefit others charged with meeting the information needs of multi-facility healthcare delivery systems, wherever they might be located.

Achievements of DHCP

The individuals who started the efforts that led to DHCP originally focused on meeting the needs of staff at individual medical facilities, which at that time in the late 1970s were primarily, but not exclusively, hospitals. Initially there was minimal concentration on moving data up to national databases or on office automation, and no significant effort devoted to transferring patient information among facilities as patients moved. Within facilities, the emphasis was on helping each service unit to automate its processes to achieve efficiencies over the manual system, providing reports to service managers to oversee their operations, and generating a few simple reports for top management. Reengineering was not a part of the culture at that time, although a certain amount of change was inevitable as a consequence of the automation process.

Most importantly, because VA received a budget allocation from Congress and, at that time, did little billing for its healthcare services, emphasis was not placed on financial and billing systems, unlike the private sector. Thus, VA had the opportunity to concentrate on day-to-day clinical and administrative systems to a degree not possible in other settings. This early advantage eventually created a gap in the VA information systems, which had to be addressed when the changing healthcare scene required cost-accounting systems to manage more efficiently.

Decentralized Hospital Computer Program was the first of a new wave of healthcare information systems used in the major federal healthcare delivery systems. In 1986, DoD contracted with Science Applications International Corporation (SAIC) for its Composite Health Care System (CHCS). By modifying and extending DHCP modules to meet DoD's list of requirements, SAIC developed CHCS. Although both systems have evolved since

1986, CHCS's file management system, data dictionary, and application base retain similarities with DHCP's.

In 1984, the IHS updated its existing information system. In a manner more akin to the VA's than DoD's, the IHS adopted portions of DHCP and then had their own staff modify and extend its capability to meet IHS-specific needs. For example, they added tribal information and opted to use the patient name and local chart number as the primary look-up keys instead of the patient name and social security number as in DHCP. In addition, they added new modules supporting ambulatory care activities years before the VA began to consider such applications. To keep the two systems compatible, VA and IHS staff have stayed in close communication and often attend each other's planning and major development sessions. Recently, the two organizations signed a formal memorandum of understanding to facilitate their close working relationship.

Because VA's DHCP has been developed by federal employees, the software is in the public domain and available without charge to governmental organizations and for the cost of material duplication (tapes, disks, and documentation) to any other individual or organization. Despite the fact that VA has no mechanism for providing support or enhancements specifically for non-VA settings, other institutions, most of them in healthcare, have chosen to implement portions of DHCP to address their information needs. These include private and public hospitals, clinics, and public health departments in locations around the United States. Institutions in foreign countries, including Germany, Egypt, Finland, Nigeria, Columbia, and China, have also implemented DHCP as their primary healthcare information system. Recent changes facilitate translation of applications into languages other than English, but for some time, remote non-VA DHCP sites have displayed prompts and information in the native language of their country (including Arabic, which displays on the screen from right to left).

The accomplishments of DHCP were validated through an external and competitive process. In 1995, VA was one of three organizations chosen by the Computer-based Patient Record Institute (CPRI) for recognition of the quality and extent of its implementation of the computer-based patient record, namely DHCP. Thus, non-VA institutions using DHCP not only get a bargain price but also are able to use one of the best CPR systems in production.

The Genesis of DHCP

The activities that led to DHCP began in 1977. At that time, the Veterans Health Administration consisted of 172 hospitals plus an assortment of nursing homes, domicilliaries, and outpatient clinics. Most of the care provided was on an inpatient basis. Only a few VA hospitals had any computers on site, usually for pilot laboratory systems or for research projects. Instead, computer support for VA healthcare facilities was provided at five data processing centers around the country, running batch mode support on mainframe computers.

Joseph T. (Ted) O'Neill and Martin Johnson led the group planning the automation initiative for VA health care. The group was guided by a set of principles; a subset thereof formed the basis for DHCP. The durability and quality of the DHCP product serve as testimony to the visionary nature of their early work. They conceived of a highly integrated information system based on minicomputers located at individual VA hospitals and medical centers. They proceeded to develop DHCP in a manner consistent with their principles. The software would be written in a standards-based language and designed to be independent of hardware and operating system vendors. Software development efforts would pair users closely with the programmers, making users the ones to guide software development using a rapid prototyping development method. Major software changes would be delivered in a matter of hours or days, giving those involved powerful reinforcement of their energy and creativity. The system would be built in modules, organized vertically along departments, with data being entered once and then being available to any other module. At its core would be the patient registration system.

Many of us who participated had idealistic goals for our efforts. We wanted to create a public domain system that met system needs at relatively low cost. We wanted to design an information system capable of supporting the efforts of a large consortium of healthcare facilities to develop applications in parallel and then share the best applications among all the participants without crashing into each other's application—a truly cooperative development effort. Moreover, many of us recognized that the support and participation of the private sector were essential for long-term viability. We used the analogy of the razor industry: we hoped to build a razor, providing opportunities for other parties to provide the razor blades, that is, enhancements and add-ons.

As with many new efforts, change was resisted. Initially, opposition came from individuals invested in the existing automated data processing activities and infrastructure. A group of dedicated individuals, later referred to as the "Underground Railroad," worked to continue the information initiative that had been started in 1977. Finally, in 1982, the VA Administrator Robert Nimmo established DHCP after Congress instructed the VA to proceed with a decentralized automation strategy. Although the leadership had changed, and not all of the principles that O'Neill and Johnson had articulated were implemented in DHCP, enough of them were retained to launch a successful effort. These included the following:

- Local control and responsibility for computing resources;
- Local flexibility regarding software applications;

- Commitment to using information standards;
- Vendor independence for hardware and operating systems;
- Establishment of an integrated database;
- Close user involvement in software development;
- Modular design of nationally developed software.

Shortly after DHCP was established, VA faced a major decision. Would we build each application independently or would we base applications on a file management system (FileManager) that had been written during the intervening years and that used a data dictionary? By deciding to base its healthcare information system on a data dictionary approach for application and database design, VA made what turned out to be the correct choice on this critical design issue.

At the outset, VA identified a set of tools to provide a common user interface, which would complement the conventions defined for system interactions; that is, how the user would stop a process or obtain help at any prompt. An e-mail system was selected that had also been written at the start of Underground Railroad days.

Principles Making Evolution Possible

The continuing evolution of DHCP demonstrates the long-range impact of the principles that were articulated at its onset and that guided its implementation. The DHCP experience illustrates that such principles should be explicitly established for any new system being implemented as well as for any major reengineering of existing systems.

Principle #1: Local Control of Facility-Based Computing Resources

Today, the commitment to distributed processing with computers based in each facility appears to be a mundane decision rather than a principle. In the late 1970s, however, mainframe computers located at a central processing facility still reigned supreme. (Remember that at that time, personal computers were the concerns of a fringe group of individuals viewed with disdain by those who formed the mainstream of automation.) The model at that time was to implement large, mainframe-based computing systems supported by a centrally directed staff who planned, implemented, modified, and maintained the systems that supported the users. Although this was successful for a time in meeting the needs of many businesses, this model often led to the establishment of large Automated Data Processing (ADP) staffs only minimally responsive to the needs of the people they were meant to support (Von Simson, 1990), particularly if the users were operating in a business, such as medicine, where the environment changes quickly.

Indeed, VA had been using such a centralized model. In the early 1980s, VA's top healthcare management—both at the corporate office (known as the VA Central Office) and at local medical facilities—knew that their needs were not being met. Despite assurances from central ADP staff that they would move rapidly to provide automated support to numerous VA sites using their five data processing centers and that they would have such support in place by the early 1990s, the medical leadership knew that the centralized ADP model could not meet the VA's long-term needs. And that was before the realization of just how fast changes in medical care delivery were going to accelerate.

Thus, a central principle at the heart of DHCP, as embodied in the very name itself ("Decentralized"), is that local facilities have their own computing resources. These resources come with the stipulation that the local facility meet a limited set of systemwide requirements. Beyond these requirements, the computing resources can be used to address locally determined needs. Responsibility and control of the DHCP system at each site are in the hands of local management.

Because no new staff were provided with the centrally purchased equipment, ADP staff were hired by local management from the local operating budget. Staffing levels were determined at each facility, based on local resources and on the perceived local benefits. Some facilities chose to provide more staff to implement, run, and modify their DHCP systems, while others hired fewer. Facilities could also supplement the initial central purchase of terminals and printers by using local funds. Again, some facilities chose to provide a robust computing environment for their staff, while others stayed with their initial allotment of ADP equipment. This led to an unevenness in the degree to which DHCP was—and still is—implemented and used across the system.

On the other hand, this principle, which was an early instance of "empowering" lower levels in an organization, led to an outpouring of creativity and innovation by management, ADP staff, and end users that fueled the growth, expansion, and evolution of DHCP to the benefit of VA and to all the agencies and organizations that use DHCP or its variants.

This empowerment also was responsible for VA's successful implementation of the first three components of DHCP over 3 years and in 169 sites. (Congress had instructed VA to implement commercial systems in three sites while installing DHCP in the rest.) In addition to the challenge of packaging and deploying the national software packages, this implementation required local sites to redirect or to hire ADP staff, plan for their system implementation, complete the installation, establish an application support infrastructure in each service, and train staff to use the system.

Local control of computing resources energized VA staff around the country to embrace DHCP and to ensure that the systems would meet their locally determined needs. Although facility managers embraced the system to varying degrees, resulting in an unevenness of resources being devoted to DHCP, the overall result was a more rapid implementation of DHCP throughout the VA system and a more complete commitment by VA staff to using DHCP than could have been accomplished through a centrally directed strategy.

Principle #2: Local Software Selection

From the start, applications in DHCP have been designed to be modular. The "Kernel," a core set of applications, handles the user environment, provides database tools and security features, and adapts the software to operating systems from different vendors. For patient-care applications, a core set of patient-related applications is also needed, to register the patient and to indicate the locations where treatment might be provided. In addition, VA has mandated that certain applications (such as Laboratory, Pharmacy, and Prosthetics) should be implemented at all sites due to their strategic importance for accomplishing the VA mission.

Beyond the core and mandated software, the remainder of applications (numbering more than 70 after 15 years) can be installed or not, depending upon the needs of the local facility. Thus, a site that provides no cancer treatment may choose not to install the Oncology application. Another site may decide not to implement the Order Entry application. Yet a third may not implement Social Work. The decision regarding the number and mix of applications is a local one.

This principle has advantages and disadvantages. It has served VA well during a time when (1) care was facility-based, (2) information systems were focused on meeting the needs of staffs and patients at individual facilities, and (3) all software could not be implemented and maintained at every site due to the relatively low levels of investment in information systems. This modular design has allowed facilities to pick and choose the applications most appropriate to their local needs. Some sites have had sufficient resources, in both hardware and ADP staff, to implement most of the modules, but most have chosen selective installation of nationally released application modules.

This principle has two marked disadvantages. First, the computing environment becomes less predictable at each site, because a site may or may not implement the nonmandatory modules. As applications become more interdependent, this can create problems in lost functionality or potential software errors if one module is dependent upon functions present in another module that may or may not have been installed. Over time, VA has had to increase the number of mandated modules to ensure that functions important to operating as a unified national healthcare system are supported at every site, especially as more and more care delivery is provided across VA facilities rather than within an individual one. This has reduced some of the local discretion, but occurs at a time when computing resources are more plentiful, so the resource restriction is less a factor than it was in the early implementation years.

The second disadvantage is the increased workload imposed on those responsible for running and maintaining the system at local facilities. The DHCP modules are updated independently, in a continuous stream of releases and software patches, rather than in one or two major releases each year. (The latter is the case with some commercial systems, which have all of their applications shipped and installed as a single unit.) As a result, information resources management staff at each facility have to install the updates on an ongoing basis rather than a few discrete times each year. On more than one occasion, several interrelated modules with major upgrades had to be installed simultaneously, demanding significant effort by local and national support staff.

Modules were initially designed along the lines of the disciplines or services in the medical centers—Medical Administration, Laboratory, Pharmacy, Mental Health, Social Work, Nursing, Medicine, Surgery, Fiscal, etc. This allowed deployment of DHCP over a very short time frame and helped clinical support services with the need to manage operations efficiently. As modules were brought online, the need for more integration across services became apparent.

Over time, user groups changed in membership and focus from a departmental orientation to a more interdisciplinary approach (see Principle #7 below). More interdisciplinary modules were identified and developed, including Order Entry, Discharge Summary, and Health Summary. The most popular clinical application, Health Summary, consolidates all relevant clinical information into a user-definable template that can be printed or retrieved online and even obtained from remote VA sites when a patient is treated at more than one VA facility. In fact, Health Summary was the application used during the Syracuse incident described at the beginning of this chapter. Today, essentially all new applications have an interdisciplinary focus, even when they are meeting primarily the needs of a specific department.

This principle has decreased in importance more than any other principle outlined in this chapter. It was an important factor in the first several years of DHCP implementation when the computing resources were scarce and choices had to be made to address the most critical local needs. Whether vertically oriented or horizontally integrated, the ability to select the modules that fit the local needs provided flexibility to local managers to ensure that the resources they devote to automation can be applied to those areas that yield the most return. However, the value of this flexibility decreases considerably as computing capacity increases, as applications become more interdependent, and as patterns of care change from facility-based care to the new network-based approach to healthcare delivery, leading to more and more application modules being designated as mandatory. Principles critical to the success of information systems at one point in their lifecycle may need to be reexamined at later points in time to assess whether they retain their value or should be replaced.

Principle #3: Software Flexibility—"Configuring, Modifying, and Extending"

Providing locally controlled hardware resources and choices among the software modules are two steps toward enabling local facilities to meet their needs. The third principle is comprised of three intimately related components, which together provide an essential enhancement to the modularity of DHCP software. These components provide local facilities with the vital capabilities to (1) configure national software, (2) modify national software, and (3) extend the breadth of software to meet the local needs. Without these capabilities, facilities would simply have been implementing rigid national software, even if the computing resources were distributed, and the "Decentralized" in DHCP would have been a phrase without significant substance. Moreover, without configurability, the cost of automating the VA would have increased dramatically, so this principle represents one of two design principles that helped make DHCP so cost-effective.

Veterans Administration healthcare facilities cover a wide range of sizes and functions. Facility types include hospitals, clinics, nursing homes, and domicilliaries. Even within a facility type, architectures vary from a single tower facility to a campus with over 30 buildings. Likewise, size ranges are enormous—from a 68-bed hospital in Bonham, Texas, to over a 1000-bed hospital in West Los Angeles.

The VA software was designed from its inception to be portable across sites with minimum effort and cost. The ability to configure the nationally developed software allows a single application module to meet the needs of a widely diverse range of facilities. One admission package can be used in every VA facility, because each site is able to tailor the software to encompass the variety of building arrangements, ward configurations, and room layouts, greatly reducing the initial installation and maintenance costs. When a new facility is activated or an existing one is modified, the site can make the necessary modifications in the tables to ensure that the software meets their needs.

Each site can easily "customize" the software to meet local needs instead of being forced to change the way they do business to meet a rigid, noncompromising computer program. Configuration of the software application does not require programming skills, and is often performed by a staff-level "application coordinator" chosen from within each department or within a facility (for interdisciplinary applications). By actively participating in the installation of the software in the mid-1980s, departmental staff felt a sense of ownership and pride in their implementation of DHCP, which facilitated its rapid implementation throughout VA.

This approach contrasted with implementations at non-VA institutions during that same timeframe where the software was often customized to the developing site, making portability to new sites very difficult, or where the software had to be modified by the vendor at an additional cost to adjust for the parameters necessary at each new site.

Local configurability reduced overhead at the site by decreasing the number and expertise of the staff necessary to implement DHCP and tailor the applications to the local needs, but it would not be sufficient to ensure that local needs would be met. This power is achieved by allowing sites to modify national code and by providing a framework to extend the breadth of applications in a systematic fashion. One is controversial; the other was ingenious.

The ability to modify national code is a mixed blessing, and VA has benefitted and suffered because of this capability. Over the years, we have learned to set boundaries on this activity to ensure that essential system requirements are not sacrificed, but we have continued to allow local modifications within those boundaries because of the benefit that this flexibility has provided to local sites.

With over 170 "customer sites" within VA, resources are not available to provide all the locally desired features. Priorities are set by targeting features that fulfill the needs of the largest number of sites or provide the functions necessary for VA to operate as a healthcare system. This means that there are always unmet local features within national software. Sites can either develop entirely new applications, using the extensibility capability described below, or they can make relatively modest modifications to national software.

Modifying national code is fraught with potential pitfalls, which each site must consider, because applications are interlinked and changes in one application can cause unforeseen problems in others. The source code for nationally distributed applications is provided to all sites, and the responsibility for the initiation and maintenance of modifications rests with the local facility. Sites can get into trouble through these modifications in one of several ways, and the VA has implemented policies to address these concerns.

For example, changes can be made that are not properly documented. Policy requires each site to maintain proper documentation of local modifications, and this is reviewed by the Medical Information Security Service during their periodic visits and by the Inspector General's (IG) Office as needed during episodic audits and investigations.

Changes can be so extensive that it is too time-consuming to make similar changes in subsequent versions, and the site falls behind implementing the updates of the applications. Moreover, changes can be made by staff who leave, without anyone on staff being capable of maintaining the changes. Sites are required to run a subset of modules that are deemed essential from a VA healthcare system viewpoint. New versions must be installed within a limited period of time. For nonmandatory software, older versions may no longer be supported after 6 months. This provides additional incentive to keep the changes manageable, because sites usually do not want to be left without any support for an application. At times, sites have reverted back to the standard national release, giving up their local features, because of their inability to sustain the local modifications.

Changes can be made that have unforeseen consequences in other applications, eventually threatening disaster in the system. This is the hardest to prevent and the most dangerous. If sites were to act irresponsibly and adverse impacts were to occur on patient care due to local modifications, then VA would have to prohibit such local modifications. Fortunately, such disasters are rare and have not resulted in any patient harm. With safeguards in place, the potential for such an adverse outcome remains remote. One boundary that has been set is that VA has prohibited any changes being made to a circumscribed set of applications. Applications in this set are audited by VA Security and IG staff.

In contrast to modifying nationally developed software, extensibility is the ability of a site to write new applications to meet their needs. From the beginning, VA provided a framework for local innovation by establishing a unique application naming and a database storage assignment for each site. Every site can develop new applications with the assurance that no national software would ever conflict with the local applications and data storage (although national software might overlap and replace the function over time).

In fact, any site that uses the assignments and adheres to a set of guidelines can swap locally written software with any other site and be assured that each will not "crash" into the other. Thus, the framework exists for a consortium of all participating sites, within and outside VA, to be able to leverage the skills, talents, and innovations of the programmers at all participating sites. In addition, VA has applied a process of continuous improvement to enhance the benefits from this principle. Over time, VA identified which characteristics of applications assured reliability and lowered maintenance costs and promulgated programming conventions that supported local extensibility and software exchange among sites.

Although sites are not required to develop software or to use locally developed software from other sites, DHCP provides local management with the opportunity to do so. Local choices help meet local needs and allow local control over resource allocation. To date, some sites have developed an extensive array of software. Some sites have opted not to develop any software but to use software developed at other local facilities. And some sites have chosen to run only nationally distributed and supported software on their DHCP implementation. As a result, some locally developed software has been shared among sites throughout VA; other locally developed applications have become nationally distributed and supported applications.

Software has been the driving force that delivers the benefits of automation. The DHCP principle of providing local flexibility regarding software has enabled facilities to meet their unique needs and address concerns that are not yet supported through nationally provided applications. *This flexibility has fostered the commitment of VA sites to DHCP, and stimulated and tapped the creativity and innovativeness of a broad base of personnel, in and out of the VA. All in all, it has been a critical factor in nurturing the evolution and vitality of DHCP throughout its existence.*

Principle #4: Standards-Based Development

The next two principles are closely linked, in that the adherence to standards provides part of the basis for vendor independence, which is Principle #5. From its roots in the late 1970s, DHCP was based on standards, either actual or de facto. Adherence to standards started with the use of a standard computer language as the basis of VA's healthcare automation activities. The language selected in 1977 was MUMPS (now known as M), chosen because it was one of the few computer languages to meet requirements of the American National Standards (ANS) Institute. It was also the language best suited for managing the text-based information so prevalent in health care.

Except for some medical coding schemes, such as the International Classification of Diseases (ICD) from the World Health Organization and the Current Procedural Terminology (CPT) from the American Medical Association, few other information standards existed in the late 1970s and the early 1980s. As more standards have been developed, VA staff have often been active members on the responsible committees, and VA has incorporated these standards (e.g., transport protocols and information structuring) as quickly as possible. Thus, DHCP software now uses Simple Mail Transport Protocol and X.440 for exchanging e-mail, Health Level Seven (HL7) for transferring demographic and clinical information, and National Library of Medicine's Unified Medical Language System (UMLS) for integrating medical concepts and terminology.

Work is underway by VA to explore such new areas as the Internet for use in future versions of DHCP and its successor systems. Today VA is often seen as a potential testbed for new standards, and VA staff are working with such agencies and organizations as the National Library of Medicine, Centers for Disease Control and Prevention, National Committee for Vital and Health Statistics, and American Society of Testing Materials (ASTM) to identify and implement pilot tests.

This commitment to standards works in conjunction with the next principle to help DHCP accomplish portability across a wide range of computer systems.

Principle #5: Vendor Independence

Probably the most important principle responsible for the cost-effectiveness of DHCP has been the principle of vendor independence.

This was possible from the start because of the choice of an ANS standard computer language, at a time when most languages were either proprietary or had dialects and variations across hardware manufacturers.

Hardware independence also depends on how an ANS computer language is implemented and supported by different vendors and on what proprietary extensions, if any, are used in the application code that is written. To sustain the hardware independence, VA took two actions. First, VA staff wrote a "shell," which insulates the VA applications from different implementations of MUMPS language. This "shell" selects a small set of applications, identified when the system was first installed, that are unique to the particular vendor that is being used. These are the only applications that need to be changed if the system is moved to a different MUMPS vendor's system.

Second, VA established guidelines and conventions—essentially its own internal standards—that precluded the use of vendor-dependent functions in application code that is intended for national distribution. The result is that DHCP software runs without change across a wide array of hardware platforms (from PCs to minicomputers and mainframes) and operating systems (MS-DOS and Mac OS to UNIX, DEC VMS, and IBM MVS).

Refusing to invoke vendor-specific functions and staying with the ANS standard can involve trade-offs. Sometimes a proprietary function or extension to the MUMPS language can offer performance increases. The decision, best made during initial system design, entails several issues. What are the benefits of the performance increases made possible by proprietary enhancements? What are the savings associated over time with staying with the ANS standard? Are these savings outweighed by the loss of performance or by the increased initial cost of hardware to attain comparable performance?

By maintaining vendor independence, VA has achieved portability of DHCP across a variety of hardware platforms and operating systems. As a result, VA has been able to purchase new upgrades for its computer system based on low price and high performance, without having any significant migration costs. Because a wide array of choices can meet its needs, VA has been able to obtain competitive bids from a variety of vendors. On each major acquisition, this has saved tens of millions of dollars as compared to the price the VA would have had to pay if it were "locked in" to a specific hardware platform.

Thus, VA has been able to migrate DHCP across several platforms as technology has improved the performance at ever decreasing cost. Specifically, VA moved DHCP from Digital Equipment Corporation (DEC) 11/44 minicomputers to VAX minicomputers to clustered PCs to DEC Alpha systems. Each time, VA obtained much more powerful equipment at lower initial cost and much lower maintenance costs. Although DEC has won the majority of the bids so far, the selection was made on price and performance and could have gone to any other vendor providing a bid that included the best value. Each transition was made with minimal change to the DHCP programs.

In terms of effort and cost, migration of DHCP is in marked contrast to migrations of other systems to new hardware platforms. Often, hardware-specific or operating system-specific extensions incorporated throughout the applications delay migration from one platform to another. They may also entail added charges for modifications needed to accomplish the shift.

The combination of adherence to standards and vendor independence has resulted in DHCP being a highly portable system. For the VA, the payoff has been the ability to keep costs down over the life of the system while taking full advantage of the rapid changes in the price/performance characteristics of new technology. The trade-offs in performance have been small enough that they have been significantly outweighed by the more powerful systems that VA has been able to afford due to more competition in the bidding process because these principles were followed.

Principle #6: Integrated Database with a Data Dictionary

One of the more technical principles implemented in DHCP is the creation of an integrated database. All critical data resides in the same database in the DHCP system. Thus, once data are entered, they are available for use by any other application, provided privacy, confidentiality, or security constraints are not violated. The capability to reuse data once captured eliminates the need to reenter full name, demographic information, etc.

Once a patient is registered in DHCP, information does not need to be reentered to order an x-ray or a laboratory test for the patient. Moreover, the physician—internist, pathologist, surgeon, or radiologist—can have full access to pharmacy information or radiology reports (and at some sites, the x-ray images themselves) while reviewing and interpreting test results because the information resides on the same information system. Access is not dependent upon interfacing (connecting) different computer systems together.

The data are stored within a generalized database that is indexed with a "data dictionary," which defines characteristics for each data element. The data dictionary can include capabilities such as checking for acceptable ranges or formatting to prevent incorrect data from being stored in the database. One example of a significant cost avoidance as a result of this design is the establishment of a standardized method for representing dates in the VA database. The DHCP date field has standard methods for detecting a variety of ways that a date can be represented when typed in or printed out (7/4/93 or 9-17-1994 or January 4, 1995 or "t" to represent today's date). Because the date field has a century indicator, there will be no problem caused by the change in millennium for any portion of DHCP that might still be in use at that time. The database allows nonprogrammers to generate reports using report writing tools provided as integral parts of the system. Although common today, this was unusual for large systems in the early 1980s. Commonly, vendors needed to write new applications to generate ad hoc queries of the data residing in their databases.

Integration of the database results in a smoother interaction for the user across applications. Use of the data dictionary results in more consistent and reliable data and, when combined with general tools for report writing, can provide users with a technical means to meet many of their ad hoc data reporting needs without having to depend on programmers.

Principle #7: User-Directed Software Development— "Bottom-up Specifications"

The last principle is one that probably accounts most for the viability of DHCP over more than 15 years of use and evolution. The involvement of the users in the design and development of VA healthcare software has been a hallmark since the late 1970s. At that time, developers and end users were co-located at a medical facility to design and implement the first application modules, and all software development took place in such an environment. In a few instances, parallel development efforts occurred at different medical centers, resulting in similar applications with different approaches and sets of functions. When DHCP was established, it became apparent that the procedures at each VA medical center were different enough that relying on users from a single site could result in software that was uniquely tailored to that site. Such software might have to undergo significant modification to meet the needs of other sites.

To balance the site-specific focus that might result from development at a single site, Special Interest Users Groups (SIUGs) were formed comprised of users from several VA sites across the country and from several levels of responsibility—staff, managers, and directors—within a given department. These groups met regularly to define the functionality needed, test prototype code, debug software prior to or concurrent with its implementation at an alpha test site, review drafts of user and installation manuals, and design educational materials and training for application coordinators and users for use in the nationwide implementation. By having representatives from several sites participate, developers learned where they needed to provide site-selectable modifications that would allow facilities to easily configure applications to conform to their individual work processes, as described above.

Today, some healthcare organizations deal with heterogeneity among their facilities either by undertaking a business process reengineering effort as part of automation or by imposing a uniformity from the top down. Ten years ago, these were not options in the VA system, due both to the culture of the system, which fiercely defended facility independence, and to the wide variation in facility size and mission among VA facilities. In fact, it is not clear that such a strategy would be successful in the VA even now.

As the VA development effort grew over time, the size, number, complexity, and interrelatedness of the modules increased. Gradually, more management was instituted to oversee the development, and small groups of developers coalesced into six and later seven major development sites. These sites grew to a size where they could no longer be accommodated within existing space on the campuses of VA medical centers, and most moved off-site into commercial space. This created a gap between the development staff and their users. It was easy for developers to understand their users' needs when they could walk down the hall and see them using the computer software that had been written only hours before. It was harder to maintain this "connection" with the end user once the development efforts moved away from the medical facilities.

Thus, a trade-off was made. Better management of the development resources, increased ability to coordinate among development efforts, and an improvement in the technical quality, maintainability, and documentation of the application came at the price of increasing the distance between the developers and the environment they were hired to support, and lengthening the time between the identification of user needs and the creation of applications. Special Interest User Groups were one of the critical factors ensuring that the developers remained in touch with their users' needs and that the products they created would be adaptable to the variety of setting and business processes present in the VA organization.

Each SIUG consisted of representatives from the same or related departments and disciplines. While this was efficient because it allowed VA to rapidly develop and deploy the core computer system throughout the VA system, the "stovepipe" nature of the application modules soon became a limiting factor. The problem arose because data from applications were usually needed by staff from a wide variety of departments.

Consider pharmacy applications, for instance. Pharmacy departments needed a computer system to facilitate filling prescriptions and tracking inventory. At the same time, front-line clinicians such as physicians, nurses, and physician assistants needed to have prescription information in a way that facilitated caring for the patients and documenting that care. Identifying each medication by delivery form (e.g., pill, capsule, and intravenous) and by size (e.g., 125 mg or 250 mg) might serve a pharmacist well, but a physician documenting an allergy wants to select "penicillin" rather than review all sizes and dosage forms of penicillin stocked by the pharmacy. Similarly, because some ways of storing the data in the database might be more efficient for filling and refilling prescriptions, developers could work with the pharmacists to optimize their application by using these database structures. However, some of these structures might make it impossible to meet the need for physicians or nurses to have clinical decision support applications available. Unless the spectrum of needs for an application or database is considered at the outset, the resulting design may meet only a subset of necessary uses.

Recognizing the limitations of the vertical approach, staff began to discuss the possibility of establishing interdisciplinary teams to address clinical software issues as early as 1985. In 1988, the first interdisciplinary team was formed. In 1991, the SIUGs were replaced with a structure of interdisciplinary groups. Like the SIUGs, these groups were charged with the planning of new computer system functionality and the evolution of existing applications. They differed, however, in that they focused on meeting the needs of multiple disciplines, not of single departments. Today, virtually all computer system planning is conducted by interdisciplinary teams, whether the applications will be purchased or designed and developed. This has resulted in more robust applications that meet a broader base of user needs.

To date, VA has relied on active user participation in the identification of computer system needs and in the creation of applications developed within the VA. This has allowed VA to harness the creativity and talents of staff throughout the country to develop and improve DHCP. The specific mechanisms for obtaining user input have continuously evolved to improve the quality of the resulting applications and their ability to meet the needs of a broader base of VA users.

The Future

The principles described above have served VA well in its efforts to support the delivery of health care to veterans. However, information systems must continue to change if they are to be successful, and the principles underlying the systems need to evolve as technology progresses and the very nature of healthcare delivery systems undergoes a massive transformation. In this process, the principles serve as the guiding force to ensure that the information system achieves its goals in a manner that produces a robust and durable product that can respond to this changing environment. Principles also change more slowly than the information systems themselves, which are subject to the impact of rapidly shifting technologies. Future principles for VA information systems might include the role of reengineering at the time of automation, the integral positioning of security features throughout all systems, or the inclusion of veterans and their significant others as direct users.

Over the past 15 years, DHCP has grown to over 70 applications and has evolved to include extensive linkages with commercial products and inclusion of development in multiple computer languages. Such changes must continue at an even more rapid pace if the VA information system is to meet the needs of the Department to operate competitively in the ever-changing healthcare market. Thus, in the near future, VA is planning to incorporate such technologies as client–server architecture, graphic user interfaces, Internet access and tools, and wireless devices for clinicians to replace the "dumb terminal" architecture of DHCP. Although the external appearances of future systems may be different, at the heart of the effort will be a set of principles that may be similar to those that formed the underpinnings for DHCP in 1982.

These principles will guide VA in a direction that will ensure that future information systems continue to respond to, and even anticipate, the needs of healthcare staff and providers so that VA can achieve its mission to be a comprehensive, integrated health service serving the needs of America's veterans by providing specialized care for service connected conditions, primary care, and necessary support services.

References

Kizer KW. 1996. Transforming the veterans health care system—the "new VA." *Journal of the American Medical Association, 275(14),* 1069.

Von Simson EM. 1990 (July–August). The "centrally decentralized" IS organization. *Harvard Business Review, 68(4),* 158–162.

5
Infrastructure and Tools for Flexibility

CATHERINE N. PFEIL

Attributes of DHCP

From the beginning, development of the Decentralized Hospital Computer Program (DHCP) focused on several tenets. The system had to provide flexibility to meet the needs of information management for all services and users within the medical-care environment. It had to support the uniqueness of each VA healthcare facility (Pfeil, 1992). And it had to provide an environment for users to perform diverse sets of tasks. The same system was expected to provide flexibility for a development team geographically distributed across the country. These attributes have made the DHCP system a pioneer in the field of medical information processing, as defined by Dick and Steen (1991).

Within the Veterans' Health Administration (VHA), the system had to promote order among conflicting needs and create a consistent environment for users who worked with multiple software products. Software products had to interact with one another and with the underlying operating system and hardware in a uniform manner. Only then could a modest Information Resources Management (IRM) staff manage a universe that would eventually encompass virtually every work need of each employee. Development had to be regimented to the extent that loosely coordinated development teams could benefit from each other's products while ensuring that each product would fit neatly into the full fabric of the system.

Reconciling the needs of a runtime environment for users with the applications development environment needs of over 150 developers is the hallmark of the DHCP infrastructure. As it has evolved over the past decade, the following attributes have ensured its success.

Vendor Independence

Independence from vendor-specific implementations of the M language and from the unique features of computer hardware (e.g., CPUs, disks, CRTs, and printers) has been central to the infrastructure strategy. It has promoted an environment where all applications rely on the DHCP Kernel as the platform. This single strategy has allowed the VA to migrate its information system across a range of computer systems, taking advantage of improved processor speed and advancements without modifying application code. Avoiding systemwide reprogramming has saved the VA millions of dollars while allowing VA to take advantage of emerging hardware technology (Ivers et al., 1983; Montgomery, 1993; Munnecke, 1981; Munnecke & Kuhn, 1989; von Blanckenesee, 1989, 1991).

Adherence to Standards

Adoption of standards, both external and internal to the VA, has positioned DHCP to interact with external systems. Among the standards adopted, is the programming language M, recognized by three major standards bodies (ANSI, FIPS, and ISO) (M Technology Association, 1994). The adoption of M has been supportive of the successful migration of DHCP to multiple hardware platforms over the years. To ensure connectivity among VA systems and with external systems, the e-mail system is fully compliant with standards for message exchange. Standards are also used for data, including medical coding schemes (e.g., ICD-9 and CPT). Exchange of clinical data is promoted through the use of standards such as Health Level Seven (HL7). Financial transactions use the X.12 standard for electronic data interchange.

Consistent User Environment

Standardization of the user environment is achieved through the infrastructure. Applications are provided with a rich set of utilities to present data to and solicit input from users. Whether being prompted for a date or viewing the complexities of an entire patient record, the user is guaranteed a uniform view of DHCP. Online help is available via the same mechanisms across all applications. This consistency makes the system easier to support, simplifies training, and increases confidence because the user's expectations are consistently met.

Powerful Developer Environment

The creation and continued evolution of a rich developer environment serve to maximize productivity, promote reusability, ensure consistency, and maintain portability. A database management system, VA FileMan, has evolved, which incorporates features of hierarchical, relational, and object-oriented systems. The infrastructure provides tools for building menu trees of user options and screen-oriented lists of user actions, printing consistency, e-mail (both as a user and developer tool), scheduling of background tasks, and security. To ensure quality, the environment provides utilities to aid in testing software and online documentation to meet user needs as products evolve. Tools for editing, debugging, and distributing software products provide the developer with an environment that is independent of the underlying hardware and operating system.

Focus on Integration

Integration of data within DHCP is one of its strongest features. Redundancy of data is discouraged and allowed in only rare instances. As a result, there is only ONE patient file. All applications that refer to patients reference that file. This level of integration is supported through the VA FileMan, application programmer interfaces (APIs), and integration agreements. The single instantiation of each data element avoids the dangers of data conflicts and minimizes issues surrounding internal interfaces.

Segregation of Effort

The infrastructure segregates national software and data from those that are local. This allows innovation to flourish at individual VA facilities without risk of conflict with national priorities. The local facility "owns" its information system and can augment it with locally developed software. Using the name space and file number range assigned by the database administrator, the system infrastructure extends a full development environment to the facility while protecting national software and data from potential corruption. Thus, the local site is empowered to solve local information management problems, often resulting in products that are migrated to other facilities or evolve to become national products.

Security

The infrastructure provides security for the DHCP system. User-access privileges, auditing, and data privacy are integral to all infrastructure tools. The sensitivity of data within the medical-care setting necessitates the ability to restrict access as well as to audit allowable access. The infrastructure provides the security environment to grant access to commonly shared features such as e-mail, functional needs such as the tasks of processing prescriptions, and unique privileges granted to limited users such as the ability to authorize the dispensing of narcotics.

Connectivity

The DHCP's e-mail system offers connectivity and some unique features. Rather than using the disjointed, single-response methods used by most mail systems, it supports mail within a facility using fully threaded message chains, where each response is attached to the original message. The

MailMan utility provides the electronic link between VA facilities, serving both as a communication tool for linking users together and as a development tool for passing data.

Involvement

Involvement in the evolution of DHCP is extended to all. The infrastructure provides tools allowing any user to design files, build options, print reports, allocate privileges where authorized, and exchange ideas via e-mail. Begun as a grass roots effort, DHCP continues to derive its growth from these seeds. Support for this unique environment is facilitated by the infrastructure. It lets representatives from the various service areas within a hospital assume ownership of their applications. And through this ownership, the IRM service can extend selected system management tasks to those most directly affected by each application.

Working within their respective services, such as pharmacy or laboratory, automated data processing application coordinators (ADPACs) effectively extend IRM staff expertise into those areas. This skilled and informed body of users then provides feedback to all developers resulting in improvements to the software products, which in turn serves the other members of their service. Thus, everyone contributes to the evolution of DHCP.

Infrastructure That Facilitates Flexibility

The DHCP infrastructure is not an accident or the outcome of simple luck. It has been carefully designed to meet its multiple roles. Although the word infrastructure may suggest a rigid and fixed environment, the DHCP infrastructure provides the essential flexibility needed to respond to the complex needs of medical information management. To address the full set of tools of the DHCP infrastructure would require a book of its own. However, several can be described here to reflect the richness of this environment. The infrastructure of DHCP is often referred to as the Kernel. Yet the Kernel is actually only a part of this infrastructure.

Security

Few issues can be more important than ensuring the data supported by a medical information system are protected from unauthorized access and secured against tampering (Donaldson & Lohr, 1994). Accordingly, security is the most rigorous component of the DHCP infrastructure.

Recognition and authorization of users to gain access to the DHCP system are primary roles of the Kernel. Each user is identified by an access code and verified through a second code. No user may successfully log on without producing a pair of valid codes. These codes are encrypted using a one-way algorithm. The system has no way to reverse this encryption. Any user who forgets his codes must request IRM service to establish a new access code and destroy the old verify code. The user must then establish a new verify code to regain access to the system. At regular intervals, all users are required to change their verify codes. These methods inhibit access by unauthorized users. To detect potential penetrations, DHCP monitors all log-ons and maintains a register. The system can also trap repeated attempts against legitimate access/verify code pairs.

The DHCP uses a secure utility for users to electronically sign "documents" or orders online. Like the verify code, maintenance of the actual code is invested in the user. The algorithm to encrypt the electronic signature is different from that used for the access and verify codes. When a document is signed, selected data elements are hashed with the signature and placed on the record. Alteration of the record is readily exposed when viewed or printed because the hashed data elements appear as "garbage" or noise. The DHCP assumes that any signed document is to be uneditable, just as it would be in a traditional medical chart. Changes to the record must be filed as separate entries. This ensures that an accurate record is maintained.

The security measures of DHCP are regularly reviewed by internal and external bodies to ensure that methods are sound and appropriately applied.

Database Management: VA FileMan

The most prominent tool of the infrastructure is the database management system, VA FileMan (Davis, 1987; Pfeil & Hoye, 1993a; Timson, 1980; Winn & Hoye, 1991). Embodying the tools to define data dictionaries, to construct templates for the input and output of data, to establish relationships among files and fields, to validate data, and to audit changes to data, this database management system serves as the anchor of the DHCP system and has also enjoyed international use (El Hattab, Ibrahim, & Kadah, 1992; *FileMan Programmer's Manual*, 1995; *FileMan Technical Manual*, 1995; Giere & Moore, 1993; Kleine-Kraneburg, 1992). It presents users with an uncomplicated and direct interface, which is easily mastered by end users. For sophisticated end users, FileMan allows detailed tailoring of its constructs. Entire applications can be developed solely through the use of this tool (Davis, 1987).

With an even more robust set of features available to them, developers can use VA FileMan to establish composite keys and required fields to ensure that essential data are entered for each record of a file (*FileMan Programmer's Manual*, 1995; *FileMan User's Manual*, 1995). The data dictionary can be used to define, document, process, and constrain data. Views of data are facilitated through input, sort, and print templates. User input can be modeled using either a roll-and-scroll interface or a screen-oriented interface. When using the screen-oriented interface, the data are processed in a transaction-like manner (Pfeil & Hoye, 1993b). Editors for handling text input are another component of FileMan, again supporting the roll-and-scroll and screen-oriented environments. Complex user interfaces can be designed, which incorporate data from multiple files either using direct relationships among the files or computing a relationship on-the-fly based on the data of other fields. Complex sorts involving multiple files are readily defined and can be used as prompts for users who need to use the general logic, but also need to specify variable ranges at the time of the actual sort. Data can be exported from the DHCP database using FileMan's Export Tool, which organizes the selected data according to the format specified by the user (e.g., for Excel, SAS). And, finally, FileMan provides a full-featured API for communication with client applications (Department of Veterans Affairs, 1995; Marshall, 1994; Pfeil & Hoye, 1993a).

VA FileMan is used extensively throughout the VA. It is used to create ad hoc reports on a daily basis, by a wide range of users. Application coordinators routinely use FileMan to create modest applications which aid their services in tracking data not yet supported via national products. Within IRM, developers routinely use developer tools within FileMan to create complex applications for their site or for use across the sites within their network.

Simplifying Access to Reference Files: Multi-Term Look-Up

Many of the reference files of DHCP are so large they inhibit easy access of desired entries. This is not due to their structure, but rather to the complexity and volume of the data. The terminology of the data in these files is generally structured to meet a specific purpose, such as diagnosis coding, for example, the ICD-9. Unless the user is conversant with the schema of the data, look-up of the desired entry can be frustrating. To compensate, the Multi-Term Look-Up (MTLU) tool was created to allow users to define their own terms and phrases as look-up keys. Thus, a user may choose to use the term cancer when seeking entries for carcinoma. The MTLU allows each site to establish local aids for look-up on a wide range of files and to let users have their own terms or to use communal terms.

Electronic Mail: MailMan

The e-mail system is the most widely used tool provided by the DHCP infrastructure. Utilized directly by the entire user community, it serves as a point of focus for dialogue to define policies, publish memorandum, explore clinical issues, announce training classes, debate software design, and network with users on topics of mutual interest. The distinction of this tool, MailMan, is that it threads the user dialogue. Within a single DHCP environment, each response is attached to the original message so that all users see all responses and can easily review the entire dialogue without having to sift among several disjointed messages. This feature gives the responses their proper context and ensures that every recipient has access to the full content of the dialogue. When MailMan exchanges messages with nonthreaded mail systems, it adapts to their restrictions and forwards (or receives) only the original message or single response.

MailMan offers features that users utilize directly to manage the dialogue within their sites. Mail groups can be established to simplify addressing. Users can assign others to be surrogates to process mail for them. Messages can be secured, confidential, closed, or informational. Shared mail can be used to provide many of the features of electronic bulletin boards as a means to post policies, job offerings, training schedules, and other items of interest to all users of the DHCP system.

MailMan is also used as the vehicle to transport data between VA sites. For example, registries of selected patient groups (e.g., spinal cord injured) are updated with extracts from VA medical centers that use MailMan to send the data to the national registry. The VA's centralized processing center in Austin, Texas, is updated daily via MailMan messages containing data for financial transactions, patient statistics, and other corporate data (Dayhoff & Maloney, 1993). Comprehensive patient records are transmitted among facilities by applications that use MailMan for transport. Data exchange between the VA's administrative divisions is also supported by MailMan. For example, the Veterans Benefits Administration (VBA) defines the veteran's eligibility for health care. To establish this, VBA must request the Veterans Health Administration (VHA) to ascertain the patient's degree of disability attributed to active duty service. In exchange, the VHA must ask VBA for the financial status of the veteran to determine how to process charges for health care. The two administrations use architecturally different systems for their internal operations, yet they depend on MailMan to facilitate the exchange of information to support the veterans.

MailMan serves a valuable role in software development and distribution. Every feature described above can be affected through the MailMan APIs. As a special function, utilities allow developers to use MailMan to pick up a set of routines and transport them to another DHCP environment. When the target environment is reached, the developer can compare the message contents to the routines currently existing in that environment. Thus, the changes reflected in the received routines can be isolated and reviewed as necessary. When ready, the recipient can direct MailMan to install the new routines into the target environment to update the routine set. This mode of software distribution serves as the backbone of the patching process of DHCP. With over 70 applications and more than 155 separate computer systems at the 173 VA hospitals, the efficiency of patch distribution and installation is critical. MailMan streamlines this detailed process.

Establishing User Environments: Menus/Options, Lists, and Protocols

The DHCP offers three major tools for organizing and constraining user activities. The first and primary tool is *Menu Management*. All activities that a user is authorized to perform are accessible through his assigned menus. These are segregated into three general types:

- Common Menu: A shared menu accessible by all users, a common menu contains options to use MailMan, log off, and other basic DHCP activities.
- Primary Menu: Assigned individually to every user, a primary menu is specific to the job that the user performs. For example, a laboratory technician will have a primary menu that includes options to accession lab samples, verify results, insert controls, etc.
- Secondary Menu: Given to users as needed, a secondary menu grants access to options that are not typically part of that user's job. For instance, an application coordinator may have VA FileMan as a secondary menu option so that he can construct ad hoc reports.

Taken together, these menus grant all needed access to the user to perform the activities therein, while constraining the user from access to unauthorized activities; for example, if the user is a laboratory technician, filling prescriptions.

When a context is established, the second tool, *List Manager*, may be used (*List Manager Developer's Guide Manual, 1993*). This tool displays a screen-oriented list of activities to the user, all of which make sense in the active context. For example, if the user has selected a patient for whom to place orders, the list contains activities such as reviewing orders, requesting laboratory tests, discharging the patient, scheduling appointments, and similar patient-care activities. As the context shifts, so does the list of activities. Thus, when the activity to order lab tests is selected, the activities are based on the context of ordering lab tests for the patient already selected.

The third tool is the *Protocol Processor*. Protocols are the actions that users can take against items on the list. An action may invoke a single protocol or several linked protocols.

For each construct—menus, lists, and protocols—the developer needs to build the objects that it uses. These include templates and application code, which the control mechanism will activate. Together, these tools provide a flexible environment for developers to create work environments for users.

Background Jobs

Not all tasks in DHCP take place as the immediate response to user activity. Many are better relegated to the background to be completed as a lesser priority event. Support for this is accomplished through TaskMan, an infrastructure tool used to schedule all queued actions, such as complex data sorting, printing, and delivering e-mail. TaskMan is responsive to three distinct sources: user requests, IRM system management, and application invocation. In DHCP, users are able to monitor and affect their own tasked jobs. If a user wants to reschedule a job she queued, she can change the runtime; she can even remove the task from the queued list. In addition, IRM uses TaskMan to schedule maintenance activities in DHCP, including daily purging of outdated data, forwarding time-sensitive mail messages, and balancing the impact of the many application-specific background jobs. Applications invoke TaskMan to direct reports to printers, prepare and submit messages to MailMan, and run jobs to organize data for the next day's clinical activities. TaskMan supports flexible scheduling, allowing date, time, day of week, frequency of rescheduling, and priority to be determined. Like the other infrastructure tools, TaskMan provides an API so that developers can tailor their applications.

Servers

Complementing the use of MailMan to transport data between DHCP environments, servers can be created to process the received data directly from mail messages (Meighan, Hirz, & Curtis, 1991). Servers are actually a connection of four infrastructure components: FileMan, MailMan, TaskMan, and Options. They are activated through the receipt of a mail message, which serves as input to the server. In essence, servers are a controlled means to respond to a remote request for action. The developer of a server has several choices for action: queue up the requests for later automatic processing, process the request immediately, or hold the request until a user intervenes. All activity initiated through a server can be audited at the discretion of IRM. Security on server activation is controlled by the receiving system so that unknown or prohibited requests can be denied and recorded. The most widely known instance of a server within DHCP is the Patient Data Exchange (PDX) application (see below).

Alerts and Bulletins

A dynamic environment, health care demands that information systems be capable of reacting to events that occur throughout the care of the patient. Decentralized Hospital Computer Program uses two mechanisms to advise users of events requiring their attention. The first involves coordination between VA FileMan and MailMan. By defining an event such as entering or editing a data field, FileMan can trigger a mail message known as a bulletin. Each bulletin is sent to a mail group, including the users who need to know of the activity or event. Through this mechanism, DHCP effectively uses bulletins to track access to sensitive records, inform staff of patient admissions, and note other activities for which a mail message is sufficient.

To advise users of events that require more immediate attention, the Kernel provides the ability to issue alerts. This type of advisory is more active in that it is persistently presented to the user until the needed activity is completed. For example, when the urgent results of a laboratory test are verified, an alert will be presented to the ordering clinician (or even the entire clinical care team) until someone actually views the lab test result. Alerts can also be simply informational, though they tend to be more terse than bulletins. Importantly, alerts can be audited so that a history of significant clinical actions can be recorded.

Site Parameters for Application Tuning

Virtually every DHCP software product has parameters that can be set to influence the behavior of that product. For the infrastructure, there are parameters to define the MailMan domain, system defaults, audit parameters, and many more. Within applications, parameters can be set for consistency with the specific facility's practices, such as establishing inventory reorder levels, assigning printers for prescription labels, or defining critical values for laboratory tests.

User's Toolbox

The DHCP infrastructure acknowledges end users as responsible for the actions they perform on the system. Each user is extended privileges to manage those actions to the fullest extent possible; the infrastructure simultaneously protects the overall system and the activities of other users. Users are allowed to modify the parameters of jobs they queue to run in background; they may alter the time to run, stop actively running jobs, and modify other minor attributes of their own tasks. Users may also create their own templates to define a sequence of menu options to be executed and invoke this template to automatically step them through the sequence at any menu prompt.

Users may direct their reports to a spool device for holding and de-spool their reports when convenient, specifying the printer and even the number of copies. Alternately, they may direct the output to a MailMan message for electronic forwarding to other users. Users are granted authority to modify parameters, dictating how the system responds to them. For example, users may select a text editor and may change a menu display, verify code, and electronic signature code.

Security Keys

Security keys support segregation within menus. In simple terms, users who hold a key may view options not visible to less authorized users. This allows developers to create simplified control structures rather than duplicate menus when a few additional functions are needed. For example, a supervisor in the outpatient pharmacy can edit pharmacy site parameters, while a technician who enters prescriptions cannot. In this case, the option to edit parameters is locked with a security key. Because the supervisor holds the key, he can access the locked option. The technician never sees the option. Yet all the outpatient pharmacy staff are assigned the same primary menu.

Security keys are used in other contexts where the need to segregate employees exists. The most prevalent use of this type is the provider key, assigned to staff who are authorized to perform clinical functions such as sign patient orders electronically. Security keys are also used in clinical areas when the patient's diagnosis is sensitive, such as HIV (Pfeil, Ivey, Hoffman, & Kuhn, 1991). In such cases, the key is required before the application will allow the system to display the sensitive data to a user.

Health Level Seven for Exchanging Data with Non-DHCP Systems

The DHCP system needs to be able to exchange healthcare information with other systems. Commercial products that provide specialized functions like intensive care monitoring are used to augment DHCP. On occasion, VA needs to share data with external systems operated by other healthcare institutions, such as private or university hospitals. For data exchange, DHCP developers elected to use the standard protocol defined by HL7. As is often the case, HL7 tools were initially used to address a specific need, namely data exchange between the DHCP radiology product and commercial transcription systems.

Once the problem was resolved, HL7 was adopted as a generic solution for all of DHCP, and DHCP developers became active in national efforts to further the HL7 standard. Today this standard is the interface that DHCP can most readily support (Department of Veterans Affairs, 1995). It is important to note, however, that this standard is still evolving and interpreted variably. Thus, the use of HL7 for interfacing DHCP with an external system requires mutual agreement as to the segments to be used. To streamline the interfacing effort, VA maintains a repository of all HL7 segments approved for use with DHCP. This repository is regularly extended with new segments.

Patient Data Exchange for Exchanging Data Between DHCP Systems

One of the benefits of the VA healthcare system is that veterans can go to any VA medical facility for care. When a veteran visits another facility, the need for the patient's medical record can be promptly met by use of the

PDX. Originally considered an application, PDX has evolved to become a general tool allowing the user to request a clinical summary from any other VA. The request can specify a comprehensive profile or a report formatted to address specific needs (e.g., just outpatient prescriptions). To respond to these requests, PDX uses the features of other DHCP tools, including MailMan, FileMan, servers, and the Health Summary application.

Generic Code Sheet for Consolidation of Corporate Data

The VA reviews and processes volumes of data on workload and finances at the corporate level. MailMan is used to move the data to the central office from individual facilities, but only after the data are formatted so the received messages are ready for immediate processing. The tool used to do so is the Generic Code Sheet (GCS), a utility that bridges the gap between DHCP, where data for corporate processing are created as a byproduct of patient care, and the central systems, where batch input is required to produce national reports, pay vendors, and issue paychecks (Department of Veterans Affairs, 1995).

The GCS grew out of an interface that supported one DHCP application's need to pass data to one of the VA's many centralized systems. Its first goal was to obviate the need for keypunching, which was the method used at the time for submission of data to the corporate databases. Thus, the initial implementation simply mimicked punched cards (i.e., formatting data in 80-character records and using MailMan for actual transmission). As the utility evolved to become generic in use, it eliminated the need for manual (and often duplicate) data entry into multiple systems. Moreover, GCS improved the accuracy of VA records through the removal of duplicate data entry, which was a potential source of conflicting data. Today there is virtually no expectation that users interact with multiple systems; instead, users can focus their attention on the primary tasks they perform and all submission of data to external databases is accomplished as a byproduct via GCS.

Remote Procedure Call Broker for Client–Server Computing

During the early years of DHCP, VA based user interactions on a "roll-and-scroll" method using basic "dumb" terminals. As DHCP became more advanced, screen-oriented tools were incorporated, but still for use with character-based video terminals, that is, low-end terminals without graphical user interfaces. Today DHCP has evolved to take advantage of the client–server model of computing in which the workstation or personal computer serves as the user's tool.

To retain vendor independence, DHCP developed a brokering utility to support and manage the messaging needed for data exchange between a client workstation and the database server. The Remote Procedure Call (RPC) Broker was optimized to work in a Microsoft Windows environment and to use the component features of Borland's Delphi graphical user interface toolset. A set of functions, built as a set of data link libraries (DLLs), provides access to the client–server messaging for any Windows-based client application. This extension paves the way for incorporation of a wide variety of commercial products to be used in conjunction with DHCP.

The server side of this model is representative of all DHCP infrastructure in that it is written in the M language using VA FileMan and other Kernel utilities to manage and balance the server workload. The server side also ensures that only authorized client applications are able to query or update DHCP data.

The RPC Broker serves as the essential messaging tool to support the fetching and display of clinical images (e.g., x-rays, ECGs, and pathology slides). The Clinical Patient Record System (CPRS) relies on the broker to retrieve data for presentation to clinicians and to send data to DHCP for patient orders, which were created on the user's workstation.

Updating the DHCP Software and Data

The DHCP allows the nonstop evolution of applications and reference data files. These changes are generally developed within individual programming environments, often with each application coming from a different source. Combining these enhancements with an established production environment used by a medical center demands solid tools and strict adherence to VA's internal standards.

Each DHCP system belongs to the individual VA medical center it serves. Thus, even though most of the software and reference files are created external to the center, installation of new and enhanced products must be respectful of the customization and data that the facility has established locally. The DHCP installations do not "blow away" the previous version of the product; they evolve it.

Name-Spacing, Number Ranges, and Version Numbers

All DHCP software products, including infrastructure tools, are developed within boundaries established by the database administrator. These boundaries cover the naming of routines, options, protocols, templates, etc., according to name spaces and the defining of files within a number range. Adherence is observed through careful attention by software developers and enforced by use of DHCP's infrastructure development tools. For end users and application coordinators, VA FileMan restricts development activity to limited file number ranges established by IRM. There are benefits to developers and end users who adhere to these boundaries in that the suite of tools used to edit programs and files is optimized to work from name-space and number-range specifications. When observed, the name-spacing and file number ranges serve to isolate national software and data from that created by the facility. This isolation allows the national software and data to be updated by new versions while preserving the local additions to DHCP.

Each release of DHCP software and data is identified by a version number. All patches are tracked according to the version number of the software with a sequence number for the patch. As with name-spacing and number ranges, the infrastructure provides tools for applying and viewing software products according to version numbers.

Kernel Installation and Distribution System

Software updates are accomplished using a specialized infrastructure toolset to build, distribute, and install them. This toolset is known as the Kernel Installation and Distribution System (KIDS). During the specification or build phase of a software product, the developer identifies the routines, files, templates, options, protocols, and other application attributes that comprise the product. This identification is primarily based on name-spacing and number ranges. For each, a default as to whether the data should overwrite the existing data or merge with the existing data is established. Criteria to certify on the target system are defined as precursors to satisfy before the actual installation would be allowed to begin. Any data structure or data conversions are developed to be run as preinstallation or postinstallation activities.

When IRM initiates the KIDS process on a local system, the first event is a check to ensure that the facility DHCP system is ready to receive the upgrade. Because this involves no changes to the system, the check can be made while the system remains fully operational. When the system is determined to be ready, the KIDS process advances to a preinstallation procedure if defined by the developer. Activity at this time may include the introduction of new files or data that do not affect operation of the current system. The next step is the actual installation of the new software. This may necessitate the disabling of all or part of the DHCP system. Installing a new version of the radiology system may require disabling only the radiology application and its users. However, if the installation is for Kernel, all users are disabled because the infrastructure is being altered. Each installation makes the determination as to the extent of impact on the users. The goal is always to minimize the impact on users. Data introduced to the system via an upgrade may be merged/added to the database or may actually replace/overwrite existing data. In the latter case, a postinstallation conversion may be required to update multiple records or reestablish interrecord relationships.

For all installations, KIDS updates the online history, including version and date and other application information, so that IRM can track changes to the system over time. In addition, KIDS records installation parameters and time to complete steps for subsequent examination as needed. Should errors occur during the installation, KIDS can resume the process at the point of interruption.

Tools for Application Coordinators

Given the scope of DHCP and the vast numbers of users, IRM would be overwhelmed if they had to handle all aspects of each application. From the beginning, DHCP made automated data processing application coordinators (ADPACs) key participants, responsible for meeting the advocacy and training needs of their service area. As the infrastructure has evolved, the suite of tools for use by ADPACs has been expanded, relieving IRM of the repetitive tasks associated with supporting routine service activities.

Application coordinators enjoy a greater use of FileMan than typical end users, often developing modest applications that are purely FileMan based and require that no code be written (*FileMan User's Manual*, 1995). ADPACs can use features of menu management to make these applications available to users within their service.

Infrastructure tools enable ADPACs to manage user activities within their areas. Secure Menu Delegation can be used to invest an ADPAC with the authority to extend new options to users (*Kernel Systems Manual*, 1995). Furthermore, all applications provide a specialized menu routinely assigned to the ADPACs; these can be used to tune the application and set site parameters, allowing the ADPACs to assume responsibility for the application and all its attributes.

Scalability

Decentralized Hospital Computer Program systems support medical facilities that vary from quite small to extremely large. Yet the software that supports these facilities is the same. It is through the infrastructure that the system is tuned to adapt to the supporting hardware capabilities, the size of the user population, and the size of the patient population.

Benefits of M Technology

The choice of M (formerly known as MUMPS) as the development language and environment has been advantageous to the VA. One prime benefit is the efficiency with which M stores data. No declarations as to size of the database are needed. Rather the database simply grows as a balanced B-tree structure (Lewkowicz, 1989; Walters, 1989). No disk space is wasted on absent data; data are allotted space only when they are created. This has special importance for medical databases, as no two patient records contain the same type or volume of data. This data storage efficiency is invisible to the developer who works at a conceptual level, designing the database using VA FileMan.

FileMan contributes to the scalability by not limiting the number of entries within a file and by extending the privilege to the developer to organize the data in meaningful groupings, such that fetches of related data are optimized.

In addition, the choice of M makes it possible to provide an environment that is readily isolated by the Kernel. This independence from vendor peculiarities has allowed the VA to migrate to faster CPUs, larger disks, and a wide variety of peripherals with minimal impact. For the most part, only IRM has required training and this only for the specifics of the underlying operating system (e.g., Digital's VMS or Windows-based NT). Migration to improved hardware platforms has served to maintain a system that is able to provide response times essential for the clinical environment.

Building Blocks and Reusable Code

The DHCP makes extensive use of each software component in the infrastructure, including APIs, reference files, library functions, and control mechanisms. Applications can evolve rapidly, adding to this suite of tools by providing additional APIs and reference files. The overall result is a system that is integrated and modular. Built upon basic blocks, it makes extensive reuse of code and objects. Development is collaborative, effectively advancing each application beyond its individual contributions. For example, when the laboratory package establishes a file of blood tests, the file can be used to check drug–lab test interactions, build clinician order sets, and update inventory levels as tests are performed.

Options, lists, and protocols are introduced to the system by all applications and can be combined to offer several user profiles for meeting the facility needs. Because all are constructed from a consistent base, activities for users can be expanded without recreating them from scratch. Although no option exists in duplicate, several users can be authorized to activate it. Because the tools are consistent in form and presentation, users can help each other and adapt to new applications with minimal need for training and minimal impact on IRM and ADPACs.

The decision to build DHCP from a set of reusable components makes the transition from a traditional computing environment to a client–server environment far less painful and work-intensive. Though new APIs are needed to ensure that all data can be requested incrementally, many noninteractive APIs were available for early modeling of the client–server environment. This architectural model demands that the system behave in a more object-oriented manner, as DHCP is well positioned to do.

Performance Considerations

The DHCP infrastructure incorporates numerous parameters, which tune overall system performance. On a typical day, the number of interactive users on the system peaks in the midmorning and early afternoon, and hits its lowest level in late evening. Within this pattern, the infrastructure can contain background job activity, balance the number of active users across multiple processors, and set alarm levels on system performance. These capabilities are used in conjunction with operating system tools and commercial monitoring tools (*Kernel Toolkit Systems Manual*, 1994).

The DHCP infrastructure and all DHCP applications are continuously evaluated for performance. Through the Kernel's response-time monitoring tools, frequently executed codes can be examined to isolate areas where performance can be enhanced. Attention is paid to routines that are widely used via APIs. An improvement in a single API used frequently can leverage the entire system. The VA FileMan is regularly examined for efficiency. Because data files can be extremely large, the efficiency with which data sorts are performed has received concentrated effort. Code analysis and data sampling techniques have resulted in significant performance improvements. These improvements extend to applications where FileMan is used to manipulate large files to the degree that performance of VA FileMan has been equated to that of relational systems such as Oracle (Weis, 1995).

Much DHCP activity involves the exchange of data between systems. As the volume of exchange increases, efficiency becomes more important. MailMan addresses this need by providing a wide array of transmission protocols, from basic models like the Simple Message Transport Protocol (SMTP) to high-speed methods, which rely on Transmission Control Protocol/Internet Protocol (TCP/IP). Today DHCP is able to move data at the highest speed that the two involved systems can support (*MailMan User's Manual*, 1995; *MailMan Technical Manual*, 1995).

Database Growth

The goal of a medical information system used in a clinical environment is to retain all data for each patient in perpetuity. Neither M nor FileMan poses any restrictions, which prohibit database growth. Nonetheless, there must be disk space available for storage and hardware available to ensure rapid data access and support backup and journaling of incremental changes to the database.

Even when all clinically relevant data are retained, the need for archiving and purging of data remains. Inactive patient records are candidates for archiving. Data used by ancillary services, such as pharmacies, to meet legal requirements can be archived and purged when those requirements are met *if* the clinically relevant data have been recorded in the patient's electronic record. Thus, the infrastructure of DHCP provides the means to archive and purge data. Currently data from the ancillary services are archived. Archival capabilities for full patient records will be addressed once the clinical repository for those records is implemented.

Designing for Strategic Redundancy

Unexpected downtime for a system used for almost every patient-care activity is more than a "negative impact." It can be devastating. Medication records, laboratory results, problem lists, progress notes, patient allergies, and other vital data become inaccessible. Clearly, as the clinical record becomes more automated, the need for system reliability increases.

The DHCP began as a system designed to maximize uptime. Strategic redundancy was, and remains, a primary factor. This is achieved on four levels: application, infrastructure, operating system, and hardware.

Application Level

Although DHCP does not provide redundant database elements or user functions within a single system, many DHCP applications have taken advantage of the VA's distributed healthcare system. The PDX provides a dramatic example. In a recent natural disaster, the Northridge earthquake, patients had to be transferred to other medical centers in the Los Angeles area. Many of these patients demanded immediate attention for the severity of their situations. Unable to locate and ship the physical medical charts, the involved medical centers turned to PDX to transmit detailed patient summaries to the receiving hospitals. This ability to support patient care in the face of disaster increased the visibility of DHCP and PDX. Today PDX is widely used for all interfacility patient transfers, be they planned or emergency.

Infrastructure Level

The infrastructure of DHCP promotes uptime through its consistency. Despite different hardware and operating systems at different sites, every computer system runs the same DHCP infrastructure. The infrastructure exposes the same platform to the applications and thereby provides an environment that lets users migrate among the applications, seeing a consistent method of presentation and reporting. This presentation redundancy multiplies the user's skills in interacting with the system. Significantly, it allows training efforts to focus on the specifics of the application. Users already know the basics like inputting data, querying for help, and requesting reports.

Linked by MailMan and servers, every VA facility can serve as a fallback support system for another in the event of special needs for patient care. The platform provided by VA FileMan ensures that the structure of file systems is consistent across the VA and promotes the ability to share data without translation or conversion.

Support for sharing and redirecting output relies on redundancy within the system. DHCP's infrastructure achieves this through its device handling and background tasking. Devices can be organized into groups that function like telephone rotaries; needed reports are printed even if some printers are busy or off-line. In the event of equipment failures or altered priorities, background jobs can be redirected to run at alternate times or on different devices.

The consistency provided by the infrastructure allows VA to reorganize operationally as needed. Recently VA moved toward a consolidation of facilities such that the data from multiple DHCP systems were combined to yield a single system, which spans a singly organized, yet geographically distributed medical center. Because VA FileMan, MailMan, PDX, and other tools that operate redundantly across medical centers readily supported this effort, the reorganized VA was able to present a consistent and single view of a combined and much larger database.

Operating System Level

The choice of M as the platform upon which the DHCP infrastructure operates has also promoted redundancy. Because M systems provide a computing environment easily configured to take advantage of multiple central processing units (CPUs), the failure of any one system does not disable the entire information system. Using extended syntax, global translation tables, device maps, remotely mounted disk drives, and other configuration parameters, this type of computing environment can readily respond to hardware failure.

An M system is also capable of running multiple environments simultaneously. Thus, if needed, a single system with sufficient hardware can run the profiles of two medical centers.

Hardware Level

Since its inception, DHCP has operated on hardware platforms that incorporate redundancy. The use of multiple CPUs, dual ported disk drives, and port contention systems provides IRM with the tools to reconfigure the system rapidly in the event of virtually any type of failure. Advances in hardware technology are rapidly incorporated into the architecture so that DHCP has increased its fault tolerance to make unscheduled downtime a rare event.

Infrastructure capabilities have enabled DHCP to migrate across major hardware upgrades over the past decade. Beginning on a platform of PDP-

11 computer systems, DHCP has evolved to a platform of Microsoft Windows NT servers operating on 486 and Pentium systems and VMS operating on DEC Alpha systems (von Blanckensee, 1989). The IRM staff successfully migrated their local systems to these more advanced platforms while limiting downtime from a few hours to a few days in the most extensive case. Significantly, users did not alter their interaction with the new system in any way, aside from commenting that it seemed to be much faster after the conversion was complete.

This extreme change of platform demands a much longer transition time for systems that have not made such an investment in the infrastructure. The dollars saved on avoiding retraining and reprogramming are measured in millions (von Blanckensee, 1991). The dollars saved in clinician time not being wasted waiting for the system to be back up are estimated to be equally impressive.

The interconnection of VA facilities over a wide area network (WAN) is also redundant. The WAN used provides dynamic routing and state-of-the-art messaging support. MailMan serves as the gateway to the WAN, isolating DHCP from the WAN just as Kernel isolates the applications from the underlying operating system. Thus, when VA upgraded its WAN and changed vendors, the DHCP system required only minor adjustments to MailMan and no other reprogramming for applications or infrastructure.

Building for Robustness

No system can guarantee that unforeseen events will not occur. How the system responds to such events is indicative of its robustness. The methods by which DHCP prevents and recovers from misadventure are embedded in its infrastructure.

Prevention

The first step toward a robust system is anticipating and preventing potentially disruptive situations. For DHCP, the first line of defense is the active data dictionary provided by VA FileMan. Through the definition of data structures, the developer is able to create rules for acceptance of user input. Created at the field, file, and template (i.e., view) levels, these rules guide the user as to expected input and issue helpful instruction when unacceptable input is attempted (Davis, 1987; Timson, 1980; Winn & Hoye, 1991).

By controlling access, the Kernel plays an important role. The Kernel gives users access to authorized functions, blocks users from disallowed functions, and even prevents users from being aware of functions for which they are not authorized. Through the use of menus, keys, protocols and lists, users are cocooned in an environment that gives them the confidence to work without fear of "hurting" the system.

Detection

No matter how carefully a system is designed and built, untoward events will occur. Detection of such events is essential in a healthcare system. Validation, review, and audit functions are needed to ensure that patient records are accurate.

Validation of data is best accomplished at the time of data entry. Trapping invalid data as close as possible to the point of entry provides the best opportunity to prevent propagation of the error. FileMan validates the form of data at the time of input through the input template. To determine whether data are wrong in context, FileMan provides utilities to examine the database and identify problem areas. The underlying M system supplies tools to examine the structure of the database to identify when the physical linkage of the database may be corrupt. When a problem is located, IRM can use the matched tools to repair the damage.

Auditing of DHCP data is constant, selective, and strategic. Monitoring is ongoing for users empowered to modify programs and data structures. Logs are kept of user access to patients with designated sensitivity such as HIV. Externally originated messages requesting patient data are conditionally trapped for manual intervention. User sign-ons are recorded.

Auditing of other data is activated depending on the need to monitor activity. The use of specific options and the activity of designated users can be tracked for later review. Intervention in the event of repeated attempts of an unauthorized user to access the system is also built into the infrastructure of DHCP.

Alerts and Bulletins are triggered when the system calculates that advisories are needed. The system automatically monitors itself to sense when critical levels of overall performance are reached and alerts IRM so that action can be taken before irrecoverable levels are attained. Processes exist to scan the M programs to detect variance from expected contents, thus detecting changes to code.

Recovery

It is possible to detect and recover from situations without significant impact to the user or the database. Upon the initiation of each user session, the Kernel wraps the user in an error recovery envelope sensitive to conditions signaling that program code has failed. If an error occurs, the Kernel traps the error and smoothly exits the user from the option or protocol being executed. Rather than being bumped off the system, the user is returned to the operational level above the level where the error condition was detected. The fact of the error is recorded for later examination by IRM or a developer who is able to use the Kernel's features to recreate the situation of the error and eventually affect a fix to the problem (*Kernel Systems Manual*, 1995; *Kernel Technical Manual*, 1995).

For tasks running in the background, the Kernel follows much the same procedure, trapping and recording the events and facts of the error and utilizing the same mechanisms to fix the problem. When the fix is done, the task is rescheduled for completion.

Of all the infrastructure tools, MailMan is perhaps the most persistent in ensuring that the user's intentions are met. Failure to deliver mail will result in saving the message in a queue for later attempts. Every time a new message is forwarded to the same destination for which a previous failure has been queued, MailMan will attempt to forward all the mail for that destination. Information Resources Management regularly examines the queues and intervenes to resolve continued impediments.

DHCP: Tensile Strength and Flexibility

The DHCP's infrastructure provides the tensile strength to secure the information system and the flexibility to allow it to adapt to the divergent data and user community of the healthcare environment. It is independent of the operating system and hardware base, allowing the VA to migrate to new technology without a need to reprogram applications. It promotes a concordant environment for users who easily transition across a suite of applications. It advances development of software through a rich suite of software tools, which allow each developer to be highly productive and more focused on the needs of the users. It is the core of DHCP from which all applications take their form.

References

Davis RG. 1987. *FileMan: A user manual.* Bethesda, MD: National Association of VA Physicians.

Dayhoff RE, & Maloney DL. 1993. Exchange of Veterans Affairs medical data using national and local networks. *Annals New York Academy of Sciences, 670,* 50–66.

Department of Veterans Affairs. 1995 (August). *Decentralized Hospital Computer System* (Monograph). Washington, DC: Author.

Department of Veterans Affairs. 1995. *FileMan programmer's manual, version 21.0.* Washington, DC: Author.

Department of Veterans Affairs. 1995. *FileMan technical manual, version 21.0.* Washington, DC: Author.

Department of Veterans Affairs. 1995. *FileMan user's manual, version 21.0.* Washington, DC: Author.

Department of Veterans Affairs. 1995. *Kernel systems manual, version 8.0.* Washington, DC: Author.

Department of Veterans Affairs. 1995. *Kernel technical manual, version 8.0.* Washington, DC: Author.

Department of Veterans Affairs. 1994. *Kernel Toolkit systems manual, version 7.3.* Washington, DC: Author.

Department of Veterans Affairs. 1993. *List Manager developer's guide manual.* Washington, DC: Author.
Department of Veterans Affairs. 1995. *MailMan user's manual, version 7.1.*
Department of Veterans Affairs. 1995. *MailMan technical manual, version 7.1.*
Dick RS, & Steen EB. 1991. *The computer-based patient record.* Washington DC: National Academy Press.
Donaldson MS, & Lohr KN. 1994. *Health data in the information age.* Washington, DC: National Academy Press.
El Hattab OH, Ibrahim AS, & Kadah MB. 1992. DHCP in Egypt. *MUMPS Computing, 22(3)*, 87–91.
Giere W, & Moore GW. 1993. Translating English into German using VA file manager. *M Computing, 1(4)*, 16–23.
Ivers MT, Timson GF, von Blanckensee HT, Whitfield G, Keltz PD, & Pfeil CN 1983. Large scale implementation of compatible hospital computer systems within the Veterans Administration. *Proceedings of the Annual Symposium on Computer Application in Medical Care, 7*, 53–56.
Kleine-Kraneburg A. 1992. The translation of VA applications from English to German. *MUMPS Computing, 22(1)*, 54–57.
Lewkowicz J. 1989. *The complete MUMPS.* New York: McGraw Hill.
Marshall R. 1994. Programming hooks 102: rules for use. *M Computing, 2(1)*, 50–52.
Meighan MJ, Hirz LJ, & Curtis C. 1991. Servers and filegrams. *MUG Quarterly, 21(3)*, 89–92.
Montgomery GV. 1993. M vital to quality health care for U.S. veterans. *M Computing, 1(4)*, 45–46.
M Technology Association. 1994. *M—The open production system (a white paper).* Silver Spring, MD: Author.
Munnecke T. 1981. Software portability considerations for multiple applications over multiple sites. *Proceedings of the Annual Symposium on Computer Application in Medical Care, 5*, 38–42.
Munnecke T, & Kuhn IM. 1989. Large scale portability of hospital information system software within the Veterans Administration. In HF Orthner & BI Blum (eds): *Implementing Health Care Information Systems.* New York: Springer-Verlag, pp. 133–148.
Pfeil CN. 1992. US department of Veterans Affairs: DHCP passes the ten-year mark. *MUMPS Computing, 22(5)*, 28–31.
Pfeil CN, & Hoye ML. 1993a. Today's FileMan. *M Computing, 1(1)*, 44–45.
Pfeil CN, & Hoye ML. 1993b. VA's newer tools bring versatility, flexibility. *M Computing, 1(2)*, 11–12.
Pfeil CN, Ivey JL, Hoffman JD, & Kuhn IM. 1991. M: Use of DHCP to provide essential information for care and management of HIV patients. *Proceedings of the Annual Symposium on Computer Applications in Medical Care, 15*, 146–149.
Timson GF. 1980. The file manager system. *Proceedings of the Annual Symposium on Computer Application in Medical Care, 4*, 1645–1649.
von Blanckensee HT. 1989. VA DHCP replacement systems: VAX clusters for the 90's. *MUG Quarterly, 19(1)*, 80–81.
von Blanckensee HT. 1991. The VA Kernel: What has it bought the VA? *M Computing, 21(3)*, 87–88.
Walters RF. 1989. *ABCs of MUMPS.* Bedford, MA: Digital Press.

Weis WG. 1995. A performance analysis of relational and hierarchical database packages based on query response. *M Computing, 3(4)*, 18–26.

Winn TK, & Hoye ML. 1991. Relational features of VA FileMan. *MUG Quarterly, 21(3)*, 46–51.

6
Structuring Databases for Data Sharing

CAMERON SCHLEHUBER

Introduction

Against great odds, the Veterans Administration (VA) has implemented and deployed a system with a common database shared across the facilities within the VA (Davis, 1987; Curtis, 1992). Moreover, VA is able to share replicated base tables of data and definitions with other domestic and foreign government agencies and with numerous private organizations as well. Such sharing across organizational boundaries is discussed elsewhere in this book.

Complexity has its price. Only about 30% of the enterprise-wide systems attempted survive. Even less likely to succeed, indeed rare, are those systems that span entire communities where management and administration are not monolithic and participation is elective. Yet the VA has succeeded, due in part to its dedicated and visionary employees. Certainly a significant contributing factor has been the architecture that supports the VA databases. Designed for flexibility, it has evolved over time, making continued growth of the Decentralized Hospital Computer Program (DHCP) possible (Davis, 1987; Curtis, 1992).

This chapter examines the technical reasons why the VA has managed a most difficult task: a superior database environment in M technology; and high demand for, use of, and feedback on a wide variety of applications.

Database Administration

The traditional role of database administrator (DBA) begins with designing the physical database based on the logical designs of data administration and the anticipated needs of end users. The role encompasses management of the database for maximum speed in performance for transactions, workflow, and reports. A key function within this traditional role is the maintenance of the most efficient use of disk space.

Today, as health care moves toward multi-facility systems and the enterprise model, the role of the DBA is evolving and becoming more complex, with DBAs functioning differently at different levels within the organization. According to Quinlan (1996), the enterprise DBA is "someone who develops and supports mission-critical, high-availability databases that not only require a high degree of security but also must sustain large transaction volumes," whereas the departmental DBA "has the ability to deliver databases and systems very quickly. New applications and databases move from design stages to production in months." As Quinlan emphasizes, the enterprise DBA needs to make use of the capabilities and advantages of the departmental DBA. This is indeed the case within the VA, with its multiple facilities and varied services. Many "departmental size" applications are needed, and have been and are being delivered within months to the enterprise level.

Database Architecture

The role of database administration in the VA is not unlike its role in other organizations. The desire is to eliminate undesired definitions, redundancies, and inconsistencies in utilities and files used by multiple applications. This can require the creation of a separate function, known as data administration, to prepare logical models for the system. In the VA, data administration was not separated out, but remained under the purview of the database administrator throughout the first decade of DHCP. Database models were defined and refined in large part by the many application developers—the content experts, analysts, and programmers—with frequent design reviews by their peers and the DBA. This resulted in an organization that was considerably "flatter" with its attendant benefits.

The nature of VA FileMan and its database permits and encourages such a flat organization. The VA FileMan encourages rapid prototyping by virtue of its automatic creation of the physical database while the user simply declares the desired entity-relationship and attribute model. Changes to the logical model automatically change the physical model, and changes to the physical model are automatically tied to the logical model. Therefore, the "model view" of the database is in fact the database. Changes to the data dictionary (DD) become the "model"; and, conversely, changes to the DD model become the "dictionary." The resulting physical and logical models can be printed and shared in reviews. More importantly, the same models and databases can be exchanged electronically during development, implementation, and production.

Available at all facilities using the DHCP software, the DD (with its inherent model) can be put in the hands of all users, thereby creating opportunities to test and develop new software. Moreover, the quality of data in the database is greatly enhanced because all users have access to the same full definitions, ensuring the highest possible integrity of the data by applying the same checks and business rules at all facilities.

The exchange of components and modules during development by selecting those parts that are under development and easily integrating them into other accounts where development in other areas is under way makes rapid, distributed development far easier. In fact, programmers at end-user facilities have equal access and capabilities to those who are responsible for initial development.

The roles in database administration have grown in scope and activity and are now divided into areas of specialization performed by several persons in technical integration.

Integration Agreements

In the early years of DHCP, some developers believed they could reference virtually any routines, dictionary, or data, no matter what package they were in. Naturally, many changes needed to be made and continue to need to be made. As modules evolved, developers of nationally supported software, as well as developers at local facilities working on local enhancements, realized that it was increasingly important to know what was "stable" and what was not. For this purpose a library of Integration Agreements (IAs) was established. The broadest and most stable Integration Agreements were the utilities already known in VA FileMan and Kernel. Added to these were other utilities, dictionary, and data references that were recognized and accepted as having long-term stability. The descriptions of these along with the various rules of use permitted anyone to use them so long as the rules of use were followed and no additional documentation of their use was required on the part of the developers using them. These first-tier IAs are referred to as "open supported." A second tier of IAs was added to the first with the difference being that any package using the specific IA had to be registered as a "subscribing package." The third tier of IA, termed "private," is reserved for one-to-one agreements between packages. Private IAs are often necessary for only a single version of a package. For IAs that extend beyond a single version, the purpose is usually to support references that are desired to be kept limited, either for reasons of security or confidentiality or because the references are likely to need to be changed in a short time.

Draft IAs may be entered by anyone who is a user of FORUM, one of the VA's large shared e-mail systems. Once entered, the IA can be "rolled-up" into a mail message that can be further discussed by the parties involved. When in final form, the IA is made "active," and the IA is no longer available to be edited further.

Over the last few years, IAs have expanded to include various messages and segments for HL7, for Remote Procedure Calls (RPCs), and for Application Program Interfaces (APIs). These are virtually all of the usage "open supported."

Some have argued against having too many IAs, believing that it is counterproductive to have to record such agreements. For the sake of simplicity, the fewer IAs, the better ... to an extent. The greatest number of associations with the fewest number of agreements is best. But there is also benefit to having a variety of "kinds" of associations, rather than all being of the same kind. And any new kind of association requires either a new agreement or a more complex agreement.

Perhaps the best "agreements" are based on "natural" law, where the "contract" appears to be fairly stable and can be discovered. The task is to create a "natural" information system environment that is fairly stable and can be discovered. The contract for gravity is a good example. As humans, we interact with other masses by the rules of gravity discovered and "written" into our nervous systems. Computers "know" the rules of gravity, for space shots, etc., from the rules we have handed them. For us, and for computers, the rules are "written down" somewhere. For gravity, there are very few rules, perhaps only one. But for every method and every attribute (e.g., the force of gravity or the gravitational constant) that anyone interacts with or has access to, there is a "contract" or "rule." Every RPC, every API, every Hyper Text Markup Language (HTML) tab, is a "contract" or part of a "contract". The more "rules" there are, the more complicated the behavior or characteristics of the system. But the benefit is in having a much richer system with far more capabilities and possibilities.

Databases in Distributed Systems

Evolution of databases in VA and the architecture of a data system can strongly influence the management of that system and its long-term cost. Technological improvements in computer processors, data storage devices, and computer networks have led to rapid changes during the past 5 to 10 years in the options available for implementing large systems. Basically, two options now exist: consolidated systems and distributed systems.

Traditionally, corporate databases often have been centered in mainframe environments that consolidate or centralize processing power (a powerful "host" computer, servicing multiple terminals) and direct access storage devices (e.g., magnetic media disk drives). With this architecture, essentially all of the computing processes occur on the host machine. There are advantages to this type of system. Administration is simplified; technical expertise can be concentrated in a single environment, and security issues are simplified. Unfortunately, many of the strengths of such a centralized model also lead to important disadvantages. Users are isolated from the data as well as from the developers, in what is perhaps the most significant disadvantage. In addition, maintenance costs increase over time, and technical stagnation occurs as a consequence of concentrating technical expertise within one administrative domain.

As the system matures, change requests become more numerous, and the resources needed to alter an increasingly complex environment raise the cost and time required to effect change. As a consequence, because budgets for support staff are always limited, the responsiveness to user needs eventually deteriorates, leading to increasing isolation of the users from the data. In the eyes of the user, the data become "their" data, not "my" data.

The alternative to a consolidated system is a distributed system architecture. With this type of system, computing processing power is placed within a network context, with direct access storage devices distributed as well. This means that the distributed system consists of an electronic network—essential to the system's operation—and a number of computers, each of which usually contains adequate direct access storage device capacity to support that computer's individual operations. The availability of fast reliable processors in desktop or small tower configurations—coupled with great improvements in local area network (LAN) and wide area network (WAN) technology—has rapidly brought the distributed system to the forefront in recent years.

Costs for development, maintenance, and expansion are significantly lower for distributed systems than mainframe systems. Typically incremental increases in distributed systems tend to be achieved by corresponding linear increases in cost, whereas the costs of increasing mainframe power tend to be exponential.

The decision whether one corporate database is enough must take into account the fact that the utilization of a database involves two distinct classes of computer processes:

- the database (the repository) itself, along with database management tools; and
- computer application(s), controlled by user(s), operating on the data.

User applications may reside as computer processes in operation on the same hardware that houses the database (the host), or on a separate machine that in turn may be located at a remote distance from the database itself. (This arrangement, which is a distributed system, is often referred to as the client-server model).

The single-site, centralized, "all things to all users" approach to corporate databases is an unrealistic solution to the need for data for decision-making and health-services research. Local communities need data that accurately reflect patient-care and management practices. The single-site solution does not foster a close relationship between the data in the database and an accurate reflection of practices.

This is not to say, however, that large database systems operating on powerful computers are never needed. When processing power requirements are great, and those computer processes utilize a common large data set, there is a role for the large data system, although the cost-effectiveness of such a system must be carefully studied.

Standard data and data dictionaries have produced many successes by readily permitting prototypes of applications—the routines, data, and dictionaries—to become the applications. As an example of how much reuse there is of existing components, the following has been found in a review of existing IAs (as well as undocumented references) found for Scheduling. Of the formally supported "class I" packages, 43 packages use over 230 scheduling routine calls or data elements, for a total of 1400 times. For example, The Hospital Location Name is accessed 190 times.

Future Database Development

Databases for the future will include object-oriented designs. This means the data and methods are stored together, hidden from direct observation by the user. It means that such "objects" will be able to "inherit" characteristics of data and behavior from already tested components. To some extent, it turns on its head the paradigm of building ever larger systems from smaller components and instead permits building smaller, more specialized components from larger existing components.

The meta DD is another aspect of a future version of the VA FileMan. The goal of the effort is to allow developers or IRM to construct views of DHCP data so that users may more easily query the database. This module provides the tools to build the views. With the meta DD, the developer defines a view that shows only meaningful fields that appear to be on a single file, thus making the navigation from the user.

Automated updates from national tables will be developed based on existing capabilities in the Kernel Installation and Distribution System (KIDS) and VA FileMan audit, as well as new tools for monitoring the complete process (change, update, and confirmation). Databases of the future will be more complex, yet easier to build and manage.

References

Curtis AC. 1992. Multi-facility integration: An approach to synchronized electronic medical records in a decentralized health care system. In KC Lun, P Degoulet, TE Piemme, & O Rienhoff (eds): *Proceedings of MedInfo 92*, pp. 138–143.
Davis RG. 1987. Conceptual foundation of DHCP. *MUG Quarterly, 17(1)*, 111–116.
Quinlan T. 1996. Time to reengineer the DBA? *Database Programming & Design, 6(3)*, 28–34.

7
Collaborating for Integration

STEVEN A. WAGNER

Importance of Standards

The importance of standards to the development of the Decentralized Hospital Computer Program (DHCP) was recognized at the time DHCP began. As a result, DHCP was developed using a standard programming language. Through the use of standards, DHCP has evolved into an integrated, portable, adaptable, extensible, and decentralized system. Standards have also helped reduce the cost of hardware and software purchased in support of DHCP. Standards are being used to promote and support collaboration with other federal government agencies and will be used to support collaboration with community healthcare facilities. Standards are considered even more important for the future of the Veterans Health Administration (VHA) than they have been to its past. The VHA believes that it must participate in the efforts to develop standards if the standards are going to support VHA's needs. Therefore, VHA is targeting those standards development efforts that it considers most critical for improving the quality of patient care and for developing a computer-based patient record.

Standards-Based Origins of DHCP

The DHCP was officially created as a program in 1982. However, the software modules that became DHCP in 1982 were primarily developed during the years 1978–1982. These software modules were developed in a decentralized environment and were based on a set of standards and principles. The key standard that formed the basis for DHCP was the M (or MUMPS as it was called in 1978) programming language (Forrey, 1988). M was, and still is, an American National Standard (ANS) programming language. In 1978, it was a brand new language that was concise and relatively easy to learn and that supported rapid development and prototyping. Because it was a standard language, it could be used on a decentralized basis by software developers located in different cities around the country. M is unusual as a programming language in that it incorporates features of a programming language, a database, and an operating system.

The importance of standards to the development of DHCP was recognized from the start. There is a wide variety of standards: international, national, federal government, and corporate/agency. Standards are referred to as internal or external, approved or de facto. International standards are those that have been approved by the International Standards Organization (ISO). National standards are those that have been approved by the American National Standards Institute (ANSI) and are referred to as ANS. Federal government standards are referred to as Federal Information Processing Standards (FIPS). Together ISO, ANSI, and FIPS standards are referred to as external standards and also as approved standards. Standards developed for use within a single corporation, government agency, or other organization are referred to as internal standards. Standards that are not approved by a standards approving organization, but that are used by a predominant sector of the market, are considered de facto standards.

Standards support the sharing of data, applications, and software tools. It was recognized early on that to develop DHCP in a decentralized environment, additional standards beyond the M programming language would be needed to ensure that individual developers wrote programs that would be portable from one site to another. Additional standards were needed because M is an extensible language and different M vendors created their own extensions, which were not part of the M standard. As a result, the principles of portability, operating system independence, and hardware independence were adopted early on by DHCP developers and incorporated into a set of internal VHA standards (Davis, 1987). A software package called the VA Kernel was developed and used as an internal standard to support these principles. The concept of a single, integrated data structure was also adopted. A database management system (DBMS) called the VA FileMan was developed and used as an internal standard to support this concept. The VA FileMan provides a standard user interface for accessing the DBMS, an active data dictionary, and a set of application programmer interfaces that software developers could use to interact with the DBMS.

Other internal standards included an e-mail application (called MailMan), that was based on a de facto standard (developed by the Department of Defense) called Simple Mail Transfer Protocol (SMTP), and a document called VA Programming Standards and Conventions, which was updated regularly and followed by all developers with minimal exceptions. This document prescribed the use of software tools such as the VA Kernel and VA FileMan and prohibited the use of those parts of the M language that would result in the development of nonportable or hardware-dependent software applications.

As DHCP has continued to develop over the past 15 years, the VA Programming Standards and Conventions document has been expanded, VA Kernel and VA FileMan have been significantly enhanced, and additional standards have been adopted (Bradley, 1989). MailMan incorporated support for the X.400 e-mail standard and an application called the Messaging System was developed to support several data exchange standards. Tables 7.1 and 7.2 list a set of standards that, in addition to the M programming language, DHCP currently supports or is considering supporting in the future. Entries with the protocol/application name in bold are currently supported by DHCP.

Advantages of Being Standards-Based

DHCP has reaped many advantages from the use of standards. Because of the extensive use of standards and the manner in which they were implemented, DHCP is an integrated, portable, adaptable, extensible, decentralized system. Standards have helped reduce costs through increased competition between vendors seeking VHA business. Standards also support collaboration within the multi-facility VHA system and between VHA and other government agencies.

The DHCP has always been an integrated system. This is due to the early adoption of three standards: the M programming language, VA FileMan, and VA Kernel. Standardizing on the M language for development of software applications supported not only the decentralized development of software applications, but also the subsequent integration of the software applications when they were brought together and installed at individual VHA facilities. The VA FileMan supported integration at the data and presentation (what the user sees) levels. The VA Kernel supported integration at the hardware and presentation levels.

The portability of DHCP applications is maintained through use of the standard M language, without extensions, and by requiring that hardware dependent functionality be provided by the VA Kernel. Adherence to these standards allows DHCP applications to run on any hardware platform and under any operating system that supports the M language. These same standards have reduced the cost of the DHCP system by allowing VHA to move from one hardware platform and operating system to another (Blanckensee, 1989). As DHCP has grown over the years, improved hardware platforms were required to support the software. The VHA has migrated DHCP across four hardware platforms so far. Likewise, the operating systems have changed. The M language itself was the first operating system under which DHCP ran. Since then DHCP has

90   System Design

TABLE 7.1. Communication protocols.

Protocol name	Description
Transmission Control Protocol/Internet Protocol (TCP/IP)	A network protocol for internetwork routing and reliable message delivery, which is endorsed by DoD and has become a de facto standard. In addition to specifying TCP/IP, Institute of Electrical and Electronic Engineers (IEEE) 802.3 (Ethernet), X.25, Frame Relay or Asynchronous Transfer Mode (ATM) should be specified to support lower level connectivity, Simple Mail Transfer Protocol (SMTP), File Transfer Protocol (FTP), and IP socket-to-socket should be specified to support upper level connectivity.
Health Level Seven (HL7) Hybrid Lower Layer Protocol	A simple point-to-point protocol endorsed by the HL7 committee.
American National Standard (ANS) X3.28	A complex point-to-point protocol that is a national standard.
Open Systems Interface (OSI)	A network protocol that is more robust than TCP/IP and that is a national (and international) standard. In addition to specifying OSI, IEEE 802.3 (Ethernet), or X.25 should be specified to support lower level connectivity and X.400 should be specified to support upper level connectivity.

TABLE 7.2. Application protocols.

Application name	Description
Health Level Seven (HL7)	An American National Standard (ANS) that is used for the exchange of clinical, text-type data within and between institutions.
Accredited Standards Committee (ASC) X12	An ANS that is used to exchange billing, insurance, and other administrative data within and between institutions via trading partner transaction sets such as 834 and 837. This standard is also referred to as Electronic Data Interchange (EDI).
Structured Query Language (ANS SQL)	An ANS approved standard that is used to send queries for information within and between institutions.
Digital Imaging and Communications in Medicine (DICOM III)	A national standard that is used to transmit radiographic images within and between institutions.
Institute of Electrical and Electronic Engineers (IEEE) P1073 Medical Information Bus (MIB)	A developing national standard that is used to interface critical-care instruments to an HIS.
American Society for Testing and Materials (ASTM) 1467	A national standard that is used to interface EEG and EMG systems to an HIS.
prENV 1064:1993	A developing European standard that is used to interface ECG systems to an HIS.

migrated across two other operating systems. Because DHCP could run on a wide range of hardware and operating systems, there was increased competition among vendors for VHA business, which resulted in lower costs to VHA.

The VA FileMan DBMS has proven to be an excellent tool for adapting DHCP to the wide variety of needs of VHA facilities. The VA FileMan, combined with the VA Programming Standards and Conventions and the VA Kernel, has provided the necessary tools to allow facilities to adapt and extend DHCP applications to local, site-specific needs (Lewis & Shull, 1990). The VA Programming Standards and Conventions prescribe the method facilities can use to adapt and extend DHCP; the VA FileMan and the VA Kernel provide the necessary tools to accomplish the needed changes or additions. The use of these standard tools has supported the development of reports, modules, and even applications by one VHA facility and their export to other VHA facilities where they could be used without further modification. The VHA has seven software development offices that develop applications for nationwide use at all VHA facilities. However, several applications that were later implemented nationally were initially developed at VHA facilities as opposed to a software development office. Examples are an application to support VA security police, an application to track time contributed by volunteers at VHA facilities, and an application to maintain information on the utilization and characteristics of buildings and rooms at a medical facility. Standards permitted decentralization of software development down to the individual facility level and allowed VHA to utilize the talents and innovation of all VHA personnel in developing DHCP software.

Role of Standards in Fostering Collaboration

The Department of Veterans Affairs (VA) and the Department of Defense (DoD) share a common goal. Both are charged with providing various services to military personnel, DoD while they are on active duty and VA once they become veterans. To accomplish this common goal, DoD and VA must share information and collaborate in a number of ways. Until recently, almost all information sharing was accomplished manually through the exchange of paper documents. Today, DoD and VA are beginning to share information electronically and explore further opportunities for electronic sharing in the future.

Sharing information electronically is greatly simplified if standards are available to support the sharing effort. As indicated in the paragraphs above, DHCP was able to share applications and software tools across all VHA facilities because of the standards that were adopted. These same standards have made possible the sharing of DHCP applications and software tools with other government agencies. Currently, DHCP applications and the software tools provided by the VA FileMan and the VA Kernel are being used by most federal government agencies that provide health care, including the Indian Health Service, Public Health Service, and Bureau of Prisons. The Composite Health Care System used by DoD is based upon DHCP and uses many of the same software tools (Geerlofs & Timson, 1989). The DHCP is also used by a number of State governments (McInnis, 1988; Russ, 1991), and a growing list of foreign governments (Werners, 1990; Schuller, 1991; Giere & Moore, 1993).

Data exchange and communication standards provide ready-made tools that define how to exchange information and assist in defining what data to exchange. Standards support cooperation and sharing of information. Electronic sharing of information with DoD will be based on standards. Both DoD and VHA are beginning to define Health Level Seven (HL7) events, messages, segments, and data elements for use in exchanging clinical information. Once initial definition efforts are complete, both parties will have a much better idea of when and what data they want to exchange. A common implementation of the HL7 standard can then be finalized. This implementation will be able to support the exchange of important clinical information when a military patient is transferred to a VHA facility, thereby supporting continuity of patient care.

The VHA has been able to leverage the DHCP Messaging System application and its support of the HL7 data exchange standard to support the exchange of clinical, text-type information. The HL7 standard is being used to exchange data with several commercial applications. One example is an interface to a single commercial application, which in turn interfaces with a wide variety of laboratory instruments. Each VHA facility has a large number of laboratory instruments from many different vendors. Each instrument utilizes its own specific interface format for exchanging information with another system. By using HL7 to communicate with the commercial application, which in turn communicates with each laboratory instrument, VHA was able to develop a single HL7 interface that will allow DHCP to receive results from almost any laboratory instrument. Another example is an interface to radiology voice-processing systems. The interface was written to allow radiologists to dictate the results of film readings. The radiologist uses the voice-processing system to request a list of radiology exams that need to be dictated (or the list of exams can be sent automatically) and to dictate and transmit the results to the DHCP system. By writing an interface specification that is robust and generic, VHA need only write its side of the interface once. Because this interface was written using a standard data exchange format, any vendor of voice-processing systems can then interface their product to DHCP using this same interface specification. Two vendors have written interfaces to DHCP using this interface specification and two more vendors have expressed interest in doing so.

The HL7 standard is also being used to exchange data between VHA facilities and corporate databases. Patient demographic, eligibility, insurance, and financial information is being transmitted daily to an income-verification database. Inpatient and outpatient treatment information is sent to a patient-care database. Future plans include transmission of clinical results, demographic, and other patient information between VHA facilities, both within and between Veterans Integrated Service Networks (VISNs). Also, a new DHCP scheduling package that is under development will use the HL7 appointment and scheduling formats to exchange patient, provider, and room scheduling information.

In utilizing the HL7 standard, VHA has found instances where the standard does not support data that VHA needs to exchange. In these instances the HL7 standard has been extended by using locally created formats. The HL7 standard supports the creation of local formats through use of a "Z" naming convention. Local formats developed by VHA will be submitted to the HL7 standards group for review and possible incorporation in future versions of the standard.

Role of Standards in Supporting Information Systems

Standards support information systems in many ways. The use of standards can improve: connectivity and communication between systems; security of systems; sharing of information between systems; data consistency and integrity; and adaptability to change by supporting open, portable, vendor-independent systems.

Although VHA has always operated in a multifacility environment, the original model for DHCP was medical center based. All patient-care services (and the information systems to support them) were controlled by the individual medical centers. Sharing of information between medical centers was limited. Recently VHA has reorganized into integrated networks of facilities where sharing of information between facilities will be a critical need. Communication and data exchange standards will be used to exchange data between network facilities. Plans are to use the Transmission Control Protocol/Internet Protocol (TCP/IP) protocol, including the File Transfer Protocol (FTP) and SMTP, as communication standards and HL7 as the primary application standard to support interfacility exchange of clinical information. DICOM III is being considered as the standard for exchanging medical images between facilities.

Now VHA is in the process of converting from terminal-based information systems to client–server systems. With the introduction of client–server software and graphical user interfaces, new standards are needed to support software development efforts. Plans are to continue use of the M language for software development on server systems and Delphi has been adopted as its standard tool for software development on client systems. Many more decisions are being made related to client–server systems, including standardizing on operating systems for clients, servers, and networks. Today a wide area network connects all VHA facilities. This network is being upgraded from X.25 packet, switching to a frame relay network.

Promoting Standards

Standards are considered even more important for VHA's future than they have been to its past. The VHA believes that it must participate in the efforts to develop standards if the standards are going to support VHA's needs. Therefore, VHA is targeting those standards development efforts that it considers most critical for improving the quality of patient care and

TABLE 7.3. Standards development groups.

Standards group	Description
Health Informatics Standards Board (HISB)	This board is responsible for coordinating healthcare standards development efforts in the United States. The VHA is a voting member of the board. One of the most important functions of this board is the elimination of duplication/overlap between standards developed by different standards development organizations.
Health Level Seven (HL7)	The HL7 is intended to be a fairly comprehensive standard for clinical, text-type data. The VHA hopes to use this standard to exchange most of the clinical information in patient records between VHA facilities and between VHA and non-VHA facilities. There are a number of clinical areas not yet covered by the standard.
American Society for Testing and Materials (ASTM)	The VHA participates on the E31 committee for health informatics. This committee develops a number of standards of particular interest to VHA related to security (privacy, authentication, etc.) and the Computer-based Patient Record.
Digital Imaging and Communications in Medicine (DICOM III)	DICOM III is a standard that is intended to support the exchange of all types of medical images. The standard is sponsored by the American College of Radiology and the National Electrical Manufacturers Association.
Institute of Electrical and Electronic Engineers (IEEE)	Two standards developed by IEEE that VHA is interested in are Medical Information Bus (for interfacing to ICU systems) and the Standard for Health Care Data Interchange, which is an object-oriented data modeling framework.
National Provider System (NPS)	The VHA is participating in the Health Care Finance Administration (HCFA) effort to assign a unique national ID to all healthcare providers, both individuals and organizations.
Accredited Standards Committee X12	The VHA is interested in the standards being developed by the X12N subcommittee for exchanging insurance, eligibility, and benefits information.

for developing a computer-based patient record. Currently, VHA representatives participate in a number of standards development groups summarized in Table 7.3.

The VHA representatives participate in standards efforts in a number of ways. They attend standards meetings where they participate in discussions, reviews of documents, and coordination efforts. They also review and comment on ballots of official standards documents and develop proposals for enhancing or otherwise modifying existing standards. The VHA is able to provide feedback to standards development organizations on the standards that it implements.

In addition to participating in existing standards efforts, VHA is trying to promote the development of standards where none yet exist. There are several key areas where standards are not yet developed or efforts are just starting.

Identification numbering systems is one key area. The Health Care Finance Administration (HCFA) is establishing a national numbering system for providers of patient-care services (both individuals and organizations). The HCFA is also establishing a national numbering system for payers. Both of these numbering systems are needed and can be used by VHA. The VHA is participating in these efforts. A patient identification numbering system is a critical system that has not yet been developed. The VHA is looking at the American Society for Testing and Materials (ASTM) standards for assistance in this area. All of these numbering systems (patient, provider, and payer) are needed to uniquely identify individuals and organizations when healthcare information is exchanged between different facilities, especially when the facilities are owned by different enterprises. Without these numbering systems, each healthcare facility must translate its identifiers to those of every other healthcare facility with which it needs to exchange data.

Several standards are needed in the area of privacy, security, and confidentiality. As VHA reorganizes into networks of facilities, including community facilities, patient information will need to be transmitted outside of the VHA telecommunications network. The security and confidentiality of information transmitted within the VHA telecommunications network can be controlled by VHA; and, as a federal government agency, VHA has been required by law to meet certain privacy requirements concerning patient information. Similar controls and requirements do not yet exist for all nonfederal facilities. The VHA is participating in initial ASTM efforts in these areas. The ASTM recently reorganized their standards efforts and established a security division to coordinate and streamline their standards efforts in the area of security. This may provide the impetus needed to rapidly advance security-related standards.

Another key area needing standards includes codes, vocabulary, and terminology. Clinical information exchanged between facilities is subject to misinterpretation and misunderstanding if the terminology and vocabulary differ from one facility to another. The VHA is implementing a clinical lexicon based upon the Unified Medical Language System (UMLS) to help cope with this problem. Although there are many coding systems for different aspects of health care, a coding system does not yet exist that specifically supports the clinician's needs. Most of the existing coding systems were designed for billing purposes. The UMLS-based clinical lexicon will support the integration of concepts and terms from various coding systems, but what is needed most is a coding system that supports the clinical concepts and terms used by the practitioners who provide health care. Some preliminary work has been done in this area by a task group of the Health Informatics Standards Planning Panel, the organization that was superseded by the Healthcare Informatics Standards Board (HISB). However, a standards development organization has yet to assume this task.

The VHA is also participating with ASTM in testing and improving an ASTM standard for computer-based patient records content (ASTM 1384). The DHCP may be used as a test bed for a number of the concepts in the ASTM standard. Other federal agencies such as the Centers for Disease Control (CDC) and the National Committee on Vital and Health Statistics have expressed an interest in possibly using DHCP as one method of testing standard data exchange formats and electronic transmission of data. The CDC is working with the HL7 standards group in an effort to develop an HL7 format to support the exchange of immunization information.

Conclusion

The VHA has used healthcare standards in its information systems since the inception of DHCP and has found the use of standards to be cost-effective and efficient. Major changes and challenges lie ahead for VHA, and standards will play a significant role in adapting to change and meeting the challenges. Participating in standards development activities provides VHA the opportunity to ensure that the standards developed meet VHA needs. Participation also results in early knowledge of new standards under development and the opportunity to plan for their implementation and use.

References

Blanckensee HT. 1989. VA DHCP replacement systems: VAX Clusters for the 90's. *MUG Quarterly, 19*, 1.

Bradley D. 1989. Healthcare computing standards lead MUMPS into the 1990s. *MUG Quarterly, 19*, 3.

Davis RG. 1987. Conceptual foundation of DHCP: The decentralized hospital computer program in the department of medicine and surgery in the Veterans Administration. *MUG Quarterly, 17*, 1.

Forrey AW. 1988. MUMPS is more than a language. *MUG Quarterly*, *17*, 4.

Geerlofs JP, & Timson G. 1989. A FileMan-based patient care management system. *MUG Quarterly*, *19*, 1.

Giere W, & Moore GW. 1993. Translating English into German using VA file manager. *M Computing*, *1*, 4.

Lewis JR, & Shull RM. 1990. Modification of the DVA DHCP software for use in a veterinary teaching hospital. *MUG Quarterly*, *20*, 1.

McInnis KA. 1988. The use of MUMPS and the Veterans Administration software in a statewide medical information network. *MUG Quarterly*, *18*, 1.

Russ DC. 1991. Western State Hospital: Implementing a MUMPS-based PC network. *Computers in Healthcare. June*. pp. 18–23.

Schuller G. 1991. An open, modular, distributed and redundant computer system for hospitals. *Computers in Healthcare. June*. pp. 24–31.

Werners M. 1990. FileMan in a foreign land: Its use in a non-medical, non-English environment. *MUG Quarterly*, *20*, 1.

8
Supporting Systems After Installation

Roy H. Swatzell, Jr. and Virginia S. Price

Introduction: A Brief History of VA's Software Presence

Prior to 1982, the use of computers in the Department of Veterans Affairs (VA) consisted mainly of centralized mainframe applications supporting payroll, finance, logistics, and the collection of aggregate management information. In 1982, drawing upon a small and dedicated group of developers and medical center workers, the healthcare portion of the VA began an effort that culminated in the deployment of computer hardware and software to every VA medical center (VAMC).

By choice, VA pursued a twofold approach, developing software from a bottom-up decentralized approach and providing VAMCs with necessary hardware through a top-down, centralized procurement. This effort came to be known as the Decentralized Hospital Computer Program (DHCP).

The DHCP laid the groundwork for the present VA healthcare information environment. In 15 years, VA evolved from automating critical workload needs to automating the patient record. Today a hybrid networked healthcare information environment provides technology solutions using in-house developed and commercial off-the-shelf products. Along the way, many issues emerged relative to installing and supporting a large-scale, integrated healthcare information network. Choices had to be made on how to deal with a number of these vital issues, as follow: defining requirements; involving users and managing user expectations; managing in-house software development processes; supporting users, systems, and applications; designing and delivering training; transitioning to include commercial off-the-shelf applications; upgrading the information and telecommunications infrastructure; and enhancing networking capability.

Over the years, VA has reached for solutions that are standards-based, cost-effective, and applicable in a number of healthcare settings, whether inpatient or outpatient, acute care or long-term care, facility-based or network-based. This chapter provides a brief historical tour and outlines the "lessons learned" during the journey, with special emphasis on issues stemming from system growth, demands for additional functionality, and the difficulties of supporting an increasingly complex environment.

Starting Out: The Early Years (1983–1985)
Design Principles

In 1983, the development approach paired programmers who had technical expertise with VA healthcare providers and administrative staff, working side by side in a VA medical center environment. This grass-roots effort enabled VA to develop software solutions that directly complemented the way VA delivered medical care. The initial DHCP software packages were selected because of their undisputed high return on investment. Critical needs for automation had been identified in the areas of admissions, discharge, transfer, scheduling, pharmacy, and laboratory. Software was developed and released to the field in phases.

Certain guiding principles were stressed. The system was to be modular, integrated, standards-based, cost-effective, extendible and customizable, portable and vendor independent, and have a common user interface. The choice to use the MUMPS programming language in itself provided a level of standardization, because a data dictionary and some database management capabilities were inherent in the language. As time went on, VA system designers created other foundational elements to implement these principles, including a database management system (VA FileMan) and a set of standard software tools that provide a uniform interface between the underlying operating system and DHCP application modules (Kernel). The Kernel addressed areas such as security management, device handling, menu management, and task management in a standard manner, relieving application developers of the need to deal with these issues.

Infrastructure

To ensure that the hardware configurations were tailored to the size of the medical center, the procurements for equipment to support the first three DHCP modules were based on a sophisticated, workload-dependent sizing algorithm. The initial selection of low-end terminals with monochrome screens proved to be cost-effective and allowed VA to develop software with substantial functionality at minimal cost to the taxpayer.

Over the 15-year life cycle of DHCP, VA has conducted four centralized procurements to provide all VA medical centers with hardware to support the growing DHCP applications. Centralized procurements have provided economies of scale and encouraged maximum market competitiveness. A strong focus on standards-based development and vendor independence enabled VA to change platforms multiple times (from PDP minicomputers to VAX minis, to 486 microcomputers, to Alpha micros) by changing only a few lines in the Kernel, rather than having to engage in major application rewrites.

During the deployment of DHCP hardware and software, VA also established a national telecommunication network and e-mail system called FORUM. Department of Veterans Affairs began to use FORUM to communicate with each other. User groups and application coordinators began to use e-mail to define necessary functional components of software applications and to share operational shortcuts and implementation approaches that had been well received at their facilities.

Implementation

During the period from 1983 to 1985, VA deployed applications that addressed critical needs for automated support in the areas of admission/discharge/transfer/scheduling functions, outpatient pharmacy, and laboratory. Medical centers were eager to install this software and were highly motivated to gain experience with using computers in the performance of their day-to-day duties. At the same time, VA management felt that deployment to all medical centers was essential to the success of the program. To satisfy this need, VA introduced a concept called *rapid prototyping*. An initial release with a minimal functionality set was developed based on user specifications and was fielded immediately, with the understanding that subsequent versions would follow periodically. The advantage of this approach over the traditional formal specifications and bid-for-product process was that VA could develop the software within its own resource limitations and expand capabilities based on user input as the users became more conversant with the technology.

The use of rapid prototyping and the low requirement for complex system testing associated with an initial system implementation meant that products could be developed and disseminated in a short period of time. As a result, VA was able to install the first wave of functionality in a year, while using traditional methods would have required several years.

The initial effort enjoyed a number of early successes, in part because of the environment that existed at the time. The VA healthcare delivery network consisted of 172 medical centers and well over 200,000 employees. Less than 1% of the staff were computer literate, and very few clinical staff, including those with appointments with local private sector medical centers, had any experience at all with computer systems. Low user expectations resulted in a maximum payoff for a minimal functionality set. Training could be directed at a fairly homogeneous population of potential users and focused at a single level of application awareness. The fact that there was no need to develop multiple levels, dependent on the user's expertise, greatly simplified the task of coordinating training needs and producing training materials.

Support

Before receiving any hardware and software, each VA medical center identified an Information Resource Management (IRM) site manager, who would be responsible for installing and supporting the equipment and the information system. As applications were fielded, the medical centers were asked to appoint application coordinators, who were trained to assist with implementation, train users, and support users after implementation.

Growth and Growing Pains (1985–1989)

Adding Applications in Program Areas

The early software development cycle was characterized by a great deal of control by the computer analyst and direct input by the users at the technical level. This enabled VA to generate a product rapidly with maximum flexibility afforded to the computer staff. However, this process did not compensate for variability from medical center to medical center in the way functions were performed. Given that the process of filing a prescription varied from center to center, many adaptations to the software were suggested.

As follow-on releases of software occurred and users became familiar with the capabilities of the technology, it became clear that formal mechanisms for gathering specifications that represented the broader user community could help reduce the coding time associated with "generalizing" the releases. This was acknowledged by the creation of Special Interest User Groups (SIUGs).

Other program areas began to see the benefits of automated support for day-to-day operations and its byproduct of accurate and standardized workload reporting. The SIUGs were established for these areas as well, and the demand for new applications began to multiply exponentially.

The number of DHCP packages increased from four in 1983, to eight in 1985, to 27 in 1987. More SIUGs were established, defining needs in program areas such as inpatient pharmacy, nursing, dietetics, engineering service, fiscal, medical records, mental health, radiology, and surgery. The functionality available in DHCP by the end of this period is depicted in Figure 8.1.

Managing Functionality Creep

The expansion of DHCP software functionality caused a dramatic increase in the amount of historical data medical centers were required to keep on hand. As database growth stressed available disk space, medical centers

102 System Design

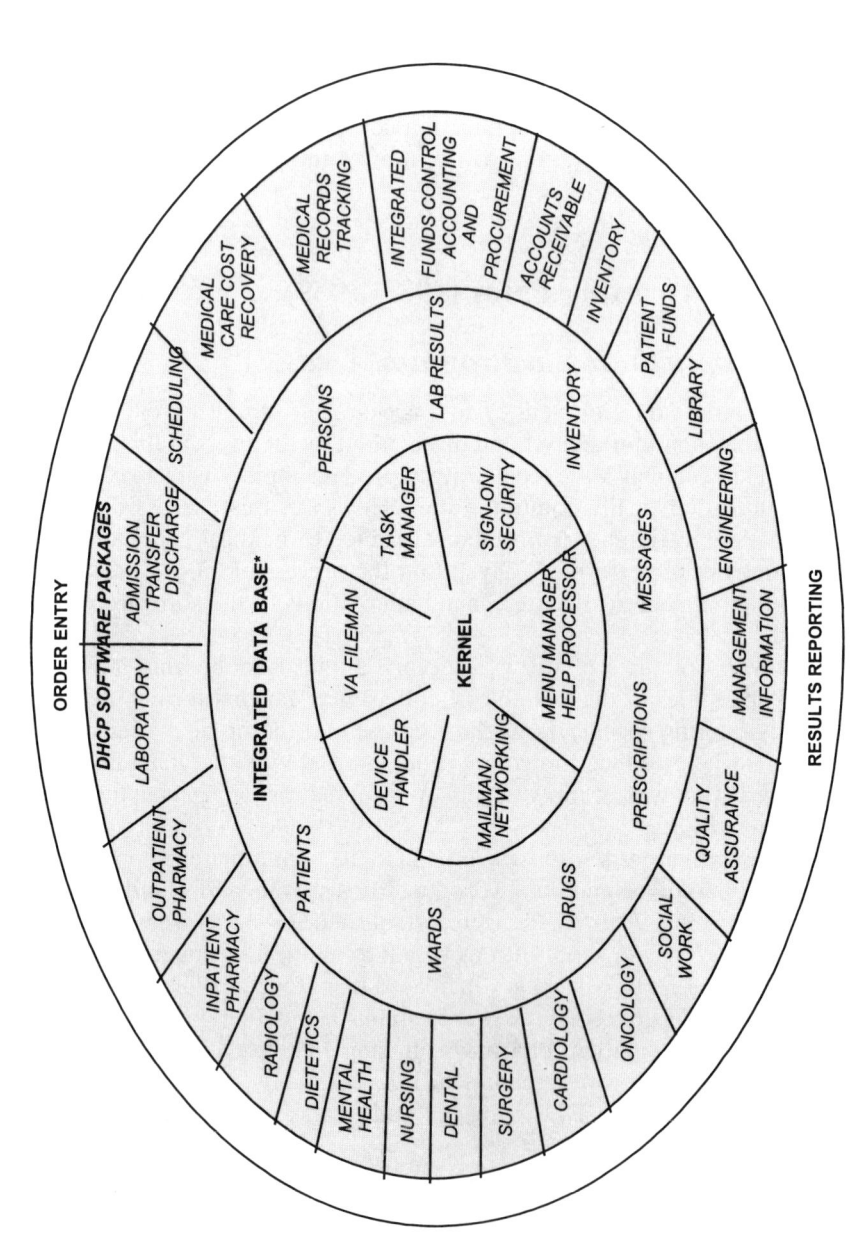

FIGURE 8.1. Decentralized Hospital Computer Program Functionality, 1989.

began to demand hardware and software solutions for managing longitudinal data.

In the mid-1980s, when disk storage was relatively expensive, solutions tended to focus on archiving and purging data. In the early 1990s, when disk drives became less costly and systems often offered built-in redundancies, the focus shifted to retaining data longer and thus providing more complete information for clinical decision making and for a comprehensive electronic patient record.

Disk storage was not the only component of the computer system that enjoyed a price reduction. During the course of 15 years, VA cycled through four hardware platforms (Digital PDPs, Digital VAXs, a 486-PC platform, and Digital Alphas). With each hardware platform, the processing capacity increased, while costs decreased markedly with the transition from VAXs to PCs. With lower costs for the hardware and disk base, the medical centers found it easier to accommodate "functionality creep" on a temporary basis.

However, the success enjoyed by DHCP soon began to stress the development and support structures at the Information Systems Centers and the medical centers. As users became more familiar with computer capabilities, they wanted more "bells and whistles" and more complex functionality. Additionally, as the DHCP information set increased, more uses for DHCP data were identified and requested. These new requests began to place serious demands on development and support resources, as well as increase the computer resource consumption and create extra demands for disk space. The process of increasing demand for additional features came to be referred to as "functionality creep."

The overhead associated with a maturing software process—verification, systems testing, documentation, and integration—also began to extend the product delivery timelines. It soon was apparent that managing the expansion of functionality beyond the critical mass necessary to meeting basic needs was becoming a key element in containing cost and effectively administering development and support resources. Three categories of demands on resources were especially critical.

User-Driven Demands

Users were showering development resources with requests for enhancements to existing software and for new and more complex packages requiring integration among system elements, such as order entry, and quality assessment. Extracts of DHCP data were also requested to feed external decision support systems.

Mission-Driven Demands

Mission-driven demands included mandates from Congress and oversight groups for VA to explore certain directions in automation and to supply automated solutions to answer identified material weaknesses in program areas. Special missions like Desert Storm ranked as number one priorities.

Agency-Driven Demands

Systems that served the entire agency began to request DHCP to feed off data for payroll and supply activities. These requests varied from writing extract routines to full-scale system implementations, with all the attending interface work. Because DHCP is public domain software, answering Freedom of Information Act (FOIA) requests for DHCP routines and implementation assistance also took its toll. Portions of the DHCP system are running in Egypt, Finland, and Germany. Sharing activities with other government agencies, such as the Department of Defense, Indian Health Service, and the Federal Bureau of Prisons, also complicated the process of determining development priorities.

Standards

During this period, the DHCP product continued to mature, local VA medical centers began to develop automated solutions for local needs, and the Agency began to integrate more commercial solutions into the healthcare information environment. Inevitably, VA began to see the needs for standards to complement those inherent in MUMPS (now renamed M), VA FileMan, and Kernel.

New standards were drafted for internal DHCP development; a Data Base Administrator (DBA) was appointed, and DBA integration agreements were developed for file sharing and package integration. As a result of DHCP programming standards, packages developed to address local needs at one VA medical center could be ported to another VA medical center without significant modifications. Standards were also developed to meet newly emerging needs to verify software and provide documentation to users. Methodologies for assuring that all national software and documentation adhered to standards soon followed.

Security

Because the security of patient information was always a prime concern, a program was developed early on to protect the physical security of DHCP terminals and to control internal access to DHCP information. Standards and guidelines were developed and provided to sites, along with assistance in training and performing on-site evaluations of information security. Teams were formed of technical security specialists, regional information security officers, and customer support personnel; these teams visited VA medical centers, resolved ad hoc security questions, and assisted with contingency planning and risk analyses. The security program had to address security aspects associated with new developments, such as automating the patient record, providing electronic signatures for chart entries, implementing new information and networking technologies, and data sharing.

Training

The demand for training became a high-profile issue during this time. Staff rotation and turnover took their toll on computer literate staff. Continual training was needed as more and more employees began to use DHCP applications.

As the software expanded into a user population that could not be easily released to attend face-to-face training sessions, VA sought innovative computer-assisted training modalities. Using a software product called ClassMan, sites could dial into a test account at a support center and view a session while listening to audio instruction on a conference call. The session was instructor-facilitated, including interactive questioning. The VA also developed videotapes that provided a narrated and visually enhanced tour of the DHCP software for the user to view at leisure. These Computer Assisted Training (CAT) tapes showed users interacting with software modules while a trainer explained the various options. The CAT tapes were prepared for many of the early DHCP applications to address the continued education and inservice training needs of VA staff.

Customer Support

As functionality expanded and products matured, VA turned its attention to structuring support services, establishing regional centers to address software development and other customer needs. Initially, computer specialists were selected for support positions. However, it soon became apparent that candidates possessing firsthand knowledge gained from experience in a VA medical center had excellent potential as support representatives. A computer language could be taught in a short time, but there was just no substitute for years of experience in a VA medical center environment.

Hiring Automatic Data Processing Application Coordinators (ADPACs) and others with functional experience in VA medical centers changed the face of support. The result was a breed of generalists who were well rounded, able to work together in small teams to solve problems. Readily accessible to the medical centers, they even repaired malfunctioning software and developed custom programs and code enhancements requested by sites. In part because their catchment areas were typically from 18 to 25 sites, regional support staff were able to build rapport and establish a relationship of trust with the IRM shops they supported.

Customer Feedback

In a decentralized healthcare system, the process of communicating problems, desired enhancements, bugs, and fixes among support staff and developers can be quite challenging. For the VA, matters were compounded because the implementation was a nationwide effort. Workdays did not coincide, and time zone differences impeded verbal communication. To meet this challenge, DHCP developed three systems and made them accessible to users over the VA communication network, FORUM, for requesting and distributing enhancements and reporting problems, as described below.

The Electronic Error and Enhancement Reporting (E3R) system started out as an enhancement reporting system and an error reporting system, but it was soon discovered that the information to document a Request for Enhancement and an Error Report differed so markedly that a single

TABLE 8.1. Sample E3R (request for enhancement).

E3R Report Number: 7497   **Suspense Date**: JAN 13, 1996
Priority: Future Enhancement
Status: Assigned to Package/Developer
Submitted By: Jane Doe   **Date Submitted**: DEC 14, 1995
Location: VAMC
Phone numbers:
Responsible Package: ORDER ENTRY/RESULTS REPORTING (UNKNOWN)

Description: We are in desperate need of a legible discharge instruction sheet for the patient that pulls from many packages. This sheet needs to include the following fields:

- Treated for   (text entry field)
- Discharge to
- Discharge Appointments from DHCP
- Medications from DHCP
- Instructions   (text entry field)
- Diet (text entry)
- Activity (text entry)
- Equipment   (text entry)
- Contact if emergency (text entry) though could be primary care provider in DHCP
- Other follow-up Services like VNA with phone numbers (text field)
- Patient signature that instructions were understood

This would be an important quality enhancement. We spend so much time and money treating patients and it would be great to have a concise, legible instruction sheet.

Impact: This would result in better care delivery and clearer instructions to patients. It would also help to meet JCAHO requirements for patient instructions and documentation of receipt by patient. It would save the nurses lots of time, and be good for the chart, too.

Recommendation: Implement above changes.

format would not suffice. The current E3R system is a structured reporting approach that identifies the module, the desired functionality, and proposed solutions if they are readily apparent. Table 8.1 shows an example of an E3R to request a desired new functionality. The user fills in a template with the desired functionality, and the request is forwarded to the appropriate developers and/or prioritization groups.

National On-Line Information System (NOIS) was specifically designed for managing "bugs" and problems reported by sites to customer support services. By preventing multiple support specialists from working on the same problem, the system increases staff productivity. Sites also receive a more consistent response to a given problem, because there is only one support specialist working on that problem with the development staff. The system can combine problems that initially appeared to be different into a single problem and allow for comments by more than one specialist as a problem is refined and referred to other specialist(s) for input. Entries can be made by computer specialists at the VAMC level, as well as customer service support specialists.

In addition to tracking problems assigned to developers, NOIS is used for continuous quality improvement studies. Sites are called on a random basis to determine the quality of the support and are asked for suggestions to refine the support delivery process.

Patch Module

Designed to deal with errors introduced when sites entered fixes manually into production accounts, the Patch Module sends code to the site. The site then has the capability to automatically install the code in their test account, verify that there is no negative impact, and finally automatically install the code in the production account.

Together, the E3R module, the NOIS system, and the Patch Module provided a powerful, automated, feedback loop for communicating problems and desired enhancements to the developers, sending back software patches to the sites, and making comprehensive information on reported "bugs" and "fixes" available to all system users.

Reaching for the Gold Ring (1989–1996)

The DHCP continued to grow and expand to meet user needs and to take advantage of technological advances (Figure 8.2). A more sophisticated user community began to demand more generic solutions and more powerful tools. Users asked for clinical data repositories, research access, increased compute power, intuitive user interfaces, and online help similar to what applications on their home computers provided.

Drawing on the richness of the DHCP database, users began to see infinite possibilities of assembling data in different ways and views to sup-

"Our hero tiptoes delicately across the narrow path of shifting paradigms alert to the dangers associated with change...."

FIGURE 8.2. The perils of planning.

port management and clinical decision making. They envisioned new levels of clinical functionality—obtaining clinical data from other treatment settings to provide more complete views of their patient, and handing off data to other systems to satisfy national requirements and research needs.

Newly redefined, functionality became more interdisciplinary in nature. Quality assurance staff had valuable suggestions for designers of nursing applications, while medical records specialists, nurses, pharmacists, and physicians were all interested parties in developing the order entry function. With many of these users expressing conflicting expectations, it soon became apparent that the user group structure needed a major redesign.

In 1988, VA released a Strategic Information Systems Plan (SISP), which defined a strategic direction for expansion of VA information technologies. This direction included the following features/capabilities for the evolving DHCP:

- broadbased information integration;
- data accessibility for every workstation;
- on demand data roll-up capability;
- integrated Commercial Off-the-Shelf (COTS) applications;
- formal and unbiased buy/build decision process;
- full and adequate investment in training;
- consistently fast system response times.

The SISP also identified two broad information priorities that were to set the direction for all VA development activities:

- Highest priority: Applications that integrate patient data for clinical care.
- Second highest priority: Applications that integrate information for management decision support.

Acting on SISP recommendations, an Information Resources Advisory Council (IRAC) was established to prioritize functionality needs for VA. Separate groups were formed to prioritize functionality needed for clinical, management, and information technology. Cross-functional groups called Expert Panels were formed to work with developers on specific functional applications.

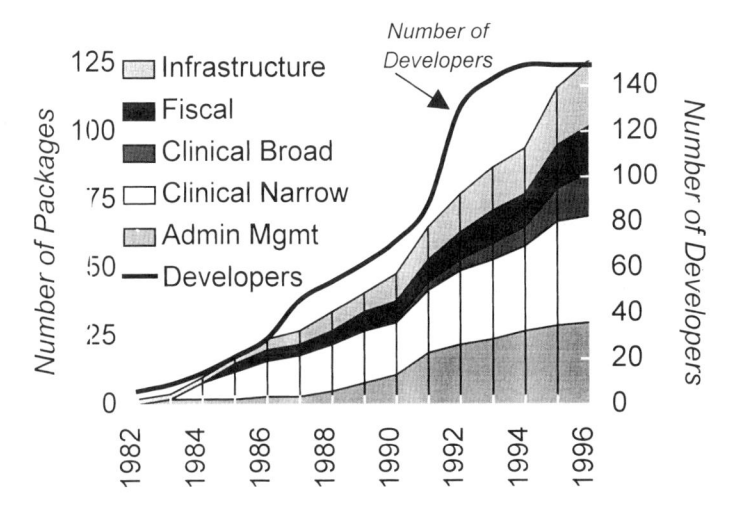

FIGURE 8.3. Growth, 1982–1996.

However, prioritizing requests for additional functionality using just these basic guidelines proved to be inadequate. The volume of work far exceeded the resource pool, and multiple initiatives were identified as "number 1" priorities. By 1991, pending requests for new functionality and E3R enhancements reached a total of 4353. Clearly, development within existing resources could not keep pace with user expectations, and a new method of prioritizing requests and managing user expectations was needed (Figure 8.3).

In Pursuit of the Clinical Record

As mandated by the priorities set forth in the SISP, VA development focused on expanding the information available to clinicians. An application called Health Summary provided much of what is traditionally found in an automated clinical record. Predefined building blocks of health information could be assembled in a summary to meet the needs of a specific clinic or healthcare provider. In many VA medical centers, clinics printed out Health Summaries and gave them to healthcare providers prior to scheduled patient visits. The information was remarkably complete, accurate, and many times was available in advance of the medical chart. Provider response was immediate and enthusiastic. Then VA developed an application called Patient Data Exchange that allowed VA medical centers to send the Health Summary to other VA medical centers, and the race for the clinical record was on.

The computer-based patient record (CPR) is DHCP's long-term goal, and all clinical development is oriented toward that goal. The excellent progress made so far is reflected by the 1995 award given DHCP by the Computer-based Patient Record Institute. The DHCP was one of three honorees recognized for "Excellence in the Implementation of Computer-based Patient Record Systems."

Today VA continues to work toward the CPR through projects integrating images (e.g., x-rays, pathology slides, endoscopy views, and CAT scans) into the DHCP database, expanding support for intelligent workstations, and furnishing treatment providers with clinically relevant information from treatment episodes at other VA facilities.

Standards and Commercial Applications

To address the growing demands for added functionality, VA instituted the Hybrid Open Systems Technology (HOST) program to explore and evaluate innovative Commercial Off-The-Shelf (COTS) technologies of possible use to the VA healthcare network. The HOST program was charged with formally testing commercial technologies that offered potential value-added benefits and integrating these technologies with proven components of VA's existing DHCP.

Accomplishing these goals required that significant development resources be assigned to develop interfaces for COTS products and that a first-line support interface be designed for COTS and DHCP products. A long-term solution, this will allow both sets of products to exchange information in a seamless manner. As a solution, it requires compliance with maturing open systems standards and substantial changes to the DHCP architecture.

Efforts are underway. In 1993, VA developed the first standards-based bidirectional text and image linkage with a Picture Archiving and Communication Support (PACS) radiology system. Currently, VA is examining a technical architecture that will make it easier for VA medical centers to enrich their base DHCP systems by incorporating commercial applications that are able to use industry standard messaging protocols (e.g., HL7) as their interface to DHCP.

Work to evolve and develop the VA's approach to standards continues. Changing the paradigm to "open" DHCP so that the VA system can more easily incorporate commercial technologies will require significant changes to VA standards, as well as a new commitment of resources and enhanced skill sets for VA developers and support staff.

Communicating with Users

By 1989, the FORUM system had expanded to allow access to the National Patch Module, National VA Interlibrary Loan system, tracking systems for facility profiles, applications running at each facility, an ADP equipment inventory, and software development plans. Gateways to other systems were established, and users with appropriate permission could access systems such as PDQ, an online system for oncology treatment protocols. Department of Veterans Affairs Medical Centers could also browse an online catalog of software developed by other VA facilities and request software routines that met their local needs.

Today, FORUM supports over 25,000 active users (both VA and non-VA users) with 13,000 sign-ons per day. This extensive communications network has proved to be an extremely valuable asset for VA, especially since the addition of network mail and gateway support. Users can now send mail messages through FORUM to internal mail systems at individual facilities and also access the Internet.

Changing the Face of Support

As more and more complex DHCP packages were released, it became more and more difficult to provide support with only minimal increases in support staff. Support managers had to look for better ways to utilize available resources. To continue established support levels, staff had to become specialists in addition to being generalists. Support teams were formed to assure coverage, backups, and crosstraining. Staff began to work together, across team boundaries, to solve more complex support problems. The complete menu of support services is shown in Table 8.2.

In 1991, two landmark decisions were made concerning support. Support coverage was extended to 24 hours per day, 7 days per week for emergencies, and Help Desks were established to help triage calls from VAMCs.

Due to the growing reliance of medical centers on their DHCP systems, around-the-clock support coverage became mandatory. Backups were becoming lengthy and were consuming the time formerly used for installing new software and patches. Installation of new applications was also becoming more complex, and a growing number of sites felt that it was critical to have experienced support staff available afterhours.

Help Desks were established to streamline call-handling functions and assist with workload management. Help Desk staff received, logged, and dispatched incoming calls from the field and served as a firstline interface between medical centers and VA support centers. Support staff could then concentrate on solutions to customer problems without having to directly answer new calls from medical centers. Workload could be distributed more evenly among staff, and more complex calls could be referred to senior staff.

In 1995, VA further streamlined support delivery by aligning all Customer Service activities under a single manager. National teams were established to provide support for all sites. Currently there are five national application support teams (Clinical 1, Clinical 2, Fiscal Management, Medical Center Management, and Patient Information Management) and three technical support teams (486, AXP, and LAN Systems).

TABLE 8.2. Support services provided to VA users.

MENU OF SUPPORT SERVICES Veterans Health Administration

A. Operations Support
   1. Medical Center Support
   2. E-mail
   3. Consultation/Emergency Call-Back
   4. ADP Planning
   5. Communications
   6. Conferences
B. Application Support
   1. Problem Analysis and Resolution
   2. Documentation/Reference Materials Library
   3. Consultation/Emergency Call-Back
   4. Software Distribution and Testing
   5. Routine and Urgent Patch Support
   6. Implementation Planning
   7. Emergency Fixes
   8. Limited Non-VA Support
   9. Custom Programming
C. System Support/Telecommunications
   1. Operating System Support
   2. Database Repair
   3. Configuration Systems Analysis
   4. Performance Analysis
   5. Communication Network Support
   6. Regional/National Mail Support
   7. System Tuning
   8. Failure Recovery and Contingency Planning
   9. ADP System Integrity Consultation/Guidance
   10. Hardware Support
   11. Optional Services
       a. Microcomputer Support
       b. Office Automation Support
D. Administrative/Management Support
   1. Management Support
   2. Information Resource Management Planning
   3. Quality Management
   4. Information Dissemination
   5. Emergency Site Management Support
   6. VA Management and Headquarters Support
E. Training Support
   1. National Training Program Support
   2. Development of Training Materials

Benefits derived from a national team approach include uniformity of service, economy of scale, and overall improved service to customers. Customer Service is able to provide a consistent level of support to all sites because regional barriers have been eliminated, along with the perceived disparity of service between regions. Despite the streamlining and improved efficiency, some sites still long for the "good old days" when they had a one-on-one, more personal relationship with their customer service representative.

Lessons Learned

You Cannot Neglect Strategic Thinking/Planning

Before the VA healthcare network achieved a high level of integration and competition for automated resources, it was far easier to accommodate the phenomenon of "functionality creep." Interesting ideas and personal projects could be included in the workflow without heavy impacts on product delivery timelines or support functions. As the information environment became complex, it became apparent that planning, providing information to sites, and managing processes are required to deal with the interdependencies between applications, training, equipment needs, and infrastructure requirements. To deal with these competing demands for resources, VA has instituted a strategic planning process which includes the following key elements:

- Identifying business needs;
- Mapping technology solutions to these needs;
- Determining priority and cost-effectiveness;
- Linking the functionality to infrastructure requirements;
- Disseminating information concerning application delivery; equipment needs, such as workstation profiles; and infrastructure standards and requirements as far in advance of implementation as possible so sites can factor this into their own planning cycles.

Build on a Good Foundation

The guiding principles upon which DHCP was built have stood the test of time. Involving users in the development process and using standards to the maximum extent possible have enabled VA to install an integrated healthcare information system at all VA medical facilities in an efficient and cost-effective manner. Platform and application changes can be made across the entire healthcare network with a minimal impact and without custom programming. These principles have served VA well, and future enhancements will honor these same principles.

Focus Your Focus Groups

Involving users at the grass-roots level has always been an important consideration in developing DHCP and has helped to build a healthcare environment that is robust and satisfies the needs of those who are interfacing directly with our veteran patients, whether scheduling appointments or delivering direct care.

114   System Design

This personalized development approach has resulted in a larger than average percentage of users who passionately identify with DHCP as being "their system." Highly knowledgeable users can also be very inventive in their concepts of how automation can be used in the healthcare environment. Finding the balance between the needs of these "power users" and the general user community can become a fine art. Arriving at a process that has continuity and results in a clearly stated and easily understood end product can also be a formidable task. The following are suggestions distilled from the VA experience dealing with user groups:

- Identify clearly the product wanted from the focus group.
- Facilitate the process to arrive at a clear specification.
- Manage user expectations.
- Simplify. Use the 80/20 rule—80% of the users will use 20% of the functionality.
- Document, document, document!

People Issues Are Paramount

In the early days of DHCP, VA's information environment was virtually a clean slate. Developers, implementers, and VA medical center staff had few expectations and a strong sense of adventure and accomplishment. Over the years, the pendulum has swung in both directions. Resource constraints have at times caused VA to move out in the "new frontiers" arena at the cost of the existing structure. We can attest to the fact that the lowly task of maintaining an ongoing information environment must be handled with the same care as the flashy new frontiers. During the 15 years of DHCP, we have found that attention to the following areas will result in a steady and productive implementation.

People Move More Comfortably in the Field of the Familiar

Investments in usability testing, user and systems training, good written documentation, online help, and an attentive support staff pay big dividends in user acceptance and perception of quality, both in terms of product and service.

Implementation Issues Are Just as Important as Good Software Solutions

Preparing staff for the introduction of new applications, and especially new technologies, is extremely important and will make or break the reputation of an organization. Familiarization campaigns, adequate initial and follow-on training and support (both on-site and at a corporate level) are essential to establishing an environment where users are comfortable and frustration levels are low.

Do Not Give in to the Tyranny of the Immediate or Be Unduly Sidetracked by New Opportunities

Balance new ventures with maintenance activities, and make sure that development and support staff skills are kept up to date. The automation field is basically a service-oriented business. Identifying needs and providing solutions are the motivating force behind most of the work we do. Our workers are our tools to accomplish our goals, and not much can be accomplished with dull tools. It is easy to neglect recurring requirements for systems and application maintenance, support structures, infrastructure upgrades, and ongoing training when faced with ever expanding vistas of new opportunities and gratified users.

Looking Forward

Increasingly, health care is no longer delivered primarily by freestanding healthcare institutions, but rather by integrated healthcare delivery networks that cover the entire continuum of care, from in-home services to tertiary care.

To meet the challenge, VA is in the process of combining facilities and increasing its emphasis on outpatient care. The VA's information systems are moving forward to meet changing information needs as well. This quantum leap into the next generation of information systems calls for a broad range of accurate information for clinical decision making and managed care, and a sophisticated ability to share integrated patient information within the VA healthcare environment and across service providers. The DHCP is developing an information architecture that will facilitate integration with COTS products, and is developing text, voice, and image teleconsulting capabilities that will be critical to VA's future ability to participate in healthcare information networks and to extend primary care access points.

9
Privacy and Security Protections for Healthcare Information

GAIL BELLES

Challenges of Automation

Many challenges are associated with transforming manual data collection systems into integrated healthcare delivery systems, including those related to the ability to process data securely and protect the privacy of each individual's record in a distributed healthcare database. Federal laws and regulations mandating security and privacy protections continue to evolve, although at a much slower pace than the technology deployed to collect, process, integrate, and electronically store data. Mainframes, personal computers, networks, and telecommunication links are widely used to facilitate the delivery of health care.

Thus, ensuring the availability, integrity, and confidentiality of data is paramount. Healthcare personnel rely on system managers to keep systems operational and on the database to be accurate; patients expect their privacy to be protected. These interdependencies are inherent with automation, but they are particularly significant in healthcare settings, such as those in the Veterans Health Administration (VHA), also referred to as VA, where treatment decisions are made using data maintained in automated systems.

Recipients of VA services furnish personal data to healthcare providers to assist in selecting and administering treatment plans and to determine their eligibility for services. These data are collected and stored in systems accessible to and used by many VA employees over the course of care; it is not uncommon for large numbers of personnel to have access to a patient's record because the provider team can consist of physicians, nurses, residents, therapists, social workers, and the like. Patients entrust personal data to VA employees, with the expectation their data will be used for their care. It is the patient's responsibility to disclose health information pertinent to his care and to inform his provider when the information given should not be included in the written record or divulged to other health professionals or family members. It is VA's responsibility to implement measures that ensure the data are accurately entered into the system, accessed only by individuals with a need to know, and not misused or disclosed to other parties without patient consent.

Computer applications play a major role in the privacy and security protections afforded to personal data maintained in healthcare databases in the VA and elsewhere. Applications must contain controls to restrict or limit access to menus, options, and data to those individuals who have a need for them, as well as controls to prevent inaccurate recording of entries in the healthcare database and to identify users and track actions within the system so that inconsistencies can be identified and investigated.

Legal and ethical issues have been raised and continue to surface as VA systems are implemented and used. Issues range from the inappropriate use of e-mail and whether e-mail on VA systems is public or private, to the legality of electronic monitoring of user activity on VA systems and whether individual identifiers such as full social security number or date of birth can be displayed during database lookup operations. The resolution of these issues requires a comprehensive approach, beginning with the promulgation of policies to address activities in the automated environment. The approach should include training for all users that stresses practices consistent with policies, assurance from software developers that technical controls are incorporated to meet the intent of policies in place, and enforcement of policies by managers and supervisors. The effectiveness of the controls depends on multiple factors, including system management, legal issues, quality assurance, and internal and management controls.

To protect and secure its data and systems, VA selects and uses technical, administrative, and physical security controls and safeguards. Technical controls include password management, menu and option management, security key management, audits, and electronic signatures. Administrative controls address management issues such as security policy, training and awareness, establishing and terminating user accounts, contingency planning, incident handling, and personnel screening. Physical controls protect the systems, buildings, and related infrastructure such as power, heating and air conditioning, and telecommunications.

This chapter discusses the implementation and use of technical, administrative, and physical security controls that work conjointly to provide security and privacy in the VA healthcare systems. It is important to note that while these controls are effective in the VA's current environment, they may not be applicable in all healthcare environments.

The National Institute of Standards and Technology (NIST) Special Publication 800-12, *An Introduction to Computer Security: NIST Handbook* (Guttman & Roback, 1995), was used as a reference throughout this chapter; it is a valuable reference tool for understanding computer security needs and for developing a sound approach to the selection of appropriate security controls.

Common Threats to Automated Systems

Certain threats are typically encountered when automated information systems are in place. A partial listing includes errors and omissions, fraud and theft, sabotage, loss of support services (power, communications, and water), hackers, computer viruses, and threats to personal privacy. Damages inflicted range from errors, which affect the integrity of the database, to fires, which can destroy the entire computer center. It is difficult to estimate actual losses. Many are never discovered; others are not reported for fear of unfavorable publicity or punitive action. Placing a dollar value on the data collected and maintained in a database is a nontrivial task indeed.

Publications on computer security identify the most common source of intentional disruption as authorized individuals performing unauthorized activities (Cohen, 1995). Referred to as "insider attacks," these incidents result from human errors, accidents, and omissions. Typically, insider attacks account for 75% of the threats to information security, physical disruption accounts for 20%, and outsider attacks on systems for only 5% of the threat (Madron, 1992). According to a senior manager in Ernst & Young's IS auditing and security practice, "Sloppy procedures around terminations and transfers of employees can result in circumvention of even the best security controls" (Panettieri, 1995).

Responsible computer security management can lessen and even eliminate security incidents and system disruptions by putting in place a plan with the following elements:

- Implementable and enforceable policies;
- Proper training of users;
- Routine review and monitoring to evaluate the appropriateness of access;
- Viable contingency plans.

Automated Information Systems Security Program Structure

The Automated Information Systems security program for the Department of Veterans Affairs (VA) is implemented at several levels. Each level contributes to the overall security program by providing expertise, authority, and resources. At the top level of the VA, policy that forms the foundation of VA's AIS security program is developed and disseminated based on current federal laws, regulations, and national requirements. Each VA element is required to implement the policies, regulations, and guidance developed at the top level of the VA, adapting those policies to secure the operation of automated systems within their specific environments. This includes decisions about the selection and installation of security controls, daily administration of the security program, evaluation of threats and vulnerabilities, and responses to security incidents.

The Veterans Benefits Administration (VBA) chief information officer (CIO) is responsible for the VBA AIS security program. Oversight and daily management of the program are delegated to the VBA information security officer, from the business and technology integration staff. The VBA enterprise-wide network is an integrated AIS consisting of multiple applications installed on multiple and dissimilar computing platforms, interconnected by wide area and local area networks. To ensure that security is maintained in this environment, the VBA considers security to be a separate but parallel architecture integral to all applications and computing platforms. Through an interagency agreement with the General Services Administration (GSA) and a follow-on contract with a private vendor, the VBA has developed a comprehensive policy, a security architecture for the modernized environment, and a computer-based security awareness training program for all field facilities (Department of Veterans Affairs, 1996).

The Veterans Health Administration's AIS security program pertains to all VHA elements and encompasses all computer systems operating therein. Managing the complex scope, size, and organizational diversity of the AIS security effort is a major challenge to VHA. The VHA AIS security program designs, develops, and implements national policies and a comprehensive AIS security training program. Field support is provided by the AIS security program through the development of automated risk analysis tools, facility level policy guidelines, site visits, system vulnerability studies, incident reporting, and distribution of antivirus software. Automated Information Systems security training, awareness materials, and security alerts are distributed to the field. The Medical Information Security Service (MISS) provides oversight and administers the organization-wide AIS security program for VHA. The director of MISS serves as the VHA information security officer. This security service maintains active liaisons with other AIS security staffs, within VA and with other federal government agencies such as the Indian Health Service, Department of Defense, Department of Health and Human Services, and National Institute of Standards and Technology.

Computer Security Program Elements

The goal of computer security is to protect an organization's valuable assets by selecting and implementing safeguards (Guttman & Roback, 1995). Unfortunately, computer security is often viewed as a hindrance to productivity because of poorly selected controls and bothersome rules and procedures imposed on users, managers, and systems. To the contrary, the selection and implementation of appropriate policies, procedures, and controls are intended to protect valuable assets. In the VA, computer security supports the mission by helping to improve the services provided to our nation's veterans.

The selection of computer security controls requires an evaluation of the benefits to be derived versus the direct and indirect costs of implementing a given control. Implementation of the controls can have an impact on system performance, morale, and retraining requirements. Countermeasures that exceed the cost of the resource to be protected obviously are not cost-effective.

When computer systems are linked to external systems, computer security needs extend beyond organizational boundaries. Management must be knowledgeable about the controls and practices protecting those external systems and require assurances that security is adequate to protect their own organization's data and resources. Management must also require that remote users observe the security policies and practices governing the systems they access.

Many organizations continuously add new and improved technologies to their automated systems and change the configuration of systems. This introduces the possibility of obliterating security controls in place in the legacy system. Each time a change is made to the system, a reassessment of threats must be accomplished to ensure that system security has not been compromised. In addition, policies and procedures may need to be revised or updated to reflect the changes to the technology or the environment.

Implementing an effective computer security program places explicit security responsibilities and accountabilities on all users of computer systems. Development of the components requires the involvement and assistance of many personnel throughout the organization, in the functional areas of audits, software development, physical security, contingency planning, quality assurance, personnel, training, systems management, physical plant, and procurement. In sum, computer security requires a comprehensive approach that extends throughout the organization.

Federal Laws and Regulations

The VA's AIS security program is guided by a number of federal laws and regulations enacted to provide for security and privacy protections in government information systems. Pertinent laws include the Privacy Act of 1974, Electronic Communications Privacy Act of 1986, Computer Fraud and Abuse Act of 1986, Computer Security Act of 1987, Computer Matching and Privacy Protection Act of 1988, and the Freedom of Information Act. Bills such as the Medical Records Confidentiality Act of 1995 (US Senate) were introduced to address the agency use of computers to collect, maintain, use, and disseminate personal information, directly affecting the privacy of individuals.

Federal regulations have been published mandating the protection of computers, the information they process, and related technology resources. Specific security requirements are found in Office of Management and Budget Circular A-130 and its Appendices (Office of Management and

Budget, 1993), Federal Information Resource Management Regulations (FIRMR) and Federal Information Processing Standards (FIPS). Federal managers are responsible for compliance with applicable requirements prescribed in these laws and regulations.

Technical Controls

Technical protective mechanisms built into hardware, software, and communications systems can protect against most types of computer crime and abuse (Baker, 1991) and preserve data integrity. A number of technical controls have been implemented by the VA to protect the security and integrity of data, programs, systems, and telecommunications networks, as described below.

Program/Application Security

A set of Decentralized Hospital Computer Program (DHCP) software utility programs, known as the Kernel, provides an interface between operating systems, DHCP application packages, and users. A key component of Kernel is Sign-on and Security Tools; its features address user security, menu management, audits, and package integrity. The security modules function to restrict access to the DHCP computer system to authorized users and to restrict authorized users to those tasks they need to perform their jobs. In addition, they provide monitoring of user actions, selected changes to the database, and changes to programs. Through these features, the Kernel provides systemwide protection of all data on a DHCP system (Department of Veterans Affairs, 1995).

Access Control

User security is the cornerstone of Kernel's system security features. Identification and authentication are accomplished as a two-step procedure, which validates the user's identification through password queries and, if a set number of failed log-in attempts has been exceeded, locks a user ID. Audit logging of failed access attempts is activated when the device is locked. Password security implemented in Kernel mitigates the problem of easy-to-guess passwords, reduces the damage done by stolen passwords, makes brute force attempts to break into systems more difficult, prevents reuse of codes, and provides technical protection of the password file.

Logical access control is provided through use of menus, options, and security keys. Access authority (e.g., read, write, and delete) is incorporated into the VA FileMan database management system to govern data input and reporting. The use of menus and options provides a constrained user interface that restricts access to specific functions and prohibits users from viewing or altering data not essential to their tasks. Security keys are used to lock options that provide specialized or supervisory access.

Audit Trails

Well-designed computer operating systems and healthcare applications provide the capability to record various system security-related events. This record of events is typically stored in system log files. By using reporting tools, systems and security administrators can discover when system and software changes have been made, when users have accessed the system, and in some cases, which commands have been entered and which functions have been performed. These logging capabilities and associated reporting tools are valuable assets for preserving patient confidentiality. (Miller, 1996)

There are three important considerations for recording events in audit logs: (1) make sure the system log files are protected from unauthorized access and modification; (2) set the logs to record only useful events; and (3) regularly review and analyze the logs. Events that should be considered for recording include the following: changes to or attempts to access the system audit logs, unauthorized access attempts, log-on and log-off activities, and changes or additions to user files and access privileges. Logging and analyzing security relevant events can provide an organization with the capability to detect security violations, system performance problems, and flaws in applications.

Audit features of the Kernel make it possible to monitor a wide range of computing activity. Several audit logs are maintained within the Kernel system. System access, option and server usage, and database management system audits can be initiated to address security needs and help ensure system integrity.

Package Integrity

Several mechanisms exist within Kernel to ensure the integrity of DHCP programs. The output for a package-specific integrity checker will list the routines included in the package and indicate whether the check sum on record matches that of the routine in its current state. Variances between the record and the current routine are flagged on the display. These tools give system managers the capability to ensure that program code has not been exposed to accidental or malicious alteration or destruction.

Electronic Signature Codes

A primary aspect of security in many DHCP packages involves the use of electronic signatures. Healthcare personnel authorized to make entries in a hard copy paper medical record are also authorized to use electronic signatures to authenticate information in the DHCP patient medical record documents, except for those pertaining to Schedule II drugs. Each person at a VA facility who is authorized to make entries in a hard copy paper record is provided with both a personal computer code and a DHCP code. The code serves as the "signature" of the provider and is affixed to computer-based medical record documents such as, but not limited to, treatment orders, reports of histories and physical examinations, progress notes, radiology reports, operation reports, and discharge summaries to authenticate and/or countersign medical record entries (Department of Veterans Affairs, 1991). In compliance with federal laws and regulations, electronic signatures on medical record documents are subject to prescribed procedural and personnel controls and practices.

File Access Security

File access security provides mechanisms to control user access to VA FileMan files in DHCP applications. System managers can control access to the database files for any user on the system, granting (or denying) access by adding (or removing) files from a user's accessible file list. This degree of control over a user's file access enhances security for VA FileMan files, but requires a high degree of system administration to maintain user access to the files they need to perform their duties.

Operating System Utilities

Operating system utilities allow VA system managers to protect the security and integrity of computer systems. These utilities include Access Control Lists (ACLs), global "classes," integrity checker utilities, database repair utilities, and various error log utilities. Proper management and administration of access control lists and regular monitoring and review of error logs are critical to providing security protections at the operating system level.

Encryption

Currently VA is embarking on projects that will help meet the needs of healthcare providers by allowing healthcare facilities to transmit patient data across networks and utilize Internet and World Wide Web resources. These initiatives give privacy and security issues high priority. Currently several applications transmitting patient data among facilities include limited encryption capabilities to secure password files. The integration of encryption methodologies into the VA's network infrastructure is being considered as a means of supporting data confidentiality and integrity and of providing sound user authentication and access controls.

Database Backups/Testing

"The most common business interruptions are caused by power outages, network failures, and other computer stoppages. Without proper back-up procedures these temporary problems create long-term business delays" (Bernstein, 1995). The VA computer system managers are required to perform journaling and complete system backups on a regular basis. Additional requirements include periodic testing to ensure that system backups are actually useable, and storage of backups outside the computer room in a physically and environmentally controlled area that protects against heat, cold, harmful magnetic fields, and unauthorized access.

Integrated Data Communications Utility Security

A primary source of wide area communications services for the VA, the Integrated Data Communications Utility provides a variety of network services, including packet switched, direct point-to-point, and frame relay data transmission between VA facilities throughout the 50 states, the District of Columbia, Puerto Rico, and the Republic of the Philippines. The IDCU also provides a limited number of gateways to non-VA locations, including connections to the Internet, FTS2000 Network, and Indian Health Service. Connections to US Army hospitals are proposed. Designed to operate at a high performance level, the IDCU currently connects over 500 user service points and supports access to over 15,000 user ports.

The IDCU provides extensive network security management functions, including support for specified security value-added services, by a network service center (NSC), which operates 24 hours a day, 7 days a week, and 365 days a year. The ICDU security management function is controlled and managed by the VA and implemented by a team consisting of VA staff and IDCU contractor personnel. This management function has overall responsibility and authority for security on the IDCU, overseeing and enforcing IDCU security policies and procedures. Each administration has a designated security component with the responsibility and authority to address local matters.

The IDCU security mechanisms and procedures are intended to protect the network resources and to minimize the opportunity for unauthorized use from inside and outside the VA. The primary security protection method used on the IDCU is access control, which is accomplished by the IDCU's Access Control System (ACS) and by approved firewall systems. Other functional security elements include enforcing compliance with security policy and procedures, defining and establishing closed user groups, reviewing and making recommendations on network configuration implementation, managing the ACS, managing gateway access controls and firewall systems, managing trusted host network interfaces, and overseeing the general security of the network.

The primary tool of the security management function is the contractor-provided Security Data Base System (SDBS). The SDBS automatically records and reports on security-related parameters and events. Management can access, monitor, and query this database to examine and analyze data communications activities. All IDCU gateway interfaces are challenged by either the IDCU's ACS or an approved firewall system. An annual security review and assessment plan by independent contractors addresses IDCU security features, identifies security vulnerabilities, contains a vulnerability assessment, and makes recommendations for IDCU security improvements. The IDUC contractor provides a contingency and system recovery plan, updated annually.

VA/Internet Gateways

In November 1993, the first national VA/Internet Gateway was established at the Washington Information Resources Management Field Office (IRMFO) in Silver Spring, Maryland. Services currently supported through this gateway include e-mail, telnet, File Transfer Protocol (FTP), gopher, Archie, Domain Name Service (DNS), and piloting of World Wide Web (WWW) access at some locations. Wider deployment of WWW access is anticipated following the completion of network communications upgrades in 1996. A number of VA facilities have also established direct Internet connections through university affiliations or commercial providers.

Internet connections are secured through the use of a firewall system providing components and configurations that reduce security risks. Firewall services are based on the levels of the following: security required to protect systems and data, services that will be permitted or denied, performance and user friendliness required, and ability to handle the threats inherent with Internet connectivity.

Individual requests for Internet accounts are subject to an approval process to ensure that all connections are configured in a manner providing adequate security protections for VA's local and wide area networks and resources. In addition, VA has established the Internet Management Review Board (IMRB) to develop and maintain an overall plan for the deployment of VA/Internet Gateways and to formulate policy related to their use, security, funding, and general operation. The IMRB ensures that policies and plans are developed and that Internet gateways provide a communications medium that will benefit the VA community by improving the services available for data and information exchange, while concurrently providing security to protect all VA host systems.

Network Security

In the VA as elsewhere, local area networks (LANs) have become the modus operandi for providing distributed data, applications, and communication services. Prior to the deployment of LANs, healthcare information was created, stored, and processed on a centralized system in the mainframe environment. Also centralized were the security controls required to maintain the confidentiality of data, prevent unauthorized modification or destruction of data and applications, control user access, and provide physical access control. The LAN environment differs. Because data are distributed over the LAN throughout the organization, the security controls required to protect the data must also be distributed. Backing up critical data and planning for contingencies become the user's responsibility. The paradigm shifts from limiting user access based on the need-to-know concept to allowing unrestricted access unless there is justification for refusing access.

Practices must be adopted that provide for adequate security in the networked environment. Users require training to back up critical data files regularly and store the backups in a secure location away from the device on which the files reside. Users also need training to recover erased files, to erase files permanently, to scan for viruses, to change their passwords on the network, and to promptly remove sensitive documents from shared printers and directories. Security training and awareness programs for LAN environments must emphasize user accountability for actions and user responsibility for security. Policies and procedures must define and support these requirements. Dial-up protection mechanisms may be needed to prevent unauthorized access to or intrusions upon the network. The LAN system administrator must perform rigorous reviews of usage reports and audit logs, as well as implement controls to protect the servers, directories, and files.

Virus Protection

Because of the lack of controls and the freedom with which users can share and modify software, personal computers (PCs) are prone to attacks by viruses. Personal Computer users must be constantly aware to detect and contain potential threats and recover from any damages (National Institute of Standards and Technology, 1994). Without proper training on how and when to use virus detection packages, users may misapply the technology, undermining its effectiveness.

The VA maintains a central contract for virus detection software, providing protection for all of its PCs. A VA-operated bulletin board distributes the software to organizational elements from a VA-operated bulletin board system for dissemination via LAN or installation on individual PCs. The software can be configured to scan drives when the PC is turned on, or it can be activated to scan diskettes before they are executed. Users are trained in the proper use of the software and provided with procedures to follow if viruses are detected. Virus alerts are communicated to all organizational elements through the VA's national e-mail capabilities.

Vulnerability Testing

During the security evaluation process, a carefully orchestrated attempt is made to bypass a system's security features. Known as vulnerability testing, this step in the field evaluation can focus future security design efforts and user implementation, if explicitly organized to do so. Vulnerability testing is an integral element of system security assurance and risk-assessment methodology. Conducted regularly without announcement, it can be a valuable management tool.

Within VA, vulnerability testing is used to identify security weaknesses. Any weakness detected is communicated to system managers for immediate corrective action, and follow-up testing is conducted to ascertain that the problem has been corrected. The process minimizes opportunities for system intrusions, and thereby enhances system security.

Administrative Controls

To be effective, a security program requires written policies and procedures for risk analysis, review and certification, incident handling, and contingency planning. Personnel hiring practices and training and awareness programs are also critical. In sum, no matter how many technical controls are in place, there is no substitute for careful management of personnel, procedures, and controls (General Services Administration, 1992).

Computer Security Policies

Policies can be written when an organization establishes a computer security program or when specific issues arise, such as Internet access, contingency planning, and use of e-mail. These policies define the organization's strategy for protecting their technical and information resources; they assign responsibilities, guide employee behavior, and address consequences for noncompliance. To be effective, policies must be communicated throughout the organization. As Guttman and Roback (1995, pp. 33–43) state, an organization's policy is the vehicle for emphasizing management's commitment to computer security and making clear their expectations for employee performance, behavior, and accountability.

Risk Management

"Risk management is the process of assessing risk, taking steps to reduce risk to an acceptable level and maintaining that level of risk" (Guttman & Roback, 1995, p. 59). It encompasses risk assessment and risk mitigation.

The risk assessment evaluates physical protection, supporting utilities, exposures to natural disasters, hardware reliability, protections for sensitive data and from errors and fraud, and adequacy of contingency and backup plans. Results are summarized, and specific recommendations are made to address areas of noncompliance and reduce specific loss exposure. Through a contract with a private vendor, an automated risk-assessment tool was developed for VA. The assessment is qualitative, focusing on the functional attributes of a system, and results in the recommendation of specific measures to offset threats. The automated tool allows organizational elements to assess single- or multiuser systems and aggregates system data to assess the entire organizational element. In addition to the automated assessment, VA disseminated specially developed checklists to use in performing risk assessments.

Risk mitigation involves selecting and implementing security controls to reduce risks to acceptable levels. To evaluate the effectiveness of the controls, a periodic assessment and reanalysis of risk must be performed.

When security controls have been implemented in a computer system, security certification and accreditation are in order. Certification is the formal testing of security safeguards to determine whether they meet requirements and specifications. Accreditation, a formality unique to the federal government, is the authorization for system operation and an explicit acceptance of risk. Guidance for computer security certification and accreditation is provided in FIPS Publication 102 (National Bureau of Standards, 1983).

Background Investigations/Suitability Determinations

The federal personnel manual mandates that certain positions be assigned sensitivity levels based on information security criteria, such as computer-related responsibilities, types of data to which the individual has access, a reasonable analysis of the risk, and principles of responsible management. Positions designated as low, moderate, or high risk based on these criteria require that the incumbent be subjected to a background check conducted by the office of personnel management. The depth of the background check is commensurate with the sensitivity level designation of the position.

Position Descriptions

Specific information security responsibilities are annotated in position descriptions for VA employees. They describe the employee's obligation to protect sensitive data and to use such data and information only in the execution of their official duties. Supervisors communicate specific information security responsibilities to their subordinates as well as the consequences of noncompliance.

Training and Awareness Programs

The Computer Security Act of 1987 mandates that all federal agencies provide information security training for staff and other users of federal computer systems. Both initial training and annual awareness programs are required. In addition to describing the risks to security and privacy and the importance of safeguards, these programs focus on the specific responsibilities of the individuals being trained and the penalties imposed for violation of security policies, procedures, and controls. Within VA, computer security training programs are conducted as part of new employee orientations, in workshop formats, and as audio teleconferences. Awareness materials are disseminated throughout the organization to keep employees apprised of computer security program requirements, changes to policies, and technological changes impacting computer security.

User Account Management

Procedures to manage user accounts are important in preventing unauthorized access to systems. An organization's security policy should set out user account decisions, such as who can access the system, how often user accounts are reviewed, how and when user accounts are terminated, and what use (including personal) each account holder can make of the system.

To control user access, VA practices require written access requests signed by a management official. In addition, all users sign the access notification form identifying their responsibilities as users of the system. Procedures are in place for reviewing changes in user status such as departmental transfers, terminations, or separations.

Computer Security Specifications in Acquisitions

The NIST Special Publication 800-4 (1992) provides guidance to federal agencies for including computer security requirements in federal information processing (FIP) procurements. Contract specifications include the following: delineation of computer security responsibilities; who can introduce hardware and software onto the system and under what circumstances; clauses to prevent the contractor from disclosing protected information; clauses to address the return or destruction of data/information; security training requirements; personnel screening; physical protection of computer systems for contracts where work will be performed at the contractor location; and security functions to be incorporated or bought with applications, operating systems, and hardware.

Incorporating these specifications into procurement contracts ensures that the contracts will meet agency missions, protect federal assets, and provide for protection of individual rights.

Contingency Plan Development and Testing

In the VA, contingency plans are developed within each organizational element to assure that critical mission and business functions can be continued in the event that information technology support is interrupted. Events contributing to the loss or interruption of computer operations can include power outages, hardware failures, viruses, fires, and storms. In a distributed environment, coordination of plan development and testing is critical to ensure that all dependencies on systems, applications, and other resources have been addressed.

The development of contingency plans begins with assessing threats and their consequences. Once it has been decided which activities and resources are most vital to accomplishing the organization's mission, controls are implemented to prevent contingencies or disasters. Procedures should be developed to test and retest the plan on a regular basis; the process of updating and revising the plan must be ongoing to reflect changes in personnel, assignments, system operations and equipment, and organizational missions and functions. In addition to detailing individual roles and responsibilities, each plan itemizes three categories of actions that must be taken: at the onset of an event, to limit the level of damage, loss, or compromise of assets; to restore critical functions; and to reestablish normal operations.

To help organizational elements address the full range of issues, VA developed and disseminated standardized contingency plan development templates. The templates include formats for analyzing the criticality and dependencies of programs and applications, identifying the critical data maintained and utilized by each functional component, identifying the work flow within each functional component, and documenting contingency plan tests and the effectiveness of the contingency plan during testing of specific components. Detailed guidance is also provided in the FIPS Guidelines for ADP Contingency Planning (National Bureau of Standards, 1981).

Computer Security Incident Handling

A computer security incident handling capability targets resolving computer security problems in a way that is efficient and cost-effective. Combined with policies that require centralized reporting of security incidents, this capability can reduce waste and duplication while providing a better posture against potentially devastating threats. It enables a proactive approach to computer security, combining reactive capabilities with active steps to prevent future incidents from occurring (Wack, 1991).

The VA policy requires organizational elements to track and report annually on major security incidents that have occurred within their units. Analysis of these reports enables the VA, at the departmental and organizational levels, to identify where improved security controls, further training, and additional policies and operational procedures are needed. When properly managed, the computer security incident reporting process alerts the organization to potential security threats and needed security controls. The end result for the organization can be a stronger computer security program.

Benefits of the internal reporting process can be enhanced by an awareness of security incidents and potential threats discovered and reported by other government agencies and the private sector. The VHA is a member of the Forum of Incident Response and Security Teams (FIRST), a voluntary effort among governmental and private organizations to help combat and prevent computer and network security problems. The FIRST provides a forum where its members can share current threat and vulnerability information, solve problems of common concern to the global information security community, and plan future strategies and technical countermeasures. The FIRST includes teams from government, industry, computer manufacturers, and academia. Information from this group that is relevant to the VA systems environment is disseminated throughout the VA using an alerts notification system.

Program Compliance Assessments

The VA's AIS security programs are subject to review for compliance with federal laws and regulations and with departmental policies and standards. Conducted cyclically, internal assessment determines whether the security program actively addresses key program elements. External reviews are conducted by oversight organizations such as the VA Inspector General, General Accounting Office (GAO), and JCAHO.

In 1994, the JCAHO introduced a performance improvement framework for the information management function. As detailed in their accreditation manual for hospitals (Joint Commission on Accreditation of Healthcare Organizations, 1995), the objectives for of the framework are as follow: more timely and easier access to complete information throughout the organization; improved data accuracy; balance of proper levels of security versus ease of access; use of aggregate data; improving efficiency through information-related process redesign; and greater collaboration and information sharing to enhance patient care. The framework sets explicit standards for an organization's information security program and implementation of security controls to protect healthcare data.

Physical Controls

Physical security addresses the environment of the computer system and telecommunications equipment areas, such as construction features, physical access controls, and other measures used to protect computer resources from theft, vandalism, deliberate corruption of data, accidents, loss of utilities, and natural hazards. Considerations include adequate air conditioning, fire protection, and access to back-up or emergency power supplies; strict control of access to computer rooms, telephone closets, and media storage areas; and maintenance of system console logs, error logs, and downtime logs. Equipment and software inventories need to be reviewed regularly and updated when changes are made to ensure accurate accountability and compliance with software licensing agreements. Implementing a procedure for removal or loan of equipment to employees assists in maintaining an accurate inventory of equipment and locations.

A security measure frequently overlooked is the removal of sensitive data from equipment that is sent for repair, transferred to another facility, or released for disposal. Sensitive data must be removed or rendered unusable before equipment is released. Additionally, equipment that has been returned after repair must be scanned for viruses before it is put back into operation.

The Future of Security and Privacy in Health Care

"The ideal security solution is either hardware or software; it incorporates smart-card technology or biometrics, or a little of both; it has single sign-on capabilities or tiered access; it boasts enterprise-wide functionality and is impenetrable to viruses, hackers and disgruntled employees; it requires no training to use; and its cost to purchase and maintain is acceptable to management. Obviously, this solution does not yet exist, but as each aspect of information security moves forward, this system is closer to becoming a reality" (Crane, 1996).

While we await this solution, vendors are focusing their development efforts on network access control products for securing networks, firewalls, encryption technology, virus scanning technologies, and authentication products such as tokens, smart-cards, and biometrics. Security issues will be prevalent in the technologies that provide ubiquitous communications, pervasive networks, intelligent devices, and wireless services.

While we ponder the benefits of utilizing these technologies to improve or enhance the delivery of healthcare, one thing remains obvious: security managers will be faced with security and privacy challenges well into the future.

References

Baker RH. 1991. *Computer security handbook*, 2nd ed. Pennsylvania: TAB Professional and Reference Books.

Bernstein DS. 1995 (May/June). Recovery planning a to z. *InfoSecurity News, 6(3)*, 27.

Cohen FB. 1995. *Protection and security on the information superhighway*. New York: John Wiley & Sons, Inc.

Crane E. 1996 (January/February). Tomorrow's technology: Promising a safe new world. *InfoSecurity News, 7(1)*, 24.

Department of Veterans Affairs. 1996 (February 21). *Veterans Benefits Administration Automated Information Systems (AIS) security program*. Washington, DC: Author.

Department of Veterans Affairs. 1991 (July 17). *DHCP [Decentralized Hospital Computer Program] electronic signatures on medical record documents*. Directive 10-91-074. Washington, DC: Author.

Department of Veterans Affairs. 1995 (July). *Kernel security tools manual, version 8.0*. Washington, DC: Author.

General Services Administration. 1992 (September). Security and privacy protection of Federal Information Processing (FIP) resources. *Federal Information Resources Management Regulation (FIRMR) Bulletin C-22*. Washington, DC: Author.

Guttman B, & Roback E. 1995. *An introduction to computer security: NIST [National Institute of Standards and Technology] handbook*. Special publication 800-12. Washington, DC: US Government Printing Office.

Joint Commission on Accreditation of Healthcare Organizations (JCAHO). 1995. *Automated CAMH (Comprehensive accreditation manual for hospitals)*, release 1.0. Oakbrook, IL: Author.

Madron TW. 1992. *Network security in the '90s: issues and solutions for managers*. New York: John Wiley & Sons, Inc, p. 10.

Miller DW. 1996 (January/February). Using system logs to protect confidentiality. *IN Confidence, 4(1)*, 1–3.

National Bureau of Standards. 1981. *Guidelines for ADP contingency planning. Federal Information Processing Standards (FIPS) Publication 87*. Washington, DC: Author.

National Bureau of Standards. 1983 (September). *Guideline for computer security certification and accreditation. Federal Information Processing Standards (FIPS) Publication 102*. Washington, DC: Author.

Guttman B. 1992 (March). *Computer security considerations in federal procurements: A guide for procurement initiators, contracting officers, and computer security officials*. National Institute of Standards and Technology (NIST) Special Publication 800-4. Washington, DC: US Government Printing Office.

National Institute of Standards and Technology. 1994. *Guideline for the analysis of local area network security. Federal Information Processing Standards (FIPS) Publication 191*. Washington, DC: Author.

Office of Management and Budget (OMB). 1993 (June 25). *Management of federal information resources. Circular A-130, revised*. Washington, DC: Author.

Panettieri JC. 1995 (November 27). Security. *Information Week*, 32–40.

US Senate. 1995 (October 24). *Medical Records Confidentiality Act of 1995*, S.1360. Introduced by Senators Bennett, Dole, Kassebaum, Kennedy & Leahy.

Wack J. 1991 (September 24). *Establishing a Computer Security Incident Response Capability* (CSIRC). Washington DC: US Department of Commerce.

Section 3
Clinical Information Technology

Chapter 10
Integrating A Clinical System
  Rusty W. Andrus   137

Chapter 11
Capturing Data in Ambulatory-Care Settings
  Susan H. Fenton   148

Chapter 12
Moving to Clinical Workstations
  Sharon Carmen Chávez Mobley   155

Chapter 13
Controlled Representation in Patient Records and Healthcare Delivery Systems
  Kenric W. Hammond   164

Chapter 14
Clinical Decision Support in a Distributed Environment
  Curtis L. Anderson   183

Chapter 15
The CARE Decision Support System
  Douglas K. Martin   203

10
Integrating A Clinical System

RUSTY W. ANDRUS

Introduction

As part of its mission to provide high-quality health care for America's veterans, the Department of Veterans Affairs (VA) has made a commitment to the development of a Computerized Patient Record System (CPRS). This electronic patient record will be part of an environment—an Automated Clinical Information System (ACIS)—which implements comprehensive acquisition, storage, processing, and display of clinical and management data. Patient data will be secure, yet accessible for use in day-to-day health care, clinical and management decision making, and education and research programs. The CPRS will be the core of a true health information system and provide the basis for its integration into the delivery of patient care.

Successful integration of clinical and administrative modules requires addressing a host of issues, only some of them technical in nature. This chapter draws upon experiences implementing Order Entry and Results Reporting at VA Medical Centers (VAMCs) to define three critical categories: user-centered design, organizational issues, and technical issues.

Historically, within the Decentralized Hospital Computer Program (DHCP), modules like laboratory, pharmacy, etc., were developed in a vertical fashion, each focusing on the requirements of the respective service. The CPRS effort integrates the data provided by each of the services into a consistent and comprehensive presentation of the patient record to the healthcare provider. This endeavor brings to the surface not only technical issues such as standard messaging practices and event processing, but also issues such as culture change, clinician (user) acceptance, changes to medical center procedures, and implementation strategies, all of which must be addressed.

In the case of the VA, as depicted in Figure 10.1, development of a integrated CPRS builds on the foundation established by DHCP, considered by many to be an early computer-based patient record (CPR). All DHCP modules were developed and integrated initially through use of a

138 Clinical Information Technology

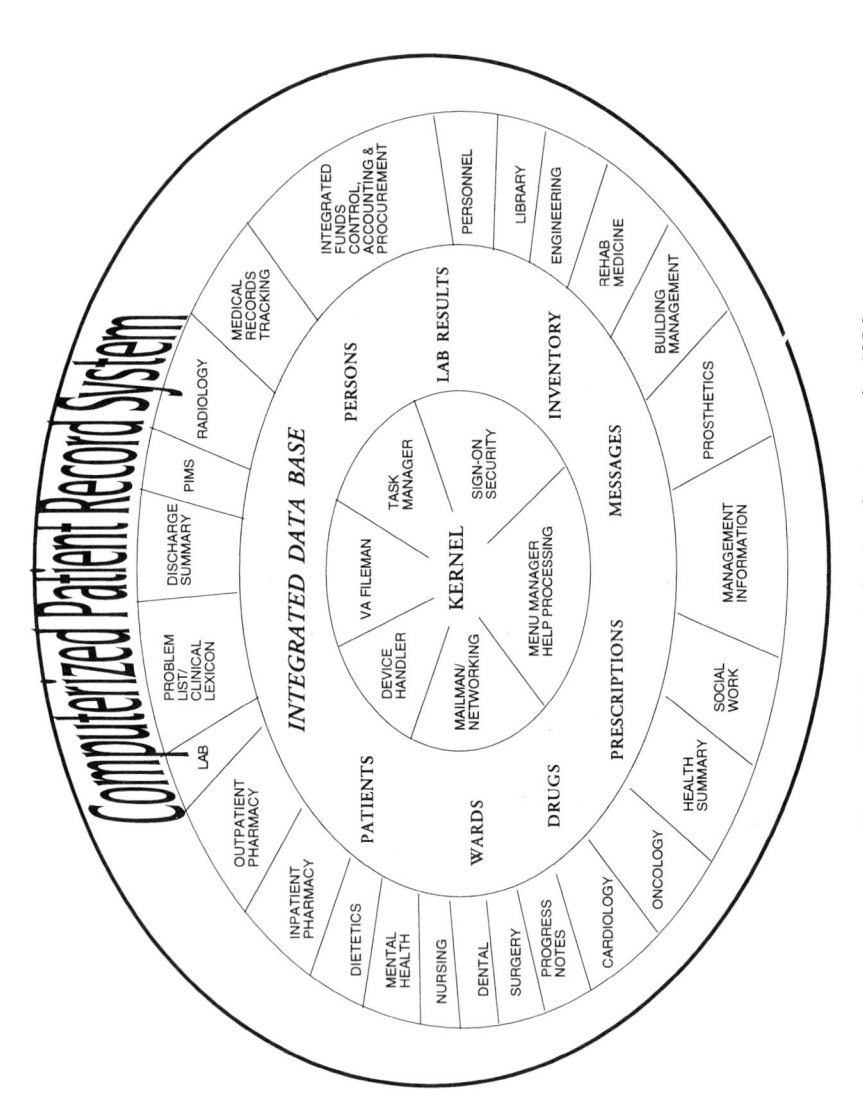

FIGURE 10.1. Computerized patient record—1996.

common database. The basic goal of the CPRS project is to derive information from complete and accurate clinical data entered at the point of patient care, and to use that information set to serve multiple clinical, administrative, financial, and quality management needs.

Current Status

The VA's DHCP is a comprehensive hospital information system that covers medical management, fiscal matters, and clinical functions. Introduced in 1982, DHCP today is the base information system for VA's medical care network, providing support to 173 hospitals, over 400 outpatient clinics, 132 nursing homes, and 39 domiciliaries. This VA-provided care reaches a large patient population, comprised of some 2.6 million veterans each year.

Each VAMC can select from over 70 integrated DHCP applications to build a flexible information system that supports its particular environment and mix of services. These applications span a broad range of clinical and administrative functionality. Clinical modules include allergies and adverse reactions, consults, dietetics, dentistry, discharge summary, health summary, immunology case registry, laboratory, medicine, mental health and psychological testing, notifications and alerts, nursing, oncology, order entry and results reporting, pharmacy, progress notes, quality management, radiology, surgery, and social work. Administrative modules include admission/discharge/transfer, billing, clinic scheduling, funds control and accounting, time and attendance, and records tracking. An ongoing development program addresses the critical clinical and administrative information needs of the medical centers. As an example, applications now in development will add medication administration, order sets, event linkages, and imaging.

All DHCP applications use a single set of tools, known as the Kernel, which incorporates user access control, menu management, device management, database management, and e-mail for data transmission and messaging. This standard approach to application development has resulted in a uniform and consistent user interface, as well as a high degree of hardware and operating system independence.

User Centered Design

Culture

It is well recognized that DHCP already captures and houses significant types of data central to the implementation of a CPR. Further progress toward a "complete" CPR component for DHCP requires that VA meet a number of major challenges. Primarily, these involve migrating from a department-centered architecture to one which adequately supports departmental and patient-centered functionality, and incrementally increasing the scope of data contained in the system to respond to major changes in the VA environment, such as primary-care initiatives and healthcare reform. This must occur, of course, while maintaining the integrity of day-to-day operations. Related tasks of imposing proportions include developing a user interface capable of effectively integrating a wealth of data for clinicians, achieving integration of commercial and VA-developed applications, and implementing standards-based exchange of data between VA medical centers and between VA and non-VA sites of care. Recognizing these challenges, VA has initiated projects to address them.

User Needs

Since the beginning of DHCP, every effort has been made to assure that the products developed met the needs of the users. Initially, groups of users were identified and named Special Interest Users Groups (SIUGs). Appropriately, the SIUGs were departmentally, or vertically, oriented. Their replacement by the Application Requirements Group (ARG) structure as mechanisms for collecting user input reflects the perceived maturity of departmental applications and the increasing need for integrated data and clinician-centered functionality (rather than service-centered). Another effect of these trends is emerging interest in modeling DHCP and planned enhancements as a way to formalize existing designs and to identify and develop the extensions necessary to implement the CPR.

Established as needed by the Information Resources Advisory Council (IRAC), the ARGs are charged with integrating and coordinating interdisciplinary and organizational concerns at all levels. Their focus is on VA's priorities:

- applications that integrate patient data for clinical care;
- applications that integrate information for management decision support.

Currently established for clinical, management, and technology integration activities, the Clinical ARG (CARG) is responsible for the clinical record and related issues. The Management ARG (MARG) represents all nonclinical areas including administrative and fiscal activities; and the Integration and Technology ARG (ITARG) represents all of the user-sensitive issues of continuity and consistency that extend beyond strictly internal aspects of any application.

Expert Panels (EPs) are created to support the ARGs in meeting these challenges. Interdisciplinary in membership, EPs provide user and content expert input into the functionality and specifications for each application or project. They are nominally transient in nature, coming in and going out of existence as ARG requirements and tasking dictate. In practice, however, many EPs like the SIUGs before them tend to be long-lived, reflecting the need for continuity in a sequence of related projects as well as the usefulness of maintaining a group with application expertise to provide long-term guidance and oversight.

Clinician Acceptance

Historically, clinician acceptance of patient-care systems has suffered from the difficulty of integrating these systems into the day-to-day routine of the clinician. For patient data to be accurate and comprehensive, they must be produced as a by-product of the care delivery process. Suppliers of data must be given sufficient return (in whatever form) to make it worth their while to contribute. After a decade of evolution, DHCP is delivering applications useful in direct-care delivery and hence attractive to clinicians. In addition to this functionality, acceptance is highly dependent upon the technological solutions to issues such as user interface, user query, medical image quality, accessibility, performance, and retrieval of longitudinal patient data. If these are not addressed to the satisfaction of clinicians, the integration of clinical modules in the CPR will not succeed.

Usability Testing

To facilitate ease of use, usability testing has been incorporated into the development process. Usability testing is the process of testing how easily a product is used by the customer, to see if a person can successfully learn and make optimal use of a product with the tools provided to accomplish a desired task. The five main usability characteristics are:

- Learnability;
- Efficiency of use once the system has been learned;
- Ability of infrequent users to return to the system without having to learn it all over again;
- Frequency and seriousness of user errors;
- Subjective user satisfaction.

In 1995, VA established a small usability testing laboratory within the development structure; its staff received training from a national usability consulting firm. The primary function of this team in its first year was to provide formative evaluation of the clinical interface being delivered with the CPRS. The development team created a series of prototypes of the important functions for the clinical record, including order entry and results reporting. The usability team within the VA created clinical scenarios, using real patient charts and clinician judgments. In a series of evaluations, VA clinicians with various levels of computer expertise performed usability tests using the scenarios. This work allowed specific improvements to individual modules and components of the clinical record. It also allowed the usability team to create schema for categorizing usability problems according to eight dimensions, such as the system not providing adequate feedback to the user to indicate that the action has been taken. Evaluations such as usability testing are increasingly recognized as an important component of clinical software development.

Organizational Issues

Management

To initiate the move to a CPRS, VA designated DHCP as the foundation of the CPR and instituting appropriate management policies. Having persons who were open to new concepts in key positions was particularly helpful. For example, the Health Information Manager (HIM) in the VA Central Office was responsible for internally approving new forms (including computerized versions) with external approval authority resting with Office of Management and Budget (OMB). No longer requiring that computerized forms exactly duplicate noncomputerized forms, the HIM insisted only that certain elements (e.g., patient identifying information) be located in roughly the same area on computer-generated forms, allowing forms to be site specific and module specific.

The transition to electronic records is forcing reconsideration of policies developed for paper-based operations. For instance, current policies require that prescriptions for controlled substances be written on VA prescription forms and maintained on file in the pharmacy. To move to a completely electronic system, agreement must be reached with internal policy bodies (e.g., professional practices groups) and external bodies (e.g., Congress and the Drug Enforcement Agency) on issues such as authorizing, authenticating, retaining, and archiving immutable representations of prescriptions, etc. Similar instances of the impact of replacement of paper records are occurring throughout DHCP.

Business Process Redesign

Introducing a new information system to solve a business problem generally requires an organization to make basic changes in its culture and how it conducts its activities. This is often termed business process redesign. A key point is that the information system is not, in itself, a complete solution. Changes in behavior and long-established practices are often necessary to take full advantage of the technology.

Redefined Roles

Today organizations that deliver health care are not resourced or prepared to absorb integrated and sophisticated information systems into their operational environments. Investment must target not only hardware and software, but also departmental management development.

Within the VAMCs, Automated Data Processing Application Coordinators (ADPACs) were originally charged with managing a particular application along with their other responsibilities. For example, a laboratory ADPAC was usually a lab technician who also managed the computer program for his/her service. As VAMCs increase their level of automation, individual modules expand and become more complex. New modules and tools, such as the Health Summary and Text Integration Utility (TIU), are introduced. In this environment, it is increasingly important to assemble key people to work together as a team. ADPACs must play newly defined roles and take on new responsibilities. Only then can implementation of an integrated CPRS succeed.

Clinical Coordinator

On the team, the Clinical Coordinator is a critically important player, bringing together services from across the facility. The coordinator needs to be familiar with many facets of the hospital's organization and operations. Although it is impossible to be an expert in every service, the coordinator can and should be respected at all levels within the hospital.

One of the chief responsibilities of the Clinical Coordinator is to "sell" computerization. This requires the ability to listen and respond to users at all levels, provide creative and effective training, communicate daily in multiple and varied forms, and make judicious use of means to accomplish an end. The coordinator must also be able to collaborate effectively with colleagues throughout the hospital. This is critical given the interdependencies created by an integrated program like CPRS. Active collaboration with Information Resource Management (IRM) and other coordinators is essential.

ADP Coordinators

As VA works toward an integrated CPRS, ADPACs can no longer be responsible for other duties. The assignment of full-time ADP coordinators within individual services is a necessity if automation is to succeed. The traditional role of the ADPAC must be reconceptualized to go beyond intraservice duties associated with specific module management to encompass hospital-wide responsibilities, including relationships with other ADPACs, hospital staff, and the IRM Service. Adoption of this philosophy by top hospital management should result in the creation of full-time ADP coordinator positions in most services, which in turn should promote accelerated acceptance and integration of automation throughout the facility.

Coordinator Council

Integrated programs such as those found in the DHCP foster increased communications and stronger working relationships among hospital services. Policies, procedures, and contingency plans can no longer be exclusive to one service because each relies so heavily upon the others in the course of daily operations and during unexpected downtime. For these reasons, it is critical that there be communication and close collegial relationships among ADP coordinators. An ADP Coordinator Council consisting of all full-time and part-time service-level ADP coordinators and selected IRM staff should meet twice monthly to:

1. Establish an arena in which open discussion and sharing of ideas related to present and future automation needs of users can occur. These needs can be prioritized and communicated to service chiefs and other managers who can address them to the appropriate committees for decision-making purposes.
2. Facilitate creative problem solving with input from those directly involved at the grass-roots level.
3. Provide a support network for new and established ADP coordinators.
4. Use the collective knowledge of all ADP coordinators so as not to "reinvent the wheel" when problems arise in any given service.
5. Use the expertise and manpower of a large group when implementing all new and enhanced modules. Rather than have one coordinator responsible for training all hospital services, train all coordinators during a council meeting and have each coordinator accept responsibility for training and implementation in his/her own service.
6. Establish a large network of cross-trained ADP experts.
7. Develop policies and procedures that are operationally sound by virtue of the fact that they have been devised, discussed, and tested by a wide range of persons and found to be effective. This is especially true of downtime procedures which, because they cross service lines, must be agreed upon collectively.
8. Establish close working relationships among council members to emphasize the interdependence of the system and reinforce the concept that "no man is an island."
9. Foster excellent time management strategy through the council's brainstorming and decision-making sessions.
10. Promote negotiation and compromise through the council. By sharing experiences and soliciting help from each other, ADP coordinators can appreciate the problems of their counterparts, develop an appreciation for the needs of the whole organization, and work toward common goals.
11. Promote dissemination of information from the IRM office.

Steering Committee and Subcommittees

Organizational development strategies can help to ensure the success of DHCP integration projects. Whatever organizational structure is chosen for integration planning and implementation, a steering committee of top management should be created to ensure cooperation from services and clinicians throughout the hospital. This committee should devise an implementation plan and appoint subcommittees to carry out the various components of implementation.

To be successful, the plan needs the support of top management and all impacted services and functional entities. The plan also needs to set timelines and specify objectives with as much precision as possible. The plan must define objectives in detail, set rigorous deadlines, and identify who is to be responsible for implementing each feature and/or module. Definitions should extend to the use of the newly implemented and integrated system. Who will enter which data and who will act on them? How will clinicians, nurses, and ward clerks be expected to interact with the module? The steering committee should exert the influence and authority of top management when needed to ensure that the subcommittees address and finalize this level of detail.

Information Resource Management Office

The IRM chief and/or a designated staff member must play a major role in planning and implementing the CPRS because of the complexities in setting it up and the magnitude of its effects on the hospital. Working closely with the Clinical Coordinator, IRM staff will act as a liaison among services and with hospital management in assessing needs.

Technical Issues

Some of the key technical issues associated with integration include standardization of terms and major concepts, event processing, and messaging formats. Each of these issues has proven to be complex; hence, we expect our solutions to them to evolve over time.

Data Content and Vocabulary Standards

Standard vocabularies and coding schemes are integral to consistency within VA medical centers and clinics, between VA facilities, and between VA facilities and external administrative and healthcare institutions. Industry standard coding systems, namely the International Classification of Diseases, 9th ed. (ICD-9-CM) and Current Procedural Terminology (CPT), are used for the reporting of diagnoses and procedures. Other nationally recognized systems are utilized where appropriate, including SNOMED, DSM-III, Diagnosis Related Groupings (DRGs), National Drug Code, Federal Stock Number, and National Stock Number.

A standard vocabulary known as the Lexicon Utility has been developed. Built from the National Library of Medicine's Unified Medical Language System (UMLS), the Lexicon Utility (LU) has mapped ICD-9-CM and DSM diagnoses, along with many other classification systems, to what is known as natural medical language. The adoption of this as a standardized reference system for clinical terminology across VAMCs will enable clinical information to be consistently recorded, transmitted, retrieved, and analyzed.

Currently VA has version 1.3 of the UMLS implemented and is committed to expanding beyond UMLS as needed to support important functions. A group of VA medical administration and medical records personnel have nominated additional terms for inclusion. The CPT codes and terms will be included with Version 2 of the Lexicon Utility. Other measures have enriched the original UMLS content as well.

The scope of any search through the LU can be customized for particular users. For example, it is possible to restrict the search, presenting only North American Nursing Diagnosis (NANDA) terms to nurses. Alternately, the search can be customized according to the UMLS-derived semantic types represented in the Clinical Lexicon. The Clinical Lexicon can be used to represent patient data elements in the Patient Care Encounter (e.g., for health maintenance) and to represent patient and pharmaceutical information in the order entry component of the CPRS. The LU is considered a VA-wide resource for naming clinically relevant data elements. Indeed, DHCP may be the first production system to implement the UMLS technology comprehensively in this manner.

Event Processing Queries and Message Formats

In the private sector, a given hospital usually has ancillary and clinical systems from a variety of vendors. Achieving any degree of integration under these circumstances is difficult at best. All DHCP modules were developed and integrated initially through the use of a common database, alleviating many of the problems known to the private sector. The decision was made that inter-DHCP module communication would occur via protocol event points and Health Level 7 (HL7) compatible arrays.

As a result of this strategy, a DHCP module can hang a protocol off another DHCP module's event protocol to obtain and process data in the HL7 arrays. A standard format for messages, HL7 was created to facilitate the development of interfaces between vendors. DHCP emulates the complex situation of the private sector by using HL7 messages to communicate between modules within the M environment. Two main kinds of interactions used by DHCP in the integration of its clinical and administrative module in support of the CPRS are:

- *Events*: Examples of events are new orders, completed results, etc. Whenever one of these events occurs, an HL7 message is created and passed to subscribers who wish to be notified.
- *Queries*: Examples of queries are requests for the narrative of a radiology exam, lab results over a given time range, etc. Application Programming Interfaces (APIs) written by the authoritative modules are used to process these queries. HL7 messages are not used in the case of queries because
  - Queries are synchronous in nature and would not need to notify other subscribers as is the case with asynchronous events.
  - The query portion of the HL7 specification is still immature.
  - Most of the modules already have or could easily create the required APIs.
  - Response time is of paramount importance in the case of a query because an interactive user is generally waiting for the results, whereas events can be processed in the background.

It should be noted that the HL7 specification is a general guideline for developing custom interfaces, not a standard in the sense of "plug and play." It stays intentionally ambiguous in many areas because the capabilities and functionalities of clinical systems are so variable.

Summary

The integration needed to implement a CPRS goes far beyond the extensive technical issues associated with integrating clinical and administrative modules. Cultural, managerial, and human factors are critically important, as the implementation of Order Entry and Results Reporting at VAMCs has demonstrated. Addressing these factors is arduous and time-consuming, yet it is also effective. An understanding of the concepts and processes involved will enable VA facilities to implement an integrated clinical and administrative system.

Acknowledgments. Staff at the Salt Lake City VAMC's Information Resource Management Field Office played critical roles in developing the Computer Patient Record System from 1994 through 1996. Their work was contained in internal unpublished documents, not cited here but reflected in the successful outcomes of the project.

11
Capturing Data in Ambulatory-Care Settings

SUSAN H. FENTON

Introduction

Like the rest of the healthcare industry, the Veterans Health Administration (VHA), also known as simply VA, has moved or is in the process of moving the majority of care delivery from inpatient to ambulatory settings. As evidenced by the fact that scheduling and registration were among the first modules in the Decentralized Hospital Computer Program (DHCP), VA has always delivered ambulatory care. However, VA has not always collected 100% of ambulatory-care data. In fact, 100% collection of ambulatory visits (one patient per facility per day) along with a VA-developed data element of stop codes, was not begun until 1984. The VA is now moving to industry standards for collecting data on 100% of ambulatory-care activity.

Background

Collectively, VA medical centers, outpatient clinics, and home health services treat over 2 million patients with 24 million ambulatory care visits each year, with a visit defined as "one patient per facility per day."

Historically, VA has collected stop codes to reflect patient activity during these ambulatory care visits. Stop codes are a VA-developed data element with a definition of "an instance of care." A visit may have one or more stop codes, with VA averaging 2.5 stop codes per visit. The approved stop code listing is maintained by the health administration service in headquarters. Stop codes have been added and deleted as requested by headquarters clinical program officials. They represent a variety of items. For instance, a nursing stop code would be used whenever a nurse interacted with a patient; a diabetes stop code would be used when a patient with diabetes was treated; an electrocardiogram (ECG) stop code would be used when an electrocardiogram was performed. Although vague and unspecific, stop codes have been used for resource allocation, costing, and performance measures. It has been almost impossible to utilize them for any type of meaningful clinical research, case mix analysis, etc.

In the early 1990s, it became apparent that VA would need to be able to compare its activity to the rest of the healthcare industry. Taken from that point of view, it quickly became apparent that stop codes were not adequate for either clinical or management purposes. In 1993, a workgroup was formed to look at the issue of outpatient workload reporting. The recommendations from the October 1993 report were as follows:

- that the VA adopt industry standards for ambulatory-care data definitions;
- that a technological solution of scannable encounter forms be utilized as a feasible interim solution (clinician workstations being the final solution);
- that 100% reporting be made mandatory.

In 1993, VHA management did not deem the projected benefits of implementing outpatient data collection to be worth the cost. This position was maintained until April 1995.

During the time between October 1993 and April 1995, modest ambulatory data capture efforts/pilots were begun by the Medical Care Cost Recovery, the office responsible for third-party billing. The goal of these efforts was to investigate several different technologies looking to maximize the ease of data capture and increase third-party collections. Technologies investigated included scannable encounter forms (mark sense and optical imaging), clinician workstations, and portable keypad computers. This work would later prove to be invaluable to the national project.

On April 26, 1995, the Under Secretary for Health signed a decision document recommending that

- The definitions from the Uniform Ambulatory Medical Care Minimum Data Set and ASTM 1384 (as applicable) should be adopted by the VA.
- The VA should require a diagnostic code, a procedural code, and provider information for positive proof that an encounter has occurred.
- The first two recommendations should be implemented as soon as necessary DHCP and Austin Automation Center programming are completed.
- The scanning equipment necessary to accomplish these recommendations should be given priority consideration for FY 95 ADP equipment funds (pending full review of Medical Care Cost Recovery funds, added by Under Secretary for Health).

Not only does this decision contain profound policy and procedural implications for VA healthcare facilities, it will dramatically change the way VHA management at all levels utilize ambulatory-care data. At the same time, VA recognizes that these minimum data elements do not meet all the ambulatory-care data needs for the next several years. Rather, these elements are a first step as the VA transitions its locus of care delivery from inpatient to ambulatory settings.

Implementation

The scope of any change in ambulatory-care data capture in the VA is enormous, given approximately 24 million ambulatory-care visits to the VA per year. Each visit has an average 2.5 encounters (formerly stop codes), for a total of almost 60 million patient encounters per year. The mandated project described above requires a minimum of three additional data elements for each ambulatory-care encounter, and is projected to increase the total number of data elements collected each year by almost 500 million. (Diagnostic codes, procedural codes, and practitioner data may each have multiple entries.)

The first step in the project involved the establishment of an implementation workgroup. Due to the scope and importance of the project, a multidisciplinary group was essential. Representatives from the chief financial and administrative offices cochair the workgroup, with high-level representation from patient services (primary care) and the chief information office. The workgroup also has medical center administrative and technical representatives, as well as training and marketing specialists.

The ambulatory data capture project is multifaceted. Below is a listing of its component parts, along with a brief description of each:

- Project Management: It is the responsibility of the cochairs to ensure that *all* critical milestones are met.
- Policy: The VA's ambulatory-care data and data collection policy must be revised to meet the new requirements in such a way that they are understandable to many people who are unfamiliar with these concepts and data definitions. (See Table 11.1 for definitions of standard terms.)
- Resources: As outlined in the fourth recommendation signed by the Under Secretary for Health, the type and amount of resources needed must be identified, designated, and reserved for this project.
- Acquisition: Hardware and software for the national technical solution must be acquired. Specific needs, including amounts, must be identified; purchasing contracts must be either identified or issued as appropriate.
- Software: Both existing DHCP and national database software (VA-written) require changes. Specifications have to be detailed and programming and testing completed by the effective date of the project.
- Marketing: The changes in different areas required for this project are enormous. It was felt to be essential to undertake an extensive marketing effort to communicate the need for changes and the methods of implementing them.

TABLE 11.1. Definitions of standard terms.

Outpatient Visit: The visit of an outpatient to one or more units or facilities located in or directed by the provider maintaining the outpatient healthcare services (clinic, physician's office, and hospital/medical center) within one calendar day.

Licensed Practitioner: An individual at any level of professional specialization who requires a public license/certification to practice the delivery of care to patients. A practitioner can also be a provider.

Nonlicensed Practitioner: An individual without a public license/certification who is supervised by a licensed/certified individual in the delivery of care to patients.

Provider: A business entity that furnishes health care to a consumer; it includes a professionally licensed practitioner who is authorized to operate a healthcare delivery facility (ASTM 1384-91). For Veterans Health Administration (VHA) purposes, a VHA medical center to include its identified divisions and satellite clinics is considered to be the business entity furnishing health care at the organizational level. Suborganizational level entities will include treatment teams and individually identified practitioners. For purposes of the National Patient Care Database (NPCD), a VHA-defined practitioner-type field will be reported together with the medical center and medical center division code.

Encounter: An instance of direct, usually face-to-face, interaction or contact, regardless of the setting, between a patient and a practitioner vested with primary responsibility for diagnosing, evaluating, and/or treating the patient's condition or providing social worker services. The practitioner exercises independent medical judgment. Encounters do not include ancillary service visits. For VHA purposes, a telephone contact between a practitioner and a patient will be considered as an encounter if the telephone contact included the appropriate elements of a face-to-face encounter, namely history and medical decision making. A patient may have multiple encounters per visit. Collateral services provided as a part of the patient's care (such as family therapy) will not be reported separately. Collateral services provided directly to the collateral (separate from the patient; for example, stress reduction skills) will be reported separately.

Ancillary Service (also known as Occasion of Service): Appearance of an outpatient in a unit of a hospital or outpatient facility to receive service(s), test(s), or procedures; it is ordinarily not counted as an encounter. These ancillary services may be the result of an encounter. For example, tests or procedures ordered as part of an encounter. A patient may have multiple occasions of service per encounter or per visit.

- Training: This is felt to be vital to the success of the project. Facility personnel need to understand the data definitions and data elements being changed, the impact upon facility processes, and technical implications locally and nationally.
- Facility Implementation: This involves the appointment of a project team at each facility to work with the national implementation group to oversee their implementation. Performance measures will be monitored to evaluate implementation progress.
- Post-implementation Support: Once the new policy is in effect, facilities will need support to address problems that naturally arise in a project of this nature.

Many obstacles encountered when capturing data in the ambulatory-care setting are not present when capturing data for inpatients. These additional obstacles arise in part from the volume of outpatient activity. Some of the large tertiary-care VA medical centers have as many as 1000 visits per day and potentially as many as 2500 encounters. These encounters are by definition abbreviated in duration with minimal supporting documentation. Thus, any solution must be efficient and easy for the ambulatory-care clinicians and administrative personnel. It must enable the collection of reasonably accurate data in a realtime mode. Given the volume of activity, delays and errors that need extensive intervention could quickly overwhelm the entire system.

The VA is also encountering resistance to change. The ambulatory-care personnel who need to collect these data now have an added responsibility. Training and marketing efforts must address these concerns and illustrate how the capture of these data will benefit the continuity of patient care, quality-improvement efforts, performance measurement, and resource management.

It is extremely difficult to ensure the integrity and validity of data with so many people entering data using different data-entry methods. By adopting national standards already reviewed and approved by different national organizations, the VA can address the initial integrity of the data. Coding and other data policies assist in this process. The software contains edits that will enforce parts of the policy. For example, only valid codes from the International Classification of Diseases, 9th ed. (ICD-9-CM), American Medical Association's Current Procedural Terminology (CPT), or the Health Care Financing Administration's Common Procedure Coding System, will be accepted. Validating the data is an ongoing, multistep process. It involves checking to ensure that all required data elements are present for each encounter, conducting a random sample clinical pertinence review of patient record documentation of the encounter against the encounter data, synchronizing the local database with the national database, and more.

There are also many different settings in which ambulatory care can be delivered, along with many types of ambulatory care. For VA, this includes hospital-based clinics, community-based clinics, outreach centers, mobile clinics, and home-based health care, among others. The care itself can range from an encounter (e.g., for an insulin check, medication check, or influenza vaccination) to ambulatory surgery (e.g., cataract removal, etc.) with a wide range of services in between. The system should enable all practitioners to report their activity accurately.

Finally, due to resource constraints, the VA must eliminate redundant data entry to the greatest extent possible. The data captured must satisfy needs of clinicians and researchers, as well as staff, in the areas of quality management, facility management, third-party billing, and resource allocation. Examples of this include enabling the clinician to designate a purpose of visit as an ongoing problem that then be stored on the patient problem list, while having the purpose of visit automatically transmitted to the billing software for claims generation, and, in addition, using the data to produce Health Plan Employer Data and Information Set (HEDIS) or HEDIS-type measures.

The overall generic process for data capture flows as shown in Figure 11.1. Patient data (e.g., demographics, previous lab tests, and medications) are presented to the clinician via the clinician workstation or a combination of the printed encounter form and health summary. The clinician then documents the reason why the patient was seen, as well as the care delivered to the patient. Again, there are several data-entry methods available, including scannable encounter forms, clerical key data entry, and clinician workstations. In addition, the clinician is required to write a progress note to support the purpose of visit and service provided; this progress note will be used later to perform data validation. When the data are entered, the clinical data repository performs specified data edits before publishing the data to software applications that have a use for them, such as integrated billing and the problem list. The data are bundled into a queue of Health Level 7 (HL7) messages and transmitted each night to the National Patient Care Database in Austin, Texas.

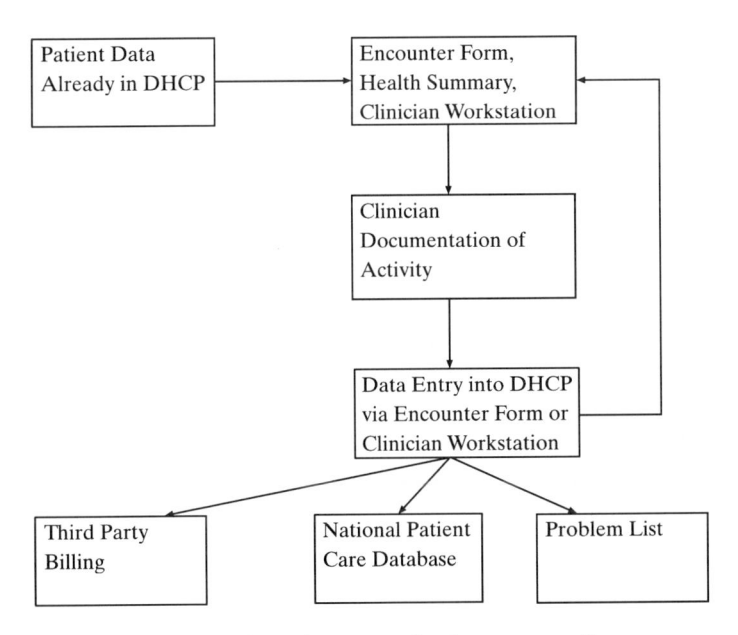

FIGURE 11.1. Generic process for data capture flows.

Future Plans

While the current functionality will be useful and provide much to VA that has been previously unavailable, the continual change in the healthcare industry will mean continued change. Items already on the drawing board include increased ability to track preventive medicine and educational counseling services, severity of illness indicators, and other health status indicators (e.g., homelessness and employment history). In addition, VA is concerned about its ability to record and track services that are not provided by VA, but may impact the health status of the patient. For example, the patient may receive a flu vaccine at a local health fair. Other features that will have to be addressed in the future range from patients entering their own data to communicating that patients via the Internet.

Ambulatory care is the present and the future. Our challenge is to collect, process, and utilize the data necessary to optimize the care of our patients.

12
Moving to Clinical Workstations

SHARON CARMEN CHÁVEZ MOBLEY

Background

The primary mission of the Veterans Health Administration (VHA) in the Department of Veterans Affairs (VA) is to deliver quality healthcare services to eligible veterans. Today VA operates the largest centrally directed healthcare system in the United States. The Decentralized Hospital Computer Program (DHCP) provides electronic information to approximately 159 medical centers as well as outpatient clinics, nursing homes, and domicilliaries.

The Decentralized Hospital Computer Program architecture began in 1982. The DHCP focus was implementation of software modules to integrate a complete hospital information system. At that time, VA decided select Massachusetts General Hospital Utility Multi-Programming System (MUMPS) as the primary programming language for DHCP and began development of integrated software modules using programmers to prototype environments. Now MUMPS is known and referred to as "M," and is an American National Standards (ANS) Institute and Federal Information Processing Standards (FIPS) standard language.

From Dumb Terminals to Clinical Workstations

Purpose

In 1995, VA made a decision to pilot a clinical workstation project to test the feasibility of using PC-based clinical workstations to enhance the quality of patient care. Moving toward clinical workstations involved

- Changing the current computing architecture from text-based to windows-based programs;
- Strengthening the network infrastructure; and
- Adding data types not currently available on the DHCP system, such as images, voice, and graphics.

The purpose of the pilot was to demonstrate whether the capital expenditure for network and PC equipment would be offset by productivity gains by healthcare providers and a general improvement in the quality of health information delivered with the new system. The pilot would also demonstrate the extent of the demand generated for maintenance and support once the transition from dumb terminals to workstations was accomplished.

A fundamental step in producing a VA computer-based patient record (CPR), the move to a clinical workstation would give healthcare providers the information they need to perform and deliver services. The pilot thus involved improving the usefulness, effectiveness, and efficiency of the DHCP system and would require major changes to the hardware and software components of the current systems.

Technology

Constructed with a graphical user interface (GUI), the clinical workstation utilizes Delphi software running on the client, while the backend remains on an M database. The Windows environment is provided by WIN95 or NT. Network protocol is TCP/IP running below the Winsock API. The physical network consists of a high speed core (ATM or FDDI); the departmental interconnected local area networks (LANS) may take one of several forms, such as 100 base T.

The VA's CPRS software modules will provide the first robust application of the PC client architecture. The Computerized Patient Record System will allow clinicians easier ordering, display, and retrieval of clinical information.

Benefits

A national systemwide upgrade to clinical workstations makes it possible to incorporate data types that cannot be currently used in tracking VA patients. Through DHCP, VHA has evolved beyond the text record of the past. The true electronic medical record will consist of graphics, sound, images, multimedia applications, and artificial intelligence. This push forward builds on current software products and network enhancements to keep the level of service to veterans equal to, or surpassing, that available in the private sector.

It is VA's goal to implement clinical workstations, and to do so by exploiting its talent pool. Better information management promises to reduce costs by reducing the number of lab tests ordered, eliminating the need to reorder hard-to-locate x-rays, and better organizing the total care that a patient receives.

The workstation approach takes advantage of the trend toward client–server computing. The VA development staff will be trained to use Delphi in the production of new software, utilizing object-oriented programming capabilities. The Hybrid Open Systems Technology (HOST) program will make third-party software tools and utility programs accessible and provide functionality in the form of reusable components. Although Delphi is currently the preferred language for development and maintenance of the presentation layer, any Microsoft Windows-based product could be used for particular application design if necessary, and work is underway to explore using Internet browsers as the graphic front-end.

All source code and production implementation will be controlled by VA. In addition, VA will work to implement better interfacility computer communication and to integrate data into central concentration points for each of the 22 Veterans Integrated Service Networks (VISNs). Internet access will be made available via the World Wide Web for professional contact to outside provider services and knowledge.

The planned workstation will offer the following features:

- Windows-based GUI, supporting quick and easy views of health summary data, detailed histories, etc.
- Innovative technology combinations in an open-system environment, providing access to data on functionally specific machines.
- Integrated access to images, data, voice annotation, dictation, phone, paging, and fax.
- Active networking infrastructure, creating virtual networks that link departmental LANs to the hospital backbone network at speeds greater than 100 mbps.
- Easy access to and creation of online help, administrative documents, and services such as Medline and Internet.
- Multimedia capabilities, making training and support functions effective and informative.
- Artificial intelligence links to commercial knowledge bases, providing automated expertise in clinical diagnosis and treatment.

Implementation

Participants and Roles

Plans identify the areas and VA personnel involved in the implementation of clinical workstations, specify the extent of effort required, and briefly describe roles and functions in telecommunications, hardware, software, support, training, evaluation, and project management.

Training

Training for technical staff will be most intense during the start-up period, when Information Resources Management (IRM) Field Office staff will need to develop expertise in the operation, support, and maintenance of

NT, GUI, Delphi, Windows dialogs, etc. Training needs will diminish as the IRM Field Office staff become more knowledgeable about the new operating system environment and ancillary issues surrounding support to the field and maintenance of key products. It would, however, be a mistake to underestimate the effort and time involved in making medical center staff computer literate; this will be the most time-consuming and the most important aspect of the training activities.

Methodology

Moving from the dumb terminal to the clinical workstations will be the single most dramatic technological change for the VA in the last decade. The paradigm shift is tremendous. Deployment to individual facilities will vary according to the size of a given site's user population, patient population, and database.

Sites will include those currently using either 486- or DEC Alpha-based systems to run DHCP. Each will test the client–server architecture. Classes will be taught in general computer literacy and Computerized Patient Record System (CPRS) software. Benefits will be measured by surveys. Vendors familiar with the VA system who have performed similar comparisons for cost-benefit studies in the past will evaluate the systems and technology used.

Resource Issues

The general areas of procurement will be wireless technology, PCs, operating system software, and various network devices. Most equipment will be depreciated on a 5-year life cycle. Maintenance contracts will be purchased only for the most critical items, such as the network operating systems (NOS). Personal computers will be purchased with "spares" onsite for quick swap out; network hubs will be handled in a similar manner. After warranties on PCs and hubs have expired, sites may choose to fix equipment inhouse or to send it to a contracted vendor. Capital expenditures will be made during the first year of the purchase; maintenance will be ongoing. When the funding is approved, the procurements can begin, conditional upon the successful implementation of equipment and software deployed by the pilots. The actual release of equipment will be keyed to the willingness and preparedness at each site. Delivery schedules will be constructed from a site survey form that will be completed by Engineering and IRM staff.

Technological Deployment

The VA's commercially provided data network will handle wide-area network (WAN) interconnectivity between VAMCs. The primary database systems (M) will be under software maintenance from the vendors. This coverage will continue at its present level, that is, 7 days by 24 hours. Personal computers currently in use will be given software upgrades and network cards if needed. Upgrading random access memory (RAM), processors, and hard disk drives will be the responsibility of the individual sites. It is assumed that each site will move desktop computer power to the areas that need it most, and redistribute equipment to match functionality necessary to perform the applications in each individual clinical area.

Technical staffing requirements will increase dramatically due to the additional hardware and software. Part of the cost-benefit analysis will include whether the benefits of the system outweigh the burden of the additional computer maintenance.

Appropriate software management tools and reliable hardware should minimize the burden of management at each VAMC. However, training the work force is an important issue to address, and local programs will need to be established if they are not already in place. Each facility will need to actively participate in training for as many administrative and clinical staff as possible.

Project Methodology

Tactical Steps

- Perform baseline measurements;
- Purchase network hubs;
- Purchase ATM core hub;
- Purchase PCs and NICs;
- Train staff on use of PCs and software;
- Construct network utilizing windows and terminal emulators;
- Transition from DOS or VMS to WinNT;
- Install SMS or similar network distribution/install software;
- Install CPRS software;
- Measure productivity gains/losses;
- Measure quality gains/losses.

Evaluation

A formal evaluation process has been set in place. User surveys and interviews will define quality issues, which tend to be subjective. Surveys/interviews will be conducted with personnel involved in the pilots, including IRM staff, ward clerks, physicians, nurses, and other clinicians. They will be asked to address such areas as the following: perceived quality of interface, user compatibility (color, icons, and sound), task compatibility (does software match activity?), workflow compatibility, consistency, sense of control, flexibility, accommodation of user skills and preferences, responsiveness, ability to handle screen construction and destruction (particularly text editing), and ease of learning/ease of use (better designed user interfaces).

The evaluation will also collect quantitative data. Random samples will be made of discharge summaries, progress notes, and lab orders, in two main categories: handwritten (i.e., paper-based) and electronic (i.e., computer-based). Samples in the latter category will be collected in two subcategories: materials composed on dumb terminals and materials created on clinical workstations.

Analysis will compare written samples to electronic versions; analysis of the electronic versions will differentiate between samples from dumb terminals and clinical workstations. For measuring workflow, timing measurements will be performed for specific tasks. Response time will likewise be measured. The results will provide measures of efficiency, for example, the speed of a patient lookup, editing records, and ordering lab tests.

Anticipated Benefits

Evaluation will attempt to validate benefits anticipated by VA as outcomes of the workstation pilot and implementation. These benefits include the following:

- Reduced idle time: less time spent waiting for information;
- Greater efficiency in operations: better healthcare delivery using fewer resources;
- Improved competitiveness: streamlined and more responsive operations;
- Better distribution of resources: employees empowered by desktop capabilities (e.g., ward clerks can do spreadsheet calculations);
- Faster response to new medical needs: easier problem solving;
- Improved product development: more comprehensive products showing more data.

New Capabilities

The above-mentioned benefits will accrue as a result of the capabilities and characteristics of the new systems approach to the clinical workstation. To begin, the PCs acquired to serve as clinical workstations are flexible as well as powerful. They can be adapted relatively easily to accommodate new software and new network-accessible hardware. Moving to client–server from standard host-based terminal servers represents significant gains in flexibility as well as the ability to run GUI software. At the same time, the client–server model provides a stable database core. New applications become standard GUI construction, with the designing of screens for new processes, and subsequent filing of gathered data. This approach guarantees interaction and activity with the central database.

Under the network model, new services such as web pages, data archives, and telephone applications become standard integrated features rather than add-on applications that vary from medical facility to medical facility.

The windows environment and its operating system allow access to more applications, developed in a great variety. Applications can be developed in faster time, with better tools, and with more synergy with commercial applications. Sophisticated windows file structures can support comprehensive help systems; these can be combined with interactive audio, visual, interactive (AVI) files to make training more robust.

Technology and the Enterprise

The capabilities provided by technology serve as an enabling infrastructure for the new networked model for VA services known as the VISN. The focus of the VISN is the provision of services to its clients. The information technologies discussed in this chapter and throughout this book support the VA as it provides those services.

The VISNs represent a new enterprise concept for VA. Technologies such as telemedicine, teleconferencing, and distance education support the enterprise, improving the quality of interactions and saving travel time by substituting video exchanges for face-to-face meetings. In the new model, work (in digital form) moves, not people. Avenues have just begun to open for telecommuting, telemedicine, and remote access.

Within individual offices and for individual personnel, technology enhances productivity by providing better messaging, alerts, and online faxes. Managers and clinicians alike can have access to more data items with higher quality and more types of data (sound, video, and images)—in sum, better information for better decisions. Online analytic processing, artificial intelligence, and decision support systems will enhance decision-making processes.

Network technologies also provide the 22 VISNs with access to resources and individuals outside VA. The Internet provides linkages to a host of online services, offering the potential for wide-ranging contacts in educational institutions, for-profit companies, and private homes. From the National Library of Medicine to specialized research databases and patient self-help groups, the Internet gives access to text-based and visual information and even sound.

New Systems, New Challenges

People are more heuristic than algorithmic in problem-solving strategies. The new workstation, with its GUI design, will allow for heuristic problem solving rather than forcing users to execute algorithmic procedures—an exercise in boredom and frustration.

The new workstation will be flexible, allowing users to take shortcuts if they are experienced and to experiment at performing functions if they have only a minimum of training. It will have robust help systems that actively provide hints or suggestions rather than passively furnish incomprehensible responses to user queries. It will require no more effort to learn than is merited by the problem at hand. Because it will be easy to learn, it will be utilized to the extent that VA will realize a return on its investment of capital resources.

With the new workstation, menus will continue to be user configured as they are at present. However, users will be able to increase menu breadth as opposed to menu depth, "flattening" items for easier selection, rather than requiring movement down a hierarchical tree. New menu styles also will lend themselves to speed key activation, context labels, and balloon style help. Icons that are concrete, familiar, and distinct will aid users in recognition and navigation through DHCP and the many robust systems accessible through Hybrid Open Systems Technology (HOST).

Conclusion

Introducing the clinical workstation into the VA environment and expanding the data sets available will help clinicians better manage patient care and make better clinical decisions. With the workstations putting information at their fingertips, VA personnel, whatever their roles, will be have data that are more powerful, diverse, and relevant to the tasks at hand. Online information will benefit critical processes. Productivity will rise, and outcomes will be more favorable. Information and technology will be used to shape the organization and the services it provides. Planning will become more innovative. The new clinical workstation environment will encourage and support creativity throughout VA, as individuals explore new options for accessing and structuring data and information to support them in their responsibilities.

The challenge is to move beyond gathering and storing volumes of data on patients and to produce meaningful information. The newer clinical workstations are capable of storing much richer and more robust data sets that can and will improve clinical, research, and business practices at every VA medical center.

The VA's clinical workstation project is one of the organization's most significant undertaking in the IRM technical operational arena in the last decade. The paradigm shift is powerful, dramatic, and awesome.

The user community will require an education in a number of areas before they can effectively take advantage of the cutting edge technology that is about to break. Staff orientation, training, and implementation of the technology will take major effort on the part of management. However, the overall benefits are worth the heavy investment in infrastructure, hardware, software, and education of personnel.

Even with the downsizing and flattening of organizations so common today, these are exciting times. Technology cannot solve all problems, but it can help identify and resolve them. With VISNs, VA is shaping the future of health care. This new model sets a precedent that does not exist elsewhere in the world.

13
Controlled Representation in Patient Records and Healthcare Delivery Systems

KENRIC W. HAMMOND

Introduction

The vitality of health care drives demand for capable information systems and encourages experimentation in the field, but its highly individualized nature requires complex systems that increase the risk of innovation. For 20 years, one or another of the best minds in healthcare informatics has predicted that either a "complete" or "paperless" electronic patient record was only about 5 years off. Presently, the year 2000 is the target of choice. Healthcare informaticians, like astronomers who build stronger telescopes to see more stars, know that the price of progress is repeated realization that completing one task merely reveals the next.

The ideal Health Care Information System (HCIS) needs to reliably record and yield valid information about millions of lives, each treated in dozens of places, by hundreds of healthcare workers. The dynamic nature of scientific knowledge introduces another, paradoxical complication: its volume doubles every 5 years, but any particular fact is likely to become obsolete within a lifetime. The diversity of organizations, arrangements, and people who participate in care adds more complexity. No wonder other wise voices express doubt that an "ideal" computer-based patient record will appear any time soon.

Programs and databases require formalization of information elements and process flows to make computing practical. Formalism is difficult to achieve when many interacting components are involved. Careful analysis may anticipate difficulties, but the most crucial insights are gained after a system is implemented. Invariably, performance failures that reveal inadequacies in information and process models drive progress. This chapter addresses the challenges and potential benefits of introducing formal symbol systems to healthcare computing. The main message is that developers must learn where and how to learn from their experiences.

Barriers and Goals

As more functionality is demanded of HCIS, current limitations become more evident. The difficulty of moving data between different healthcare computer systems is an obvious example. Despite "completeness" of any single system's records, uncertainty about what is available in another system and how it is represented impedes meaningful data exchange. This echoes the familiar lesson of success: were there no systems, interchange would not have been attempted.

A growing user community imposes another penalty. New participants bring new information requirements. Terminology that expresses concepts contributed by specialists promotes better understanding of patients, but to be communicated usefully, one group's input must be intelligible to others. Furnishing information services to a caregiver team requires vocabulary support for each member. Developing suitable vocabulary systems is challenging, but coordinating multiple vocabularies is even more so.

Here is a summary of current challenges in healthcare computing:

1. People cannot access information sources when they lack equipment.
2. Incompatible machines cannot share information.
3. Incompatible data formats render stored data unusable.
4. Information is incomplete when data acquisition fails.
5. Information is unclear when its original context is unclear.
6. Information is misleading if applied wrongly.
7. Relevant information is lost when buried in irrelevant information.
8. Relevant and correct information may not be understandable.

Items 1 through 4 concern the adequacy of physical resources and software design and are being rapidly resolved by expanding information systems in health care. Items 5 through 8 cannot be resolved by improving programs or machines. They require the HCIS to better record and transmit meaning and apply knowledge.

Preserving Meaning in Healthcare Information

Contemporary systems gain their utility from a computer's ability to route, tally, sort, and format large volumes of data rapidly. Until recently, improving the capture and reporting of data were the main concern of patient record systems. Extraction of meaning from the retrieved information is left to the practitioner.

Few, if any, production scale systems regularly process the meaning that people find in databases. However, with suitable symbols, computers that count and collate data can also process concepts, just as children who count on their fingers improve their performance when they learn the rules and terminology of mathematics. Meaning, like data, can be represented in tables of definitions, rules, and relationships. Machine-assisted decision making is accomplished by retrieving and processing datum-symbol-meaning elements using standard procedures. Data acquisition captures the meaning of an event when it reliably encodes and records an observation. In medicine, direct encoding is performed by people at points of data capture. Indirect encoding occurs when additional meaning is gleaned from the context of a recorded event.

Knowledge Processing

As current production scale HCIS become more saturated with data, attention has shifted to better use of stored information by linking data to arrays of symbolically expressed meaning. Such "knowledge bases" link observations recorded in a healthcare record system with the more constant rules, definitions, and relationships that make up general medical knowledge. Special purpose systems that "intelligently" manipulate meaning have existed for decades (Shortliffe, 1976). Their narrow focus and the user effort required to feed them data have hindered deployment on a production scale. Better ability to retrieve meaning from healthcare databases will permit more extensive use of "smart" systems. Greater use of controlled representation in HCIS may help bridge the gulf between intelligent and production systems if it provides a symbolic framework to unite the separate contributions of system developers and computer scientists.

Barriers to Building Intelligent Healthcare Information Systems

The validity and consistency of knowledge-based decisions service depends fully on the quality of the data used, which in turn depends on the fidelity of collection and encoding. The effectiveness of data-driven reminder systems in improving practitioner adherence to care standards demonstrates the power of combining knowledge with clinical information (McDonald, 1976). Knowledge adds value if applied correctly at the right time and place. When the *system* "knows" the content and context of an information session, an opportunity to enhance practice exists. Systematic control of meaning in healthcare information processes is essential for building a knowledge-aware HCIS. Because economics, not knowledge-processing capability, is the main driver of HCIS development, widespread "knowledgeable computing" will await recognition of its economic benefits.

Interaction of Knowledge, Meaning, and Process

Consider the network of information and activity depicted in Figure 13.1. Health care consists of delivering expert services to patients in a business

13. Controlled Representation in Healthcare Delivery Systems   167

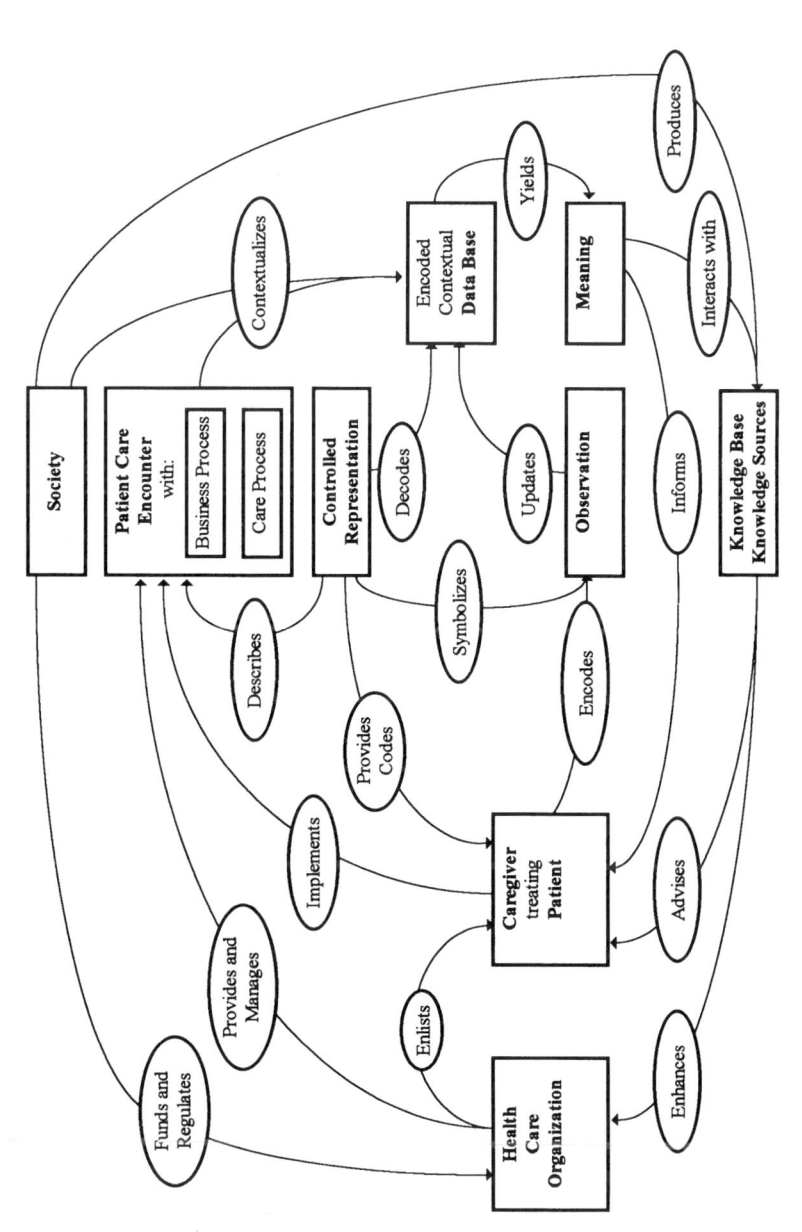

FIGURE 13.1. Information relationships in healthcare systems. Controlled representation is critical to supporting the cognitive activity of caregivers and usefully describes the business process of healthcare. Symbols give practitioners the means to encode and decode meaning. Meanings may be directly interpreted but also link to knowledge sources. Business processes confer and benefit from meaning as well. Coordination of business and cognitive tasks and representations is essential.

and social context. Its services are informed by meaning and guided by policy and knowledge. When an organization furnishes an information system to support care, business processes and caregiver mental activity are linked. A full understanding of care requires not only factual information about patients and treatment but also the business rules and scientific ideas that undergird it. Caregivers think independently as they treat patients, but when they get their information from a computer system, what they retrieve is shaped by the business processes that accompanied its capture. Currently caregivers rely upon their own fund of knowledge to interpret computer-furnished information. When systems attempt to knowledge support and even guide clinical decisions, it will be essential to know how business processes have influenced the information's meaning. Humans generally do well at resolving ambiguities that creep into HCIS data. Automated knowledge processing will be less forgiving and will need good information to succeed.

Historically, safeguarding *scientific* meaning and validity has been a concern of universities and researchers. The business side of healthcare computing has focused on practicalities: providing efficient practice support and managing finances, inventories, and employees. Figure 13.1 illustrates that the business, cognitive, and knowledge aspects in health care are interdependent. An "intelligent" system needs to interpret each of these processes to yield suitable meaning at the right time. A successful healthcare information environment must coordinate business and scientific perspectives. A suitably designed controlled representation system can bring them together.

Relevance of Controlled Representation to HCIS: Why Standards Matter

A lack of sufficient standards for representing healthcare information and processes is the most important barrier today to better use of HCIS. Carefully chosen symbol systems appropriately used to preserve meaning presents attractive opportunities, including: (1) better organization and viewing of information in growing databases, (2) transportability of information between systems, (3) access to knowledge sources, (4) enhancement of user performance by intelligent services, and (5) enhancement of organizational performance.

In knowledge-intensive computing, manipulating abstractions may be more efficient than manipulating concrete elements. Representational systems that allow abstraction (permitting a search for patients with "cardiac disease" versus requiring searching for each of the hundreds of ICD-9-CM codes for cardiac diseases), can bring new capability to the computing workspace at modest cost. Once in place, controlled terminology in the HCIS reduces the cost of maintaining and bringing new knowledge services to health care as well.

Evolution of Standards in Healthcare Computing

Historically, adoption of industrial standards have promoted progress by facilitating the sharing and reuse of intellectual effort. In HCIS, standards have evolved among the several logical layers of logic and functionality that operate in concert to support computing processes. These include hardware, communications protocols, operating systems, computer languages, applications, data, and the human–machine interface. Performance of one layer depends on the layers below. Health care itself can be described as a layered system with the patient–caregiver interaction on top. In an "informatized" care delivery system, the patient's experience of care will depend on the integrity and performance of underlying layers. To work together, functional layers must communicate by direct action or by exchanging symbolic messages. Symbolic interlayer communication between is much more flexible.

Processors, Networks, and Operating Systems: The Implementation Layer

Four decades of progressive standardization of hardware, connectors, network protocols, and operating systems have yielded an environment rich in distributed, communicating computing resources, symbolically supported by published engineering specifications. Present-day systems favor the multilocation, multiparticipant nature of health care. Standardization, driving competition, has made products more affordable, reducing the risk of building and deploying HCIS. Shared computer resources have lessen the threat of losing data; increased practitioner access sharpens demand for more clinically capable systems.

Operating systems bridge machines and applications to define the computing environment. With code that interacts with hardware and applications, operating systems transcend machine differences to allow the same programs and data to reside on many types of machines. Various operating systems, each supporting standard computer languages and installable on many platforms, were important forces promoting dissemination of HCIS. The versatility of MUMPS operating systems permitted the Department of Veterans Affairs (VA) to rapidly propagate replicas of its Decentralized Hospital Computer Program (DHCP) across a system of 172 medical centers, avoiding vendor "lock-in" and leveraging software development and maintenance resources. Microsoft Windows NT and several commercial UNIX implementations increasingly dominate the operating system marketplace. MS-DOS was instrumental in bringing computers to individuals and small healthcare organizations, and Windows NT allows MS-DOS programs to run on a full range of equipment.

Client–server architecture, especially the multiplatform "load, run, and terminate" architecture of Sun Microsystems' JAVA specification has blurred the distinction between network, processor, and operating system.

Its development was spurred by growth of the Internet and World Wide Web, which in turn were facilitated by the Transmission Control Protocol/Internet Protocol (TCP/IP) standard.

Applications: The Processing Layer

Symbolic communication between hardware and applications has freed application developers from details of machine-specific implementation and increased their productivity. Methods for storing, processing, and sharing data between applications have converged around standard specifications, affording interoperability somewhat analogous to that achieved between hardware and operating systems. Applications, of course, link processing capacity to tasks in the real world, but the interface between applications and task domains in health care is less standardized.

The first applications relevant to health care were inventory and accounting systems. Without doubt, the uniformity of procedures (accounting rules) and symbols (numbers) aided early implementation. The 1965 introduction of Medicare (with many standard business rules for health care) encouraged detailed billing and boosted the hospital financial applications market. Before long, patient registration, fundamental to billing and essential for all clinical applications, was widely established.

Hospital-Based Systems

Soon, beginning with clinical laboratories and extending to pharmacies and radiology departments, clinical support applications appeared in hospitals. Implementation followed a regular pattern: when financially advantageous, systems were established supporting well-defined business processes that could use constrained symbol systems to ease data input. Computerized support services led to placement of terminals (sometimes one for each department) on wards to retrieve data.

By the 1980s, "integrated" systems that combined financial, administrative, and departmental information in a single computing environment were marketed. The Medicare's prospective payment system of diagnosis related groups (DRGs) shaped these systems considerably. Reflecting this, in the mid 1980s the Health Care Financing Agency (HCFA) selected the International Classification of Diseases, 9th ed. Clinical Modification (ICD-9-CM) (1991) as its standard for recording discharge diagnoses for Medicare billing. Patient-care support systems lagged behind business systems. In the late 1980s growing financial pressures faced by hospitals caused integrated vendors to market "bedside" systems offering to improve the efficiency of nurses, the largest group of hospital employees. Existence of the North American Nursing Diagnosis Association (NANDA) nomenclature that provided a defined vocabulary for nursing diagnoses and care plans aided product development.

Outpatient Care Systems

Outpatient clinics and offices installed information systems later than hospitals because their capital budgets were smaller, and because their less intensive information processing needs could not justify investment until computer prices fell. Unlike hospitals, where departmental reporting systems were advantageous and feasible, clinics principally needed tools to manage billing, schedule appointments, and document care. The earliest outpatient systems were remote access billing services, but in the mid-1980s prices of stand-alone billing systems and equipment became affordable for smaller practices.

In larger clinics and health maintenance organizations (HMOs), the need to coordinate care among practitioners drove development of two key components of the patient record: progress notes and problem lists. Organizations with dictation services in place were able to put clinical notes online once storage costs fell, and the historical record of ICD-9-CM diagnoses gathered for billing served as a reasonable approximation of a problem list. Organizations that lacked dictation services or did not bill visits were slower to adopt these services. When VA received authority to bill for nonservice-related care, development of its problem list accelerated.

The Computer Stored Ambulatory Record (COSTAR) system was an early and instructive ambulatory-care information system (ACIS), developed by Barnett and others for the Harvard Community Health Plan (Barnett, Justice, & Somand, 1979). A public domain version of COSTAR has been available since the late 1970s, and it has been installed in several hundred clinics. Wider adoption did not occur because of computer expense, difficulties with data input, and a scarcity of support personnel trained in MUMPS.

To manage data flow, COSTAR introduced detailed, condition-specific forms to encode data from visits. These were filled out by the practitioner and later transcribed. In high-volume settings, encounter forms have enjoyed some popularity, but the new tasks of forms management, training, encoding, and transcription require major restructuring of workflow in a clinic. The durable contributions made by COSTAR were the encounter form, a system of codes for ambulatory care, and the lesson that implementing ambulatory-care information systems requires revision of practice patterns. The COSTAR codes are part of the National Library of Medicine's Unified Medical Language System (UMLS), and its value in representing clinical phenomena in HMO settings has begun to be recognized (Payne & Martin, 1993).

Specialized Systems

Success of computer-based clinical systems in some niche domains correlated with economic relevance and available controlled representation systems. For example, systems to administer and score psychological tests have been in use for at least 20 years, and most standard instruments are available for administration on personal computers. Few general hospital systems (VA and Boston's Beth Israel Hospital are exceptions) integrate results of these tests with general health databases. Such systems can also elicit structured health histories, but this use has not spread much to general medical practice, despite evidence of validity and cost-effectiveness. The difficulty may be due to absence of a framework for representing patient health histories in HCIS. Increased interest in administration health status questionnaires to patient populations may redirect interest to the area and encourage explicit representations of self-report information in HCIS.

Healthcare Data Interchange: A Communications Layer

By the late 1980s, enough progress in HCIS had occurred to make the matter of moving useful patient information between different systems relevant. Mobile patients and ever-changing healthcare coalitions along with greater dependence on computer information highlighted the need for machine-independent ways to send and receive data and have it end up in the right place. In the United States, the Association for Testing and Materials (ASTM) and an industry collaborative, Health Level Seven (HL7), took up this task. In Europe, the project known as Generalized Architecture for Languages, Encyclopedias, and Nomenclatures in Medicine (GALEN) addresses the same issue for the principal European languages.

The easy part of the problem was solved by specifying an ASCII-based messaging format containing the information, patient identifiers, and indications of source and destination. The HL7 standard lacks reference to a standard system for naming laboratory tests. Compliant systems that can read each other's formatted messages do not fully "understand" each other's data if they use different systems for coding test names and descriptions. A new initiative, Logical Observation Identifier Names and Codes (LOINC), is currently developing a naming system to supplement HL7 protocols. Similar initiatives will be needed to interchange other portions of computer-based records. The communication layer requires not only a common vocabulary and adherence to its definitions, but also a way to describe business practices.

Controlled Clinical Vocabularies: A Symbolic Layer for Practice

With 19th century origins, the ICD-9-CM is the most widely used vocabulary in HCIS. The ICD-9-CM was developed to classify and enumerate diseases for epidemiological purposes, but its scope and codability led to its use in billing to designate treated conditions. By custom and convenience rather than design, HCIS usually wrongly equate the ICD-9-CM code to a clinical diagnosis, but a physicians' knowledge of diseases and treatment is organized according to pathophysiology, anatomy and etiology, not the ICD-9-CM.

In contrast, the similar-appearing American Psychiatric Association *Diagnostic and Statistical Manual, 4th ed.* (DSM-IV) is studied extensively in medical schools (American Psychiatric Association, 1994). Etiologies of most psychiatric disorders are unknown, and are mostly unconfirmable by physical tests. The DSM-IV systematizes clinical observations and specifies criteria for psychiatric diagnoses. A DSM-IV code contains more explicit and implicit meaning than an ICD-9-CM diagnosis, and while used for billing, is also more suited to symbolic manipulation.

The Current Procedural Terminology (CPT) is used to designate procedures for billing and is regularly supported in HCIS (American Medical Association, 1989). Its guiding principle is to categorize in a way that discriminates cost. Different CPT are used to bill surgical incisions of different lengths, but the clinical significance is trivial. Pathologists developed the Systematic Nomenclature of Medicine (SNOMED) to assist their work (Cote, 1995). The capacity of SNOMED to describe the effects and processes of disease corresponds closely to physician's packaging of medical knowledge. Like DSM-IV, SNOMED is multiaxial, permitting encoding along multiple dimensions of procedure, disease, morphology, etiology, and anatomic location. The SNOMED codes are represented in numerous HCIS, but generally in applications that support the specialized activities of pathologists.

The COSTAR codes were mentioned earlier. James Read, with computer storage in mind, developed an extensive, practice-relevant coding system for British general practitioners (Read, 1990). The system's clinical origin interests practitioners who feel that the ICD-9-CM system restricts expressiveness. The Read system has undergone significant changes in each of its editions, and cannot be considered stable. Since 1980, the presence of HCIS has encouraged development of coding systems for nursing, among them NANDA nursing diagnoses (Carroll-Johnson, 1994), the Nursing Intervention Classifications (NIC) (McCloskey & Bulechek, 1994), and the Omaha Classification of Ambulatory Nursing Care (Martin & Scheet, 1992). Developers of the medical expert systems Quick Medical Reference (QMR) and DxPlain contributed controlled terminology systems designed explicitly for computability, but these terminologies are not systematically included in production-scale HCIS. The Medical Subject Headings (MeSH) system, developed by the National Library of Medicine for the MEDLARS bibliographic retrieval service, is an example of yet another special-purpose controlled vocabulary, built to classify and retrieve biomedical knowledge. It, too, has had little direct implementation in HCIS.

Unified Medical Language System: A Translation Layer

Emergence of multiple controlled vocabularies for medicine has paralleled the use of computer systems to manage them. Separately, each vocabulary has a distinct purpose and advocacy group, yet none maps tidily to its neighbors. They will continue to appear because even the largest vocabularies omit terms needed by new applications and users. Authoritative coding and representation systems are expensive and slow to develop because they require collective expert input and ratification by a sponsoring organization. From a distance, the ferment of vocabulary development resembles a larger system trying to describe and understand itself.

Even in their current fragmented state, information systems manage more information than any single individual can grasp. Recognizing the need to organize the body of knowledge available to biomedicine (and initially, only incidentally to HCIS), the National Library of Medicine began the visionary Unified Medical Language System (UMLS) initiative in 1986 (Humphreys & Lindberg, 1989). The project grew from the bibliographic roots of MEDLARS, a powerful biomedical knowledge retrieval system that owed much of its success to the MeSH system used to index publications.

Using MeSH as a backbone, UMLS investigators systematically mapped established vocabulary elements to a system of biomedical concepts. Content experts used powerful computing techniques to unify and interrelate meanings from separate vocabularies in biomedicine. The UMLS "metathesaurus" is available on CD-ROM disks. The 1996 version contains over 250,000 distinct concepts named by 589,400 biomedical terms from 38 source vocabularies in five world languages. Terms and their relationships are stored in relational tables that link codes from source vocabularies to definitions, synonyms, and hierarchical relationships. Each term has a semantic "type" represented in a semantic network that relates different types. In recognition of the importance of the UMLS to HCIS, the UMLS project commenced support in 1994 for clinical terminology systems by funding a cooperative research among several universities and healthcare organizations, including the VA.

Toward a Knowledge Layer in HCIS

Computing trends that aided HCIS also favored expansion of online information retrieval systems. Since 1971, Medline, supported by the National Library of Medicine, has contributed greatly to medical research and education. Its original target was the medical library, but growth of personal computing and the low-cost Grateful Med search tool have broadened its audience. In the early 1990s, publication of medical texts, journals, and bibliographic databases on compact disks increased, providing another avenue to knowledge through a computer. Some centers began to provide access to knowledge sources over HCIS networks. At the Boise VA Medical Center, Nielson developed a method to make the Scientific American Medicine textbook available through the Microsoft Windows Help engine (Nielson, Smith, Lee, & Wang, 1994).

The ability of UMLS to traverse from a clinical vocabulary to MeSH prompted other experiments that demonstrated the feasibility of implementing knowledge "gateways" in an HCIS. The Medline Button system allowed users to conduct Medline searches at any point in an HCIS session (Cimino, Johnson, Aguirre, Roderer, & Clayton, 1992). The system anticipates likely query topics (such as drug interactions when reviewing prescriptions) and offers them as default choices. At the American Lake VA, Hammond developed a system capable of extracting MeSH terms from patient data and running a Grateful Med search concurrent with an HCIS session (Hammond, 1996). The value of convenient access to the Medline knowledge that may bear on clinical decisions is not proven, but the concept is intuitively appealing. A study of Grateful Med usage in hospitals by McKibbon and Haynes demonstrated that active bibliographic searching occurred when users were trained and terminals were available (McKibbon, Haynes, Johnston, & Walker, 1991). Embedded systems that "anticipate" queries can decrease or eliminate the training requirement.

These systems also revealed an important limitation in current technology. Medline accepts external queries, but most knowledge products require using their own query interfaces. Databases that are not indexed with the MeSH vocabulary are also not accessible. In 1988, the Z39.50 applications-layer query protocol was developed to furnish client processes a common means of accessing a variety of information resources (Tomer, 1992). Z39.50 adheres to the Open Systems Interface (OSI) reference model developed by the International Standards Organization (ISO), supports the TCP/IP protocol, and formats its content in compliance with Abstract Syntax Notation, another standard that minimizes ambiguities in its implementation. CDROM publishers showed little interest in Z39.50, but the dynamic World Wide Web environment has encouraged several Internet "knowledge vendors" to build Z39.50-compatible products.

In HCIS, consistent "anticipatory" access to knowledge sources and intelligent services awaits progress in three areas: (1) encoded, standard terminology in most patient records, (2) a standard query protocol, and (3) standard indexing of knowledge sources with a vocabulary like MeSH, which improves search performance. Interest in enhancing practitioner performance with knowledge services is keen, but will be limited until the three conditions are met. One clear advantage of an open-knowledge architecture is that it permits the HCIS to furnish knowledge access without the burden of maintaining knowledge databases. Because knowledge evolves, sometimes abruptly, it is more efficient to rely on a knowledge provider to keep the database current than to rely on the HCIS itself.

Clinical practice guidelines suggest a more elaborate use of embedded knowledge access. At Boston's Beth Israel Hospital, for example, a guideline-aware function in the HCIS was demonstrated to improve adherence to recommended practice patterns for AIDS care (Safran et al., 1996). Authoritative guidelines are recommendations based on expert synthesis of current scientific knowledge. They can be far reaching, and different portions of a guideline may apply at different stages of disease management. Implementing guidelines via hard-coded rules embedded in an HCIS is feasible and has been attempted, but keeping a changing guideline (or even the list of literature citations that back it up) current is likely to be impractical for most HCIS. Storage of guideline-related knowledge source *queries* would be a more efficient way to keep the HCIS current.

An even more ambitious extension of the knowledge layer would involve "intelligent," context-aware services that recognized therapeutic goals of individual patients and inferred the treatments needed to attain them at any stage of care. These services require especially complete representation of the patient's status. Necessary features of HCIS required to support "intelligent" services are described below.

Representing the Business Process Layer

Health care consists of "real" events where patients, practitioners, organizational support, information, knowledge, planning, and treatment converge in time and space to accomplish purposeful action. The HCIS equipped to represent salient features of "real" care events may better manage them as well. Figure 13.2 illustrates the contribution of patient-specific information, knowledge, and business process to a care event. Of these three inputs, those that describe the "business" and "behavioral" aspects of healthcare processes are least well defined. Health Level 7 addresses a few event representations such as "admission," "discharge," and "enrollment" that are useful when transferring information between settings. The ASTM standards committee 1384 has done some preliminary work to elaborate an "object" model for patient records that includes semantic definitions and descriptions of expected care-related actions. Very little of this type of representation is now implemented in HCIS. Instead, most healthcare information systems represent business processes and rules concretely, through application code and specifics of deployment.

Recently healthcare quality management experts have identified the importance of understanding and improving business processes in health care. Reduction in practice and process variation is seen as essential to managing quality (Berwick, Godfrey, Blanton, & Roessner, 1990). To achieve consistent, fair, and comparable measurement of the quality of care in a large organization requires ability to define the process elements that influence quality. Improved quality is frequently invoked to justify acquiring an HCIS, but the ability of an HCIS to do so is more often promised than seen.

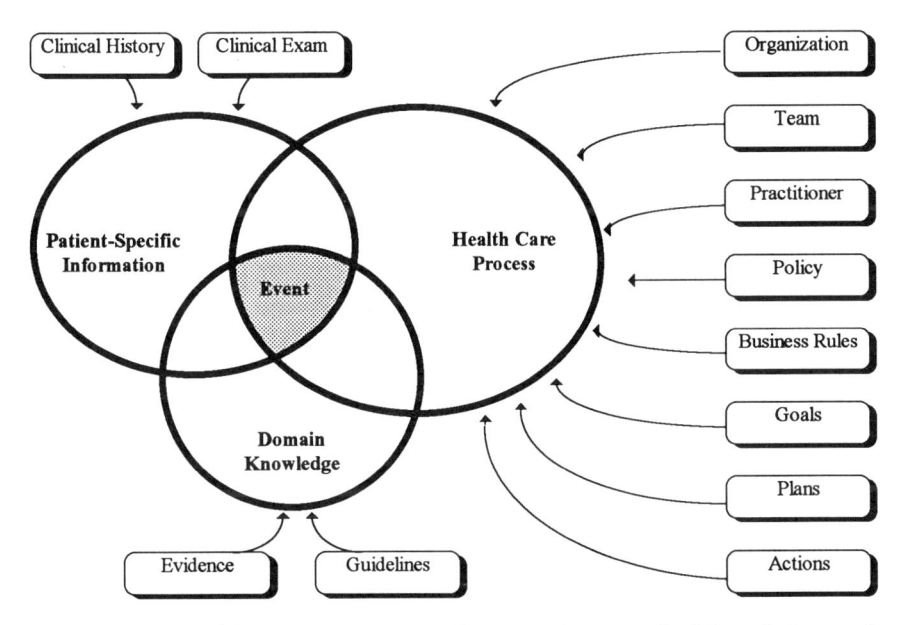

FIGURE 13.2. Healthcare events occur when a patient, medical knowledge, and business activity of a healthcare delivery organization converge. Ability of the healthcare information system to infer meaning and make predictions about care events depends on ability to reconstruct such multiply defined events. Controlled representation of inputs to care events supports modular software architecture, allowing more efficient and flexible processing.

Many effective implementations have been demonstrated, but they tend to focus tightly on specific aspects of care.

A popular organizational response to excessively narrow focus is to articulate a "vision" that flexibly guides employees to apply a quality ethic in their unfolding work. Adaptive organizations may benefit from similar capability in their software. Before this will happen, HCIS representations must support detection of opportunities to improve quality. In a recent investigation of applying computer simulations to model and reengineer ambulatory-care delivery systems, a major limitation of the HCIS was the trouble it took to retrieve commonly accepted measures of timely care from the appointment scheduling database. Straightforward questions such as "how long does it take patients to get appointments with doctors?" or "how many people are waiting in line for a particular service?" could only be reconstructed by making heavy inferences and system-specific assumptions about the significance of certain data. Far preferable would have been a system that could represent quality goals and give continuous feedback on their attainment (Hammond et al., 1996). When goal-seeking intelligent services are available in HCIS to encourage uniform care and guideline compliance, ways to represent applicable patient goals and a means to check their progress will be needed. Requirements analysis of a goal-oriented treatment planning system revealed a need to develop standard representations for patient goals, measures for goal attainment, and the treatment team (Hammond, 1995).

VA's Experience with Controlled Representation

The size of the VA's healthcare system brought it sooner than most to a realization that controlled data representation in the HCIS would serve organizational goals of providing efficient and accessible care of high uniform quality. Within a few years of its initial deployment to individual medical centers, the DHCP acquired capability to transmit hospital clinical workload data to a central processing location for budgeting and planning purposes. In evolving fashion, from 1984 to 1994, administrative data that tracked discharges, diagnoses, demographics, and appointments provided an approximation of workload. Standard ICD-9-CM codes represented diagnoses; other data elements required special definitions tightly linked to business rules. A single "visit" was defined as one or more outpatient appointments kept by a patient on a single date. Neither diagnoses nor procedures were tracked in the appointment scheduling system, but a "stop" code associated with a clinic's "type" gave a rough indication of the service given and category of patient seen. Stop codes were revised as needed to address shifting priorities, resulting in curiosities like the "stool for occult blood" code indicating that this procedure was performed during a visit.

Clinical software packages for specialty groups and packages for support departments were developed nationally and implemented locally over the same period. These served their audiences adequately for several years. Beginning in the early 1990s, a great shift to outpatient and primary care revealed major problems with integration of administrative, clinical, and support service information. Outpatient practitioners wanted problem lists and lab data from other centers. Central planners needed detailed diagnostic and procedure information for costing and billing. More need to integrate information streams appeared when VA planned satellite clinics and hospital mergers in the interest of efficiency.

Solutions that have emerged rely on controlled representation. For example, once it was decided to develop a common patient problem list, viewable by any practitioner, synchronized with the billing system and transmittable to other facilities, the need for a robust representation system became clear. The ICD-9-CM, while required for billing, was not seen as adequate to express the range of problems identified by different clinicians. Hence, a decision was made to adopt a broader vocabulary that could be mapped to ICD-9-CM. The ICD-9-CM, SNOMED, the Read Codes, and UMLS were assessed. In 1992, UMLS was selected as the source for terms in VA's problem list because of its inclusiveness, cross indexing, and national support. It was expected that VA's needs would not exactly match UMLS development, so a parallel resource was established. Known as the VA Clinical Lexicon, this resource was seeded with about 70,000 nonchemical terms from UMLS (Lincoln, Weir, Moreshead, Kolodner, & Williamson, 1994). New vocabularies (drug and laboratory test names, for example) and UMLS updates would be added.

Through 1996 organizational consolidation has continued, with formation of regional units of 7 to 12 former hospitals under a single director. Goals included realtime access to clinical information within regions, and rapid sharing between them. Concepts of a "virtual" VA allowing care in non-VA settings, and expanded billing authority including Medicare also emerged. Regional leaders were expected to improve the quality and uniformity of the direct and contracted services. Each development made it more clear that VA's hospital-centered HCIS was due for radical change.

The Clinical Lexicon was considered for supporting the terminology needs of a transformed HCIS. The Lexicon's large size made rapid modification difficult, and the need to keep legacy applications running slowed the progress of linking them to the Lexicon. The present view is that greater functionality than the Lexicon can provide is needed. Especially important is methodology to accommodate and represent business process differences across sites of care. The need to correctly normalize data representations across contexts appears to be leading to an object-oriented design that incorporates lexical and process representation capabilities and message packaging. The physical and applications architecture of VA's HCIS is changing as well, from "dumb" terminals to graphical client–server workstations. Support for interactions between VA and non-VA databases and applications places another demand for effective representation capability. At this point, a "final vocabulary solution" still seems remote. The VA's partnership with the National Library of Medicine recognizes the library's pioneering role as well as the importance of its participation in building a healthcare information infrastructure.

A Note of Caution

Adoption of standards has brought important advances to HCIS but building workable symbol systems is a huge task with many difficult problems unsolved. Clancey discourages excessive optimism, noting that controlled vocabularies symbolize what is already known, whereas uncertainty, exceptions, and discovery are frequent in medicine (Clancey, 1995). Adopting a coding system presumes commitment to maintain it as knowledge advances, and to assure that it serves its users. A wrong code may cause more harm than a missing one, so care must be taken not to misapply coding or restrict expression for the sake of efficiency. Business vocabulary is more primitive than clinical vocabulary, raising a concern whether these unmatched resources can yet work together to adequately represent healthcare events.

Implications for Large-Scale HCIS Architecture

Symbolic control facilitates communication among the functional layers of healthcare computing and delivery, but the interactions between these layers is complex. Because standards facilitate collaboration by permitting sharing and reuse of knowledge and technology, recent interest in symbol systems for HCIS is a positive development.

Progress is not equal across the conceptual layers of health care if we judge this by how ready a layer's "services" are for *practical* application. Critical mass of symbolic capability in more basic layers appears to be a prerequisite for real-world usage of higher layers. Over the 30 years HCIS have existed, a recurring theme is nondeployment of upper layers (artificial intelligence support, for example), until a breakthrough occurs below. Often enough, adopting an appropriate standard encourages the conceptual clarity that precedes an advance.

Today, larger and smaller HCIS should consider whether society, patients, and caregivers expect better performance than current systems can possibly deliver. Despite sophisticated operating systems, friendly programs and advanced input technology such as voice recognition, it is still all but impossible to move meaningful portions of a record between systems. The HCIS development community needs to recognize and communicate that the chief barrier to progress is not due to hardware or software shortages, but to deficiencies in *symbolic representation* resources. Machines and applications will not fix the problem; believing otherwise will distract attention from where it is needed.

Enormous organizations with large-scale HCIS such as the VA were the earliest to experience the need for consistent representational control at the information, business process and knowledge levels. Because of their administrative complexity, implementing controlled representation has not been a story of steady progress, and more than once a need was revealed by failure. Good innovators learn from failure and communicate the lesson to their leaders. Because the topic is more subtle, it may be harder to sell leadership on linguistic and semantic integrity than to convince them to buy equipment.

Surely many barriers will be overcome. The UMLS and activities of standards bodies such as ASTM are important assets for sharing knowledge, providing practical tools, and fostering collaboration. The VA's commitment to establishing regional patient information repositories fits its strategic plan to provide responsive, evidence- and guideline-based care of uniform quality to all patients. If it succeeds, it may provide an instructive model for other healthcare networks to follow. With sufficient advances in symbolism, processing and communication, it seems possible that the distinction between large- and small-scale healthcare information systems and organizations may disappear, forcing a new paradigm we cannot yet recognize.

References

American Medical Association. 1989. *Physicians current procedural terminology: CPT.* 4th edn. Chicago, IL: Author.

American Psychiatric Association. 1994. *Diagnostic and statistical manual of mental disorders, 4th ed.* Washington, DC: American Psychiatric Press, Inc.

Barnett GO, Justice NS, & Somand ME. 1979. COSTAR-a computer-stored medical information system for ambulatory care. *IEEE Proceedings, 67,* 1226–1237.

Berwick DM, Godfrey A, Blanton A, & Roessner J. 1990. *Curing health care: New strategies for quality improvement.* San Francisco: Jossey Bass.

Carroll-Johnson RM. 1994. *Classification of nursing diagnoses: Proceedings of the 10th conference.* North American Nursing Diagnosis Association.

Cimino JJ, Johnson SB, Aguirre A, Roderer N, & Clayton P. 1992. The MEDLINE button. *Proceedings of the Sixteenth Annual Symposium on Computer Applications in Medical Care,* pp. 81–85.

Clancey WJ. 1995. The learning process in the epistemology of medical information. *Methods of Information in Medicine, 34(1–2),* 122–130.

Cote RA, ed. 1995. *Systematized nomenclature of medicine.* 2nd edn. Skokie, IL: American College of Pathologists.

Hammond KW. 1995. Treatment planning: Implications for structure of the CPR. *Journal of the American Medical Informatics Association (Symposium Supplement),* 362–366.

Hammond KW. 1996. Systems for accessing knowledge at the point of care. In MJ Miller, KW Hammond, & MG Hile (eds): *Mental Health Computing.* New York: Springer-Verlag, pp. 304–321.

Hammond KW, Iverson SC, Nichol WP, Gabre-Kidan T, Solveson KD, Schreuder AB, O'Brien JJ, & Jeffers DC. 1996. The challenge of simulating a complex ambulatory care system. In JE Anderson & M Katzper (eds): *Simulation in the Medical Sciences.* San Diego: Society for Computer Simulation, pp. 38–44.

Health Care Financing Administration. 1991. *The International Classification of Diseases; 9th rev., Clinical Modification.* 4th edn. Washington DC: Author.

Humphreys BL, & Lindberg DAB. 1989. Building the Unified Medical Language System. *Proceedings of the 13th Annual Symposium on Computer Applications in Medical Care,* pp. 475–480.

Lincoln MJ, Weir C, Moreshead G, Kolodner R, & Williamson J. 1994. Creating and evaluating the Department of Veterans Affairs electronic medical record and national clinical lexicon. *Proceedings of the Eighteenth Symposium on Computer Applications in Medical Care,* p. 1047.

Martin K, & Scheet N. 1992. *The Omaha System.* Philadelphia: W.B. Saunders.

McCloskey JC, & Bulechek GM. 1994. *Nursing Interventions Classification (NIC): Iowa Intervention Project.* St. Louis, MO: Mosby-Year Book.

McDonald CJ. 1976. Protocol-based computer reminders, the quality of care and the non-perfectability of man. *New England Journal of Medicine, 295(24),* 1351–1355.

McKibbon KA, Haynes RB, Johnston ME, & Walker CJ. 1991. A study to enhance clinical end-user Medline search skills: Design and baseline findings. *Proceedings of the Fifteenth Annual Symposium on Computer Applications in Medical Care*, pp. 73–77.

Nielson C, Smith CS, Lee D, & Wang M. 1994. Implementation of a relational patient record with integration of educational and reference information. *Journal of the American Medical Informatics Association Symposium Supplement: Proceedings, Eighteenth Annual Symposium on Computer Applications in Medical Care*, 125–129.

Payne TH, & Martin DR. 1993. How useful is the UMLS metathesaurus in developing a controlled vocabulary for an automated problem list? *Proceedings of the 17th Annual Symposium on Computers in Medical Care*, pp. 705–709.

Read J. 1990. Read clinical classification. *British Medical Journal, 301(6742)*, 45.

Safran C, Rind DM, Davis RB, Ives D, Sands DZ, Currier J, Slack WV, Cotton DJ, & Makadon HJ. 1996. Effects of a knowledge-based electronic patient record on adherence to practice guidelines. *MD Computing, 13(1)*, 55–63.

Shortliffe EH. 1976. *Computer-based medical consultations: MYCIN*. New York: American Elsevier.

Tomer C. 1992. Information technology standards for libraries. *Journal of the American Society for Information Science, 43*, 566–570.

14
Clinical Decision Support in a Distributed Environment

CURTIS L. ANDERSON

Introduction

In a networked environment, the efficient and cost-effective provision of health care requires comprehensive clinical decision support/expert systems that can access and evaluate patient data across clinical services. Such clinical decision support systems enable clinical alerts, realtime order checking, diagnosis support, cost reduction, and research/quality assurance queries.

Decision support is the process of providing information that enables or enhances a person's decision-making capabilities. Computers are ideally suited for this task. Examples of clinical decision support information include allergy display, critical data alerts, drug/allergy notifications, suggested diagnoses, and automated standards of care.

Over the past decade, the evolution of the Department of Veterans Affairs (VA) Decentralized Hospital Computer Program (DHCP) has dramatically increased the amount of clinical data available to VA healthcare providers. This increase in information allows Veterans Affairs Medical Centers (VAMCs) to provide better patient care only if they have the resources and capabilities to monitor, evaluate, and apply that information. For example, information exists across several DHCP software packages which, when evaluated in the aggregate, can determine if a patient is suitable for blood transfusion. Working across packages, a decision support or expert system could provide the information monitoring and evaluation capabilities for screening blood transfusion orders. Monitoring and improving quality of care is a major benefit of expert systems.

The components of a well-designed expert system could also enhance quality assurance, physician resource utilization, cost control, administrative/executive decision making, and conformance with the Joint Commission on the Accreditation of Healthcare Organizations (JCAHO).

Today healthcare providers are virtually overwhelmed with information. Specialists are less aware of, or unable to recognize, the etiologies of diseases outside their domains. Pharmaceuticals are incorrectly prescribed because a physician is unaware of corresponding lab results. Lab results are erroneously indicated as abnormal because a patient is receiving conflicting medication. Postoperative antibiotics are prescribed unnecessarily, increasing costs. Patients miss meals because of staffing overlap or miscommunication. Nursing stations are overstaffed one week and understaffed the next. These are all instances where information is available to remedy the problem, yet tools are not available to filter and apply the data to make decisions. These are just a few examples where a clinical expert system could provide the information monitoring and evaluation capabilities necessary to improve patient care and reduce costs.

Ideally, the monitoring and evaluation of clinical information entering DHCP would be carried out by experts in appropriate healthcare disciplines. Conceivably, these individuals would never tire of reviewing and evaluating data, could apply those data to improve patient care (or reduce costs), and would do so consistently. In a setting of limited resources, continuous and consistent monitoring, evaluation, and application do not occur. Even with adequate resources, monitoring clinical data and their interrelations is a time-consuming and difficult task. As a consequence, much of the data's ability to improve patient care is lost. The need to develop a DHCP clinical expert system that can provide these capabilities is urgent.

This chapter describes VA plans for a system to achieve these goals. A glossary of terms is provided for the reader (see Table 14.1).

Past DHCP Clinical Decision Support

In the past, clinical decision support capabilities in DHCP were limited and most often specific to medical services and their accompanying software. Isolated examples of decision support existed in the pharmacy, order entry, allergy tracking, laboratory, and dietetics software modules. Decision support efforts outside specific service software included the Clinical Monitoring System and the Clinical Data Management System (CDMS).

Previous pharmacy clinical decision support varied between inpatient medication orders and outpatient prescriptions, but generally included checking clozapine appropriateness and displaying patient allergies and adverse reactions. Current pharmacy decision support is more integrated and generally supports the same functionality across inpatient and outpatient medications.

Clozapine is an antipsychotic medication that requires close monitoring because patients receiving it have an increased risk of mortality. Clozapine decision support occurred in the form of limiting refills, number of days' supply and maximum daily dose, and displaying white blood cell (WBC) counts resulted within the past 7 days. If no WBC has been resulted in the past 7 days, the ordering provider was prevented from placing the order for

TABLE 14.1. Glossary of terms.

Decision Support: The process of providing information that enables or enhances a person's decision-making capabilities.

Decision Support System: A computerized representation of one or more experts' knowledge. By placing an expert's knowledge and expertise on the computer, that expert is available (HL7 in computerized form) to make decisions anytime the data warrant a decision. A knowledge-based decision support system permits trained experts to develop and edit Medical Logic Modules (MLMs), bypassing the need to invoke a programmer every time a change is required or a new algorithm is mandated. In this chapter, decision support system is synonymous with expert system and knowledge-based system.

Expert System: A computerized representation of one or more experts' knowledge. In this chapter, expert system is synonymous with decision support system and knowledge-based system.

Knowledge Base: A collection of objects/modules including decision algorithms or "rules of thumb" which, when applied against data, provide the basis for making an automated decision. Knowledge-base objects (often referred to as "medical logic modules" or "frames") are usually derived from interviews with experts in the domain or subject area of the automated decision. For example, a pharmacist may aid in developing a MLM regarding drug allergic reactions, whereas a cardiologist could help create a MLM for acute myocardial infarction.

Knowledge-Based System: An expert system that uses a knowledge base and inferencing techniques to process data, solve problems, and/or provide decision support. Knowledge-based expert systems allow experts such as physicians, pharmacists, or QA nurses to personally design and test modular algorithms of their expertise. These modular algorithms are called "medical logic modules." Medical Logic Modules collectively form a knowledge base. It is from the knowledge base that an expert system derives its "expertness." In this chapter, knowledge-based system is used interchangeably with expert system and decision support system.

Knowledge Engineering: The process of creating MLMs. Knowledge engineering typically involves (1) identifying knowledge that lends itself to the expert system, (2) determining how to effectively and efficiently model that knowledge in a MLM, (3) testing and implementing the MLM, and (4) training appropriate individuals regarding the MLM's purpose and action.

Knowledge Rule/Frame: A model of expert knowledge, or, more precisely, a single, modular knowledge representation scheme that associates decision logic or a "rule of thumb" with related data. Synonymous with MLM.

Medical Logic Module: A model of expert knowledge. Synonymous with knowledge rule or frame.

Unified Medical Language System (UMLS) Metathesaurus: The UMLS Metathesaurus is a compilation of biomedical concepts and terms from several controlled vocabularies and classifications. Version 1.2 contains over 130,000 distinct concepts including representations from MeSH, ICD, SNOMED, DSM, CPT, NANDA, and others. The UMLS project is funded by the National Library of Medicine, National Institutes of Health.

clozapine. This VA-wide mandate for clozapine monitoring remains in place today.

When a medication was ordered for an outpatient, the ingredients of that medication were checked against the ingredients of all known medications the patient was receiving for adverse interactions. Ingredients were also evaluated to determine if a duplicate drug was being ordered.

In VA, medications are identified for possible adverse interactions by the Executive Committee on Therapeutic Agents (Lopez, 1990). The committee categorizes identified interactions as critical or significant and releases a short monograph regarding each interacting agent. Generally, critical interactions are those with severe consequences requiring some type of action (e.g., finding facts or contacting prescribers) to prevent serious harm. Significant interactions indicate the potential for harm is either rare or generally known and it is reasonable to expect all prescribers have taken those interactions into account.

Previous mechanisms for adverse drug interaction decision support presented the ordering provider with an interaction message, then gave prescribers the option to discontinue the current medication, discontinue the existing, active medication, discontinue both medications, or continue with the order. Whatever the response, the prescriber had to document the interaction and describe their subsequent actions online. Also during outpatient medication ordering, a list of the patient's verified drug allergies/ adverse reactions was displayed.

Each medication in the VA system is categorized into a therapeutic class. This is the same classification system used in the United States Pharmacopeia Drug Information (USPDI) publications. Outpatient medication orders were checked for duplicate therapeutic classes by comparing the class(es) of the medication being ordered against those the patient was already receiving.

In the past, inpatient medication decision support was limited to displaying the patient's verified allergies/adverse reactions and checking clozapine appropriateness. Current inpatient decision support provides for adverse drug interactions, drug/allergen interactions, and duplicate checks as well.

Laboratory decision support was available on the ordering side via the display of maximum order frequencies and identification of duplicate orders. For results display, abnormal or critical high/low values were flagged. Mechanisms for performing delta checks to establish trends or identify significant clinical variations also existed.

The dietetics software offered decision support mechanisms to facilitate patient meals. These included prompting the clinician to order a late meal tray if the requested time was past the regular meal cut-off time and notifying clinicians if a late meal tray was ordered too late for delivery. In either case, the feedback allowed clinicians to make decisions that resulted in punctual patient meals.

Mechanisms for clinical decision support in older versions of Order Entry/Results Reporting (OE/RR) were limited to the presentation of a list of patient allergies/adverse reactions at the beginning of the ordering session.

An additional component of past and present OE/RR is clinical notifications. This module generates clinical alerts regarding patient information from DHCP packages such as OE/RR, lab, Medical Administrative Service (MAS), radiology, and consults. A notification consists of one line of text displayed to the user at sign-on and "option" prompts. In addition to the clinically relevant message, notification text includes the first nine characters of the patient's name and a "quick look-up" consisting of the first character of his last name followed by the last four digits of his patient identification number. Notifications are deleted after display and/or a follow-up action is completed. The VAMCs have control over the use of notifications at their sites. Individual users can indicate that they wish to receive "non mandatory" notifications, otherwise they will not be notification recipients. This reduces nuisance alerting. Mandatory alerts, designated by a national panel of experts, are always delivered to appropriate recipients.

Notifications fall into two classes: informational and follow-up. Informational alerts are for information only and consist of a one-line message. Follow-up notifications consist of a message and an associated follow-up action triggered when the recipient selects the message.

Previously, notifications were triggered within service software, such as lab or radiology. Once triggered, notifications are sent to the DHCP Kernel alert utility, which handles the display, deletion, and follow-up actions. Notifications displayed via Kernel are termed "alerts." Recipients of a notification/alert included the provider ordering the procedure or action that triggered the alert, providers on teams that cared for the patient, or providers who included the patient on a list of patients they personally treated.

Notifications can be set up for deletion when each individual reviews the alert or takes a follow-up action. Notifications can also be deleted for all recipients when a single recipient reviews the alert or takes the necessary follow-up action. The following is a list and brief description of notifications released in past versions of OE/RR.

Abnormal Radiology Results. The alert text is variable. It typically contains an abbreviated description of the resulted procedure. It is triggered when abnormal results are entered for a radiological procedure. This alert is mandatory; it will always be sent to the clinician ordering the procedure and users on team or personal patient lists that contain the examined patient. It has a follow-up action that displays the abnormal results. The alert is deleted for each recipient after he reviews the abnormal results.

Example. JONES,EDW (JØ123): Abnormal Rad Results: CT ABDOMEN W/O CONT

Admission. The alert text is variable. It typically contains the date/time of admission. It is triggered when an inpatient is admitted. This alert is voluntary; it must be turned on for the user before it will be received. It will be delivered to users on team or personal patient lists that contain the admitted patient, if those users have this notification turned on. This alert is informational only; it only displays the alert message. It is deleted for each recipient after he reviews the alert message.

Example. GILMORE,T (G4567): Admitted on APR 22,1996@11:08:27

Consult/Request Resolution. The alert text is variable. It typically contains the service and what the service did with the consult. It is triggered when a consult/request is completed, discontinued, or denied by the specified service. This alert is voluntary; it must be turned on for the user before it will be received. The alert is delivered to the provider who originally ordered the consult/request, if he has this notification turned on. It has a follow-up action that displays the consult/request. The notification is deleted for each recipient after he reviews the consult/request.

Example. NIVEK,ALP (N8910): MEDICINE Consult—Service COMPLETED

Cosignature on Progress Notes. The alert text always appears as *Cosignature required on progress note(s)*. It is triggered when a nonsignatory provider (e.g., medical student) enters a progress note into the computer. This alert is voluntary; it must be turned on for the user before it will be received. It is delivered to the non signatory provider entering the progress note and users on team lists to which that triggering provider belongs, if the provider and users have this notification turned on. The alert has a follow-up action that displays all of the patient's progress notes that require cosignature. It prompts signatory providers to cosign them. The alert is deleted for all recipients after all progress notes requiring cosignatures for the patient are cosigned.

Example. SCHWARTZ, (S1112): Cosignature required on progress note(s)

Critical Lab Results. The alert text always appears as *Critical lab results*. It is triggered when a lab result is flagged as critical. This alert is mandatory; it will always be sent to the clinician ordering the procedure and users on team or personal patient lists that contain the examined patient. It has a follow-up action that displays all critical lab results for the patient. The alert is deleted for all recipients after all critical lab results for the patient are reviewed in detail.

Example. DEMKO,EAR (D1314): Critical lab results

Deceased Patient. The alert text is variable. It typically contains the date/time of death. It is triggered when an inpatient is discharged due to death. This alert is voluntary; it must be turned on for the user before it will be received. It will be delivered to users on team or personal patient lists that contain the deceased patient, if those users have this notification turned on. This alert is informational only; it only displays the alert message. It is deleted for each recipient after he reviews the alert message.

Example. MEILER,SA (M1516): Died while inpatient on APR 28,1996@11:08:27

Flagged Orders. The alert text always appears as *Orders needing clarification*. It is triggered when an order is flagged as needing clarification by the ordering provider. This alert is voluntary; it must be turned on for the user before it will be received. It will be delivered to the provider who entered the order and users on team lists that contain the entering provider, if the provider and those users have this notification turned on. The alert has a follow-up action that displays all active orders for the patient and allows unflagging. It is deleted for all recipients after all flagged orders for the patient are unflagged.

Example. MCCARNEY, (M1718): Orders needing clarification

New Service Consult/Request. The alert text is variable. It typically contains the service and consult time constraints. It is triggered when a consult/request is requested from the specified service. This alert is voluntary; it must be turned on for the user before it will be received. The alert is delivered to the provider indicated in the consult request dialog and providers associated with the service, if those providers have this notification turned on. It has a follow-up action that displays the consult/request order. It is deleted for each recipient after he reviews the consult/request order.

Example. GUNN,PETE (G1920): New CARD order—Within 48 hrs

Order Requires Chart Signature. The alert text always appears as *Order released-requires chart signature*. It is triggered when a nurse takes a phone or verbal order and indicates the order should not be marked as *Signed on chart*. This alert is mandatory; it will be delivered to the provider who requested the order and users on team lists that also contain the requesting provider. This alert is informational only; it only displays the alert message. It is deleted for each recipient after he reviews the alert message.

Example. JONES,EDW (J2122): Order released—requires chart signature

Order Requires Electronic Signature. The alert text always appears as *Order requires electronic signature*. It is triggered when an order is entered without an electronic signature and cannot be released to the service until it is electronically signed. This alert is mandatory; it will be delivered to the provider who requested the order and users on team lists that also contain the requesting provider. The alert has a follow-up action that displays all orders for the patient that require electronic signature. It allows signatory providers to sign the orders. This alert is deleted for all recipients after all orders for the patient that require electronic signature are signed.

Example. SMITH,BIL (S2324): Order requires electronic signature

Radiology Patient Examined. The alert text is variable. It typically contains an abbreviated description of the procedure. The alert is triggered when a radiological procedure obtains the status of EXAMINED. This alert is voluntary; it must be turned on for the user before it will be received. It will be delivered to the provider who requested the procedure and users on team lists that also contain the entering provider, if the provider and those users have this notification turned on. This alert is informational only; it only displays the alert message. It is deleted for each recipient after he reviews the alert message.

Example. SMITH,BIL (S2526): Examined: CHEST SINGLE VIEW

Radiology Request Cancel/Hold. The alert text is variable. It typically contains an abbreviated description of the procedure. The alert is triggered when a radiological procedure is canceled or held. This alert is voluntary; it must be turned on for the user before it will be received. It will be delivered to the provider who requested the procedure, if that provider has this notification turned on. This alert has a follow-up action that displays the canceled/held procedure. It is deleted for each recipient after he reviews the procedure.

Example. FISH,MILT (F2728): Rad Canceled: CHEST SINGLE VIEW

Radiology Results Verified. The alert text is variable. It typically contains an abbreviated description of the procedure. The alert is triggered when a radiological procedure is completed and verified. This alert is voluntary; it must be turned on for the user before it will be received. It will be delivered to the provider who requested the procedure and users on team or personal patient lists that contain the examined patient, if the provider and those users have this notification turned on. This alert has a follow-up action that displays the procedure's radiology report. It is deleted for each recipient after he reviews the entire report.

Example. CONNOLLY, (C2930): Rad Results: CHEST 2 VIEWS PA&LAT

Service Order Req Chart Sign. The alert text always appears as *Service order—requires chart signature*. It is triggered when an order is entered, discontinued, or edited from a service such as pharmacy, radiology, or lab. This alert is mandatory; it will be delivered to the provider who requested the order and users on team lists that also contain the requesting provider. This alert is informational only; it only displays the alert message. It is deleted for each recipient after he reviews the alert message.

Example. JONES,EDW (J3132): Service order—requires chart signature

Unscheduled Visit. The alert text is variable. It typically contains the date/time of the unscheduled visit. It is triggered when an inpatient is admitted as an unscheduled visit. This alert is voluntary; it must be turned on for the user before it will be received. It will be delivered to users on team or personal patient lists that contain the admitted patient, if those users have this notification turned on. This alert is informational only; it only displays the alert message. It is deleted for each recipient after he reviews the alert message.

Example. JACKSON,S (J3334): Unscheduled visit on APR 29,1992@ 08:59

Unsigned Progress Notes. The alert text always appears as *Unsigned progress notes*. It is triggered when a progress note is entered without an electronic signature. This alert is voluntary; it must be turned on for the user before it will be received. It is delivered to the authoring provider, if that provider has this notification turned on. The alert has a follow-up action that displays all unsigned progress notes for the patient that were authored by the recipient. It prompts for signature. The alert is deleted for the recipient after all unsigned progress notes for this patient/author are signed.

Example. RICHARDSO (R3536): Unsigned progress notes

As a precursory to distributed clinical decision support, the DHCP Clinical Monitoring System captured populations of patients meeting specified conditions across a variety of information domains. Monitors could be executed manually or in a nightly batch mode against a single patient or specified patient populations. Patients whose data met a monitor's criteria were identified for further evaluation. Examples of monitors include patients who were transferred to intensive care units within 24 hours of admission, patients with abnormal lithium laboratory results, and patients who were readmitted within 1 month and whose previous discharge was from pulmonary medicine. Monitors were created by combining and comparing one or more patient conditions. Examples of patient conditions evaluated by monitors include age, medications, laboratory results, and admission/discharge/transfer (ADT) information.

Knowledgeable users at VAMCs could edit existing monitors or add new ones. This effort required a knowledge of the M programming language, previously known as Massachusetts General Hospital Utility Multi-Programming System (MUMPS). Guidelines for monitor editing and development were provided with the software. The clinical monitoring system remains a viable option for decision support in VHA.

Developed as a prototype within VA, the CDMS provides DHCP practitioners with decision support (Andrews, 1990). The CDMS extracts patient data from legacy DHCP files, manipulates and integrates the data, then stores the data in a common data structure where they can be processed more efficiently. The CDMS approach overcame the problem of different file structures for each service-oriented software application. Knowledge rules or "frames" developed in the CDMS system were entered as records in a file and processed against the standardized data by a generic inferencing engine. Today, CDMS offers needed capabilities as a tool for specialized VA research.

Later enhancements to CDMS supported clinical reminders. The CDMS frames in the clinical reminder system are organized as "standards of care." Standards of care were developed under the direction of Charles Beauchamp, M.D., at the Durham (North Carolina) VAMC and John Demakis, M.D., at the Hines (Illinois) VAMC. Standards of care include:

- Patients receiving warfarin treatment require monitoring every 45 days.
- Patients with a diagnosis of atrial fibrillation should receive warfarin or aspirin.
- Patients with coronary artery disease need lipid levels every 12 months.
- Patients less than 1 year postmyocardial infarction should be receiving a betablocker.
- Patients with hypertension should receive instruction in weight reduction and sodium restriction as part of their antihypertensive therapy regimen at least twice yearly.
- Patients with diabetes should have a glycosylated hemoglobin every 12 months.
- Patients with diabetes should receive nutritional counseling annually.
- Patient with diabetes should have a dipstick urinalysis for protein annually.
- Patients over 30 with diabetes should have an annual eye examination by an eye specialist.
- Patients with diabetes or peripheral vascular disease should be given foot care instructions annually.
- Patients with a history of gastrointestinal bleeding who are taking nonsteroid, anti-inflammatory drugs should switch to salsalate or acetaminophen.
- Patients who smoke should receive smoking cessation counseling.
- Patients 65 or over, or in high-risk categories, should receive pneumococcal vaccine once in lifetime.

Groundwork for Present and Future Systems

There is a new direction for expanding clinical decision support methodologies in DHCP. This approach is knowledge or rule based and is derived from past experiences of the developers and lessons learned through the study of pioneers in clinical and nonclinical decision support systems (Warner, Olmstead, & Rutherford, 1972; McDonald, 1976; Clancey & Shortliffe, 1984; Barnett, Cimino, Hupp, & Hoffer, 1987; Harmon & Sawyer, 1990; Weed, 1991; Curtis, 1993; Heathfield & Wyatt, 1993; Carlson, Wallace, East, & Morris, 1995, Curtis, undated).

A key influence on DHCP clinical decision support has been the HELP System developed at the LDS Hospital in Salt Lake City, Utah (Warner, Olmstead, & Rutherford, 1972; Kuperman & Gardner, 1988; Pryor, Dupont & Clay, 1990; Rocha, Christenson, Pavia, Evans, & Gardner, 1994). HELP is a rule-based system using Bayesian logic in an If_Then_approach. A HELP rule consists of data elements, logical relationships, systemwide clinical data dictionary, and Bayesian operators. A data element represents a component of medical knowledge such as a finding, symptom, or intermediate decision. Data elements link the knowledge component with the underlying data. The data elements are then referenced in the rule's conditional logic statements. The results of the logic statements determine the rule results. Rules are evaluated when called by another rule (nesting) or an item of data is stored that is represented as a data element.

PTXT (p-text) is HELP's systemwide clinical data dictionary. PTXT is arranged hierarchically to facilitate storage, retrieval, and decision support. Clinical and demographic data are stored as eight-bit PTXT codes.

To enlist the knowledge and support of hospital clinicians, HELP used a rule editor. This allowed trained physicians to create and test their own knowledge rules. In addition, interested clinicians could participate in the process via knowledge engineering sessions. In these meetings, clinicians shared their experience and knowledge with the knowledge engineer who would fuse the clinician's input with statistical results from HELP database queries to generate knowledge rules. These rules would be tested in future knowledge engineering sessions and fine-tuned until they consistently produced valid results. With most rule editors, a knowledge engineer is useful as an intermediary between clinicians and the editor/computer representation of the clinical knowledge.

Other experiences with physicians and decision support systems indicated a general physician resistance to computer-generated patient diagnosing. Future efforts at decision support would veer away from this more controversial capability of decision support systems (Anderson, 1985).

HELP is shifting efforts to bolster decision support in ambulatory care and physician workstations. In addition, HELP developers are adopting an open systems approach with client–server architecture. Efforts are also underway to improve the clinical vocabularies of the systems (Huff, Haug, Stevens, Dupont, & Pryor, 1994).

An offshoot of HELP is the microcomputer-based Iliad decision support system. Iliad uses an expert system to solicit patient history and physical examination data. It then processes the information to derive a set of differential diagnoses sorted by percent probability. Iliad has received acceptance and acclaim as a tool for teaching medical students diagnostic skills based on physical exam and history. This approach provides significant cost saving over decision making based on expensive laboratory and imaging results. In addition to its role as a training tool, Iliad has evolved into a product that provides home counseling for consumers and specialist consultations for practitioners in rural clinics (Ben-Said, Dougherty, Altman, Anderson, Bouhaddon, & Warner, 1987; Bergeron, 1991; Hukill, Ward, Haug, Turner, & Warner, 1987; Warner et al., 1988; Cundick et al., 1989; Applied Medical Informatics, Inc., 1994).

KDS knowledge-based expert system expanded on HELP and Iliad's decision-making capabilities for commercial implementation in private hospitals. The KDS knowledge base focused on circumventing "ill-advised" orders by providing realtime feedback during the ordering process and meeting quality assurance and JCAHO needs.

Knowledge Data Systems used many of the same expert system elements as HELP and Iliad. These included a systemwide clinical vocabulary (Anderson, Hukill, Wang, Bangerter, Kattelman, & Hartmann-Voss, 1990), data elements, *If_Then_*logic, data-driven triggering mechanisms, and (like Iliad) boolean logic in rules. The approach also provided multiple destinations for rule results and preprocessing or filtering of data before expert system evaluation. Example order-related rules included suggesting the correct bleomycin medication dosing based on body surface area and prompting the orderer for cost-saving oral medications versus intravenous medications. Results/messages from these rules were sent to the terminal where the order was being placed, providing realtime feedback.

Other KDS rules were used for quality assurance (QA) purposes. Quality assurance rules included patient readmission within 30 days, bleeding time prolonged by a medication, and hysterectomy scheduled for women younger than 35. Instead of directing rule results to the user's terminal, QA rule results were sent to files where QA personnel could review and evaluate data for quality improvement or JCAHO purposes.

The KDS rule editor was based on the Apple Macintosh technology. While this editor was fairly user friendly, a knowledge engineer was required to make the links between the data elements in the rules and the actual location of data in the patient data files.

For implementing KDS rules in a hospital, a decision support/rules advisory committee was required. This committee of clinicians and administrators determined the appropriateness of each rule before it was implemented for patient care. The committee was also capable of suggesting new rules or removing existing rules if they were deemed undesirable. This approach gave the individual hospitals ownership of the knowledge base.

Local implementation required an onsite knowledge engineer. Knowledge Data Systems provided knowledge engineering training to interested individuals. Typically this included someone with clinical training and computer experience. Often these were clinicians interested in creating their own rules for research or patient-care purposes.

Additional experiences in decision support systems were gained through early efforts with VA's Clinical Data Management System (CDMS).

Present and Future Decision Support

Goals for present and future VA decision support were identified by VA staff. In part, these goals were driven by baseline requirements for order checking, a major component of OE/RR and the Computerized Patient Record System (CPRS). Order checking involves the interactive evaluation of orders for contraindications against previous orders or patient data. If the order is contraindicated, a corresponding message is displayed. Orders are checked for patient safety, relevancy, and cost containment. Order checks, notifications, and other decision support functionalities and features have been developed in VA's knowledge-based expert system.

Current order checks include:

- Allergy-drug interactions;
- Allergy-contrast media interactions;
- Display PT and PTT for fresh frozen plasma orders;
- Late diet tray ordered after cut-off time;
- Diet order after mealtime cut-off—order a late tray;
- Maximum lab order frequency;
- Lab–drug interactions;
- Duplicate orders;
- Ordered item recently resulted;
- Display total cost of an ordering session;
- Display cost of individual orderable items;
- Release requirements (cosign, ancillary review, verification, etc.);
- Items flagged for cost containment;
- Acceptable values (ranges, dosages, etc.);
- Duplicate drug class order;
- Drug–drug interactions;
- Estimated creatinine clearance if <50;
- Nonformulary orders with suggested formulary alternative;
- Order checking not available/supported;
- Recent barium study (within 2 days);
- Physical limitations for computerized tomography and magnetic resonance imaging scanners;
- Oral cholecystogram in past 7 days.

Some of these order checks are defined as expert system Medical Logic Modules (MLMs). Medical logic module is a term synonymous with rule or frame. It is derived from the Arden Syntax, a standard for sharing MLMs and knowledge bases among decision support developers (ASTM Committee E-31, 1992; Jenders et al., 1995). The DHCP decision support system will support the Arden Syntax and will share knowledge bases with other nonprofit entities.

Goals for a CPRS decision support system include the following:

- Use and acceptance by clinicians;
- User-friendly editor for modifying and expanding the knowledge base;
- VA-wide "clinician friendly" vocabulary;
- Comprehensive access to patient and nonpatient data;
- Support for generic data evaluation;
- Continuation of legacy decision support functionality;
- Management tools.

A key to clinician acceptance is the speed with which decisions and/or order check results are delivered. Several mechanisms can be used to reduce processing time. One mechanism splits order checks into two categories, those allowing/requiring local VAMC modification that lend themselves to MLMs and non-MLM checks that require no local modification. Non-MLM checks typically can be most efficiently processed by encoding them in the software. An example non-MLM check is duplicate order evaluation. An MLM example is checking for lab–drug interactions. Another efficiency mechanism involves compiling MLMs into M code that has been optimized for data retrieval processing efficiency. Additional processing improvement schemes are being explored.

The philosophy of development for the CPRS decision support system is to place the right information in the clinician's hands at the right moment with the right tools for acting on that information. In order checking, this means the checks are presented in the appropriate context of the ordering session. For example, an order check regarding drug–drug interactions will be triggered at the end of the ordering session. Triggered at the conclusion of the ordering session, this order check would not interact with orders discontinued earlier in the process. This reduces false positives. Other order checks (such as a drug–allergy interactions) will be evaluated when the drug is selected for potential ordering.

Almost all order checks allow the ordering clinician to override the alert. (In some cases, congressional mandates or VA central office policy forces clinicians to take certain actions or inactions.) Some order checks require the clinician to enter a reason for the override. The order check message and override reason are stored as specialized progress notes in the patient record, are sent with the order to the service filling the order (pharmacy, lab, etc.), and can be reviewed in the future.

Another feature to facilitate physician use and acceptance of MLMs will be the MLM/knowledge rule editor. Like the HELP, Iliad, and KDS rule/frame editors, this tool will encourages providers at medical centers to create their own MLMs. Currently CPRS includes 22 MLMs. Installed nationally, these MLMs can be reviewed and modified to meet a local VAMC's unique patient population or need. To supplement national MLMs, a clinician could design, develop, and implement local MLMs deemed by their local VAMC as clinically significant or useful for research. In the future, CPRS will contain a mechanism for the collection, review, and inclusion of locally developed MLMs into a national clearinghouse of MLMs. The VA medical centers could then download relevant MLMs from the clearinghouse and implement them with local modifications.

Medical logic modules typically include (1) references to data elements that will be monitored and evaluated, (2) a statement expressing logic applied to the data, and (3) some message or action associated with the fulfillment of the logic. For example, a MLM screening for male patients over 65 who smoke would include the elements sex, age, and patient history. The MLM's logic would be set up to check patient data for a sex of "male," an age over "65," and a history of "smoking." When a patient is identified as fulfilling all of these requirements, a message such as PATIENT IS AT HIGH RISK FOR HEART DISEASE, would be displayed or some form of action would be triggered. In another example, a MLM may monitor lithium medication dosages and lithium lab results, notifying providers when they are out of balance.

To support a national knowledge base and promote physician authoring of MLMs, a VA-wide, "clinician friendly" vocabulary is required. The DHCP clinical lexicon has been chosen for this task. Initially seeded with about 70,000 terms from the National Library of Medicine's Unified Medical Language System (UMLS) Metathesaurus, the clinical lexicon is continually expanded to support additional clinical terminologies. Terms in the UMLS Metathesaurus and clinical lexicon are universally recognizable (e.g., acetaminophen and serum creatinine). This property enables healthcare experts to easily identify the correct terms for MLM development.

When creating or modifying a MLM, the user works with clinician-friendly clinical lexicon terms that are linked to the underlying data elements. Linking data elements with the actual data source is accomplished via the metadata dictionary.

The metadata dictionary links data sources with discrete medical concepts from a wide variety of clinical areas including the following: orders, problems, vital signs, labs, medications, allergies, ADT events, clinic schedules, radiology/nuclear medicine procedures, CPRS expert system rule results, and virtually all data in HL7 messages (HL7 Working Group, 1994). The metadata dictionary is currently able to extract data from legacy DHCP files, the longitudinal patient record, order dialogs, and HL7 messages. It will be extended to extract data from the Clinical Information Resource Network's (CIRN's) Clinical Object Repository.

The object-orientedness of metadata dictionary will enable the CPRS expert system to support future ad hoc and compiled/boilerplate queries. Ad hoc queries will empower CPRS users to conduct on-the-fly rule creation and processing across one patient or population(s) of patients. Compiled queries will be accessible to DHCP software modules, which perform data "gophering" tasks. Some compiled queries will be available to CPRS users via a "clinical calculator." Another planned feature will allow

VAMCs to create compiled queries to capture and stuff complex data into spreadsheets for monitoring cancer patients or for actuating clinical reminders.

To support this kind of functionality, the clinical lexicon must contain all terms used in the MLMs. For example, if a MLM checks for a laboratory serum glucose result, serum glucose (or a synonym) must be represented in the lexicon. In the future, the clinical lexicon will be a repository of all clinical terms that could be used by MLMs. This scenario requires the definition of MLM data elements in the clinical lexicon and their linkage to data sources via the metadata dictionary.

An expanding knowledge base of MLMs requires comprehensive access to patient and nonpatient data. As indicated, the metadata dictionary can access information in legacy DHCP files, the longitudinal patient record, HL7 message content, ordering dialogs, and, potentially, sources outside DHCP.

Many MLMs in CPRS will access, filter, and make decisions based on patient data from several different services (e.g., lab blood culture and pharmacy antibiotics). To allow data comparison across packages, it is most efficient to store that data in one standardized, easily accessible, patient-oriented location. The Decentralized Hospital Computer Program is expanding a visit-oriented, longitudinal patient record that will contain most relevant clinical data. The CIRN Clinical Object Repository could supplement DHCP's longitudinal patient record as a source of clinical data for CPRS processing.

Data used in MLM data elements are collected automatically, without direct input from users. These automated mechanisms include data-driven decision making, decision-driven data acquisition, and time-driven decision making/data acquisition. The CPRS will use an event-driven model for interpackage communication and messaging. This event model will initiate data-driven MLM processing. For example, when a new lab order is placed, OE/RR will post an event and a related HL7-formatted message. Other events and processes could subscribe to this event. The subscribing entities evaluate the content of the HL7 message and process it to meet their needs. In this example, the lab system is a subscriber. Lab identifies the order as new and responds back to OE/RR acknowledging receipt of the message. The lab software processes the information to initiate collection, testing, and resulting. The results are also posted via an event and HL7 message.

The expert system would be another subscriber to these events. It evaluates the HL7 message for MLM data element content as mapped by the metadata dictionary. If found, the data are processed against the logic and conditions in the MLM. The expert system and OE/RR could subscribe to the lab result event. Order Entry/Results Reporting would update the status of the order, and the expert system would process the message content for MLM relevancy. During the ordering/resulting process, additional events are posted and evaluated. As data flow through these events and HL7 messages, those data would drive or initiate MLM processing and decision making.

Decision-driven data acquisition would occur when the results of a MLM initiates the collection of data. The DHCP clinical expert system would accomplish this by supporting methods for collecting data from the legacy patient files. If part or all of a MLM becomes "true" and one or more data elements for that MLM look at legacy files for data, the system will go to extract routines for those files and attempt to retrieve the data within the parameters identified in the MLM. The MLM decision "drives" this collection of data.

The system also has the capability to be triggered by a chronological event. Medical Logic Modules processing can be triggered nightly, hourly, or at a certain date/time. For example, a MLM can be created to remind providers to order a weekly lab procedure.

To satisfy the desire to make the CPRS system a generic tool for all data evaluation needs, the system provides Application Programmer Interfaces (APIs) and allows foreground, background, or batch processing. This modular approach will enable users to trigger and process MLMs in a mode that best suits their information needs. Most decision support needs involve obtaining and evaluating data and applying some mathematical or logical algorithm against those data to generate a value or conclusion. With its combined API list and MLM editor, the expert system will enable DHCP software developers and VAMC providers to author and process MLMs to best fit their needs. Current users include Order Checking, Clinical Notifications, and a clinical calculator. Future uses are health maintenance order sets (e.g., an order set for diabetics), patient questionnaire/history taking, and clinical practice guidelines.

In contrast with clinical notifications of the past, which were triggered by encoded algorithms, current clinical notifications are triggered by MLMs. This gives VAMCs an option for modifying trigger logic. A clinical calculator aids clinicians by determining estimated creatinine clearance and performing other calculations. New notifications include:

- Medications Nearing Expiration;
- STAT Results Available;
- Patient Discharged;
- HBHC Patient Admitted;
- NPO Diet for More Than 72 Hours;
- Order Requires Cosignature;
- Unlinked Provider (non-Attending, Team), Order;
- Order Results Available;
- Exam Canceled Due to Inadequate Patient Preparation;
- Test/Procedure Canceled Due to Inadequate Specimen (Preparation);
- Consult/Request Canceled/Held;
- STAT Order;

- Unverified Pharmacy Order;
- MSRA, TB, etc. Results;
- Patient Transferred from Psychiatry to Another Unit;
- Lab Order Cancelled;
- DNR Order Nearing Expiration;
- New Order Placed;
- Orders Unverified by Nurse;
- STAT Imaging Request;
- Urgent Imaging Request.

Nonencoded notification triggers were accomplished by using the knowledge editor to create an MLM for each notification, as shown in Figure 14.1. The data elements of these notifications were linked to components of the HL7 messages passed between OE/RR and the filler packages (lab, radiology, pharmacy, etc.). The system was set up to intercept and evaluate HL7 messages for the MLM's data elements and conditions. If a data element was found and conditions were met, the results would be sent to a notification processing algorithm for dissemination to the appropriate recipients. In this manner, notification triggers were removed from M code and embedded in the MLM where they could easily be modified.

In the future, Health Maintenance Order Sets will present predefined order sets based on a patient's problem list or other criteria. The CPRS will also evolve to support patient or problem-centric questionnaires with the line of questioning determined by expert system processes and branching logic. The VHA requirements for Clinical Practice Guidelines could be met with appropriate MLMs and knowledge bases (Bailey et al., 1995).

Utilities to manage, support, and audit the CPRS expert system perform tasks such as knowledge base integration, dictionary editing, file clean-up and optimization, and decision/integrity audits.

```
Rule: STAT RESULTS AVAILABLE (ACTIVE)

---------------------------------- Data Elements -----------------------------
HL7 OBR FINAL RESULTS [OBR FINAL RESULT] **DATA DRIVEN**
HL7 OBX FINAL RESULTS [OBX FINAL RESULT] **DATA DRIVEN**
HL7 ORC STAT ORDER [ORC STAT] **DATA DRIVEN**
HL7 OBR STAT ORDER [OBR STAT] **DATA DRIVEN**

---------------------------------- Relations ---------------------------------

If -> (1) (OBR FINAL RESULT AND ORC STAT) OR (OBX FINAL RESULT AND OBR STAT)
Then
    -> Send Notification: STAT RESULTS
    -> Notification Message: STAT results: |LOCAL ORDERABLE ITEM TEXT|
```

FIGURE 14.1. Example of a clinical notification Medical Logic Module (MLM) displayed in low-level data element format.

Summary

The VA's CPRS decision support system was developed as a robust, modular knowledge-based system, which can function in realtime as well as background modes in a distributed environment. A generic utility for meeting a wide range of data collection and evaluation needs, the system design drew from previous experiences developing academic, commercial, and governmental decision support systems. It provides a strong foundation for the growing decision support needs of our nation's veterans.

Acknowledgments. The DHCP Clinical Decision Support project would not have been possible without the vision and guidance of Gordon Moreshead, the insight and skill of Richard Spivey, and the knowledge and expertise of Dr. Michael Lincoln. Other VA staff, not named here, made invaluable contributions by serving on expert panels, preparing white papers, and otherwise helping to define key issues.

References

Anderson CL. 1985. *Techniques for physician review of patient history data in the HELP computer system.* Master's thesis, University of Utah Department of Medical Biophysics and Computing, Salt Lake City, UT.

Anderson CL, Hukill M, Wang W, Bangerter B, Kattelman, & Hartmann-Voss K. 1990. A practical approach to structuring data in an integrated expert system. *14th Annual SCAMC Proceedings*, pp. 599–603.

Andrews RD. 1990. A common data structure for complex clinical data. *MUG Quarterly, 20(1)*, 49–53.

ASTM Committee E-31. 1992. *Standard specification for defining and sharing modular health knowledge bases (Arden syntax for medical logic systems).* ASTM E 1460-92. New York: NY.

Bailey WC, et al. 1995 (April 22). *Clinical decision making aids: Position statement.* Durhan, NC: VHA Quality Management Institute and Education Center.

Barnett GO, Cimino JJ, Hupp JA, & Hoffer EP. 1987. DXplain: An evolving diagnostic decision support system. *Journal of the American Medical Association, 258(1)*, 67–74.

Ben-Said M, Dougherty N, Altman S, Anderson CL, Bouhaddou O, & Warner HR. 1987. KESS: Knowledge engineering support system. *11th Annual SCAMC Proceedings*, pp. 56–59.

Bergeron B. 1991. Iliad: A diagnostic consultant and patient simulator. *MD Computing, 8(1)*, 46.

Carlson D, Wallace J, East TD, & Morris AH. 1995. Verification and validation algorithms for data used in critical care decision support systems. *19th Annual SCAMC Proceedings*, pp. 188–192.

Clancey WJ, & Shortliffe EH. 1984. *Readings in artificial intelligence.* Reading, MA: Addison-Wesley.

Cundick R, Turner C, Lincoln M, Buchanan J, Anderson CL, Warner HR Jr, & Bouhaddou O. 1989. Iliad as a patient case simulator to teach medical problem solving. *13th Annual SCAMC Proceedings*, pp. 902–906.

Curtis AC. Undated. *Knowledge-based applications in an imperfect world: Database decision support systems in ambulatory care.* Tucson, AZ: Indian Health Service.

Curtis AC. 1993. *YRULER expert system framework user's guide.* Tucson, AZ: Indian Health Service.

Harmon P, & Sawyer B. 1990. *Creating expert systems.* New York: Wiley and Sons.

Heathfield HA, & Wyatt J. 1993. Philosophies for the design and development of clinical decision support systems. *Methods of Information in Medicine, 32(1)*, 1–8.

HL7 Working Group. 1994 (December 1). *Health level seven interface standards version 2.2.* Ann Avbor, MI: Author.

Huff SM, Haug PJ, Stevens LE, Dupont RC, & Pryor TA. 1994. HELP the next generation. *18th Annual SCAMC Proceedings*, pp. 271–275.

Hukill MJ, Ward KM, Haug PJ, Turner CW, & Warner HR. 1987. HELP decision support on the Macintosh. *11th Annual SCAMC Proceedings*, pp. 155–157.

Applied Medical Informatics, Inc. 1994. *Iliad user manual.* Salt Lake City, UT: Author.

Jenders RA, Hripcsak G, Sideli RV, et al. 1995. Medical decision support: Experience with implementing the Arden Syntax at the Columbia-Presbyterian medical center. *19th Annual SCAMC Proceedings*, pp. 169–173.

Kuperman GJ, & Gardner RM. 1988. *The HELP system.* Salt Lake City, UT: LDS Hospital Department of Biophysics.

Lopez JR. 1990. *Significant drug interactions.* Washington, DC: Department of Veterans Affairs Executive Committee on Therapeutic Agents.

McDonald CJ. 1976. Protocol-based computer reminders. *New England Journal of Medicine, 295*, 1351–1355.

Pryor TA, Dupont R, & Clay J. 1990. A MLM based order entry system. *14th Annual SCAMC Proceedings*, pp. 579–583.

Rocha BHSC, Christenson JC, Pavia A, Evans RS, & Gardner RM. 1994. Computerized detection of nosocomial infections in newborns. *18th Annual SCAMC Proceedings*, pp. 684–688.

Warner HR, Haug P, Bouhaddou O, Lincoln M, Warner HR Jr, Sorenson D, Willamson JW, & Fan C. 1988. Iliad: An expert consultant to teach differential diagnosis. *12th Annual SCAMC Proceedings*, pp. 371–376.

Warner HR, Olmstead CM, & Rutherford BD. 1972. HELP: A program for medical decision-making. *Computers and Biomedical Research, 5*, 65–74.

Weed LL. *Knowledge coupling.* 1991. New York: Springer-Verlag.

15
The CARE Decision Support System

Douglas K. Martin

Introduction

In the virtual revolution called healthcare reform, a similar and parallel revolution in clinical information system design is critical. It is not enough to capture and provide access to data describing the clinical encounter, health status, and patient outcomes. We also need systems that improve clinician attention to important clinical events and give managers insights into how to provide the highest quality of care for the least cost.

The creators of a handful of legacy systems (McDonald, Blevins, Tierney, & Martin, 1988; Pryor, 1988; Stead & Hammond, 1988; Whiting-O'Keefe, Whiting, & Henke, 1988) recognized the potential of clinical decision support systems two decades ago. These systems have continued to grow, reflecting the clarity of vision and the soundness of concept with which they were built. Their failure to be widely applied outside the academic settings that spawned them likewise reflects the forces underlying the delivery of care in the past.

Today reform is leveling the playing field, and market forces are targeting efficiency and cost. As a result of these trends, practice guidelines have emerged, and unprecedented demands for decision support systems are inevitable.

In search of a single national solution that addresses these market trends, VA has responded by examining how its own healthcare facilities have individually dealt with these needs. This search has led the agency to base its national strategy taken by the VA's Richard L. Roudebush Medical Center in Indianapolis, also known as the Indianapolis Veterans Affairs Medical Center (VAMC). What follows is a detailed account of how this innovative medical center successfully rehosted a well-known legacy system, the RMRS, and its associated decision support component, the CDSS, within its own information system environment to meet critical clinical, research, and management needs.

Bringing Decision Support into the VA

The Components

The CDSS was developed by the Regenstrief Institute for Health Care in the 1970s under the direction of Dr. Clement McDonald (McDonald, 1981; McDonald, Blevins, Chamness, & Haas, 1989). A rule-based language sporting an English-like, user-friendly syntax, the system is an integral part of RMRS (McDonald, Blevins, Tierney, & Martin, 1988; McDonald, Tierney, Overhage, Martin, & Wilson, 1992). Regenstrief Medical Record System and consequently CDSS are a blend of VAX Basic and Assembly language routines built upon a locally developed, proprietary database management system. Layered around and operating upon a rich repository of highly encoded clinical data derived from a variety of sources and represented in a common internal format, CDSS was originally designed to be a decision-support tool, generating protocol-based clinical reminders and a sophisticated database query tool, capable of performing complex database queries across selected cohorts of patients.

CARE Decision Support System was demonstrated in a number of controlled clinical trials to impact clinician behavior by increasing compliance with health maintenance protocols (McDonald, 1976a; McDonald, Murray, Jeris, Bhargava, Seeger, & Blevins, 1977; McDonald, Wilson, & McCabe, 1980; McDonald et al., 1984; Tierney, Hui, & McDonald, 1986) and improving the monitoring of therapeutic interventions (McDonald, 1976b).

The DHCP is the hospital information system (Department of Veterans Affairs, 1995) developed by the Veterans Health Administration in the early 1980s and currently deployed in all Veterans Affairs Medical Centers (VAMCs). Written in the M programming language formerly known as Massachusetts General Hospital Utility Multi-Programming System (MUMPS), DHCP includes support for ancillary services such as clinical laboratory, pharmacy, bed control, and patient scheduling. Recently DHCP began to include clinical decision support capabilities, although these remain limited by the lack of a formalized approach and integration across data sources. Decentralized Hospital Computer Program provides vehicles for the delivery of decision support through sophisticated e-mail and alert generation systems. It is readily extendible, both directly, through modification of its source code (which is in the public domain), and indirectly, through modification of the data dictionary definitions of DHCP's database management system, VA FileMan.

In 1987, we introduced RMRS into the Indianapolis VAMC through a software interface (Martin, 1992) enabling realtime transfer of DHCP-based clinical data to RMRS running on a dedicated VAX minicomputer. This provided access to RMRS functionality with minimal development effort, but the lack of integration with DHCP limited its potential. In 1991, the Department of Medical Informatics at the Indianapolis VAMC began development of a DHCP-based RMRS written entirely in M; rehosting CDSS in this new environment would provide much needed patient-based clinical decision support. Through the design process, we balanced competing needs and made the necessary compromises to

- maintain a high degree of backward compatibility with the large number of CARE protocols accumulated over the years;
- extend the system's range, providing broader decision support capabilities;
- create a more modular and user-friendly protocol authoring and support environment;
- interface seamlessly with the host information system; and
- comply as much as possible with VA's specifications for software development under DHCP.

In June 1993, after 2 years of intense development effort, we introduced the DHCP-based RMRS into production. The crowning achievement was a greatly augmented decision support system, which is the primary focus of this chapter.

The Clinical Repository

At the heart of the DHCP-based RMRS is a repository housing patient-based clinical data derived from multiple sources. The database was implemented as an M global, a native M data structure that houses persistent data, is self-indexing, and is typically implemented internally as a balanced tree structure. Closely paralleling the structure of the original, the database receives clinical data in realtime from DHCP subsystems such as the laboratory, pharmacy, and patient scheduling modules. Thus, each module must have the ability to signal when a data event has occurred.

Where possible, this was accomplished by modifying the source file's data dictionary to automatically create an entry in an event queue. Supported by DHCP's file management system, this capability obviates the need to maintain software patches as new versions of the DHCP software are installed. In cases where DHCP packages bypass the standard data access methods in updating their respective databases, the native DHCP software must be modified directly and changes propagated manually through subsequent generations of the software.

A second major source of clinical data is through manual data entry, which enables the capture of clinical data not present in DHCP. For example, structured encounter forms enable the capture of encounter-related information such as problem lists, immunizations, and vital signs. Other data sources for manual entry include traditional free-text reports such as radiographic interpretations. In such cases, salient features of the report are translated by trained personnel into a controlled vocabulary and entered into the clinical repository in abbreviated form.

Regardless of the data source, all data events are submitted to an event queue. An event processor periodically examines the queue, identifies the data source for each event, and invokes the respective data extraction routine. The data extraction routine formats the data of interest into an ASTM E1238-compliant message (Martin, 1992) and forwards this information to the message server which, in turn, performs any necessary translations and stores the data in the repository. The result is a broad array of clinical data derived from multiple and often dissimilar data sources, in a common format within a single repository, optimized for rapid retrieval. This creates an infrastructure well suited to decision-support applications; it is this infrastructure upon which CDSS is built.

Overview of CARE Syntax

The CDSS operates on entity lists. Entity lists are time-ordered tuples extracted from the clinical database or in some cases from an alternate data source (e.g., a native DHCP file). Each tuple has several components: date, time, entity type, source, and value. Tuple values may be any combination of numeric, free-text, or encoded data types. Entity types and any encoded values stored under them all reference a single, controlled clinical vocabulary called the Entity Dictionary. For storage in the clinical database and for the formats in which data are rendered, RMRS and CDSS rely on the Entity Dictionary for

- representation of coded results and
- identification of the types and properties of observations.

The core construct of the CARE language is the *conditional statement*, formed in the traditional IF–THEN–ELSE syntax common to most high-level computer and query languages. The general form of the conditional statement is:

IF <conditional clause> **THEN** <actions>|<conditional statement>
[**ELSE** <actions>|<conditional statement>]

A *conditional clause* is of the form:

[<reduction>][<transform>][<selector>] <entity list> [<precriteria>]
<verb> [<postcriteria>]

Conditional clauses manipulate entity lists and produce entity lists that are subsets of the original. An entity list may be constructed de novo by a direct reference to an entity defined in the Entity Dictionary, or it may be an entity list previously defined by the protocol. If *precriteria* are present, entities are screened by the specified criteria that may contain a variety of Boolean, arithmetic, and temporal operators. Next, when present, *selectors* (such as FIRST, LAST, ALL, ANY), *transforms* (such as INTERVAL, CHANGE), and *reductions* (such as AVERAGE, MINIMUM,

MAXIMUM) are applied in that sequence. The resulting list is screened by any postcriteria specified. Composite conditional clauses may be created by connecting individual clauses by the AND and OR operators. In the case of two conditional clauses connected by the OR operator, if the first clause evaluates to true (i.e., the entity list meets all of the applied criteria), the second clause is not evaluated. Similarly, if the first of two conditional clauses connected by AND is false, the second clause is never evaluated. This is important in determining which entity list is active after the evaluation of a composite conditional clause, since the last entity list created becomes the default list upon which subsequent actions operate.

Actions enable CDSS to interact with the external environment, performing such tasks as generating clinical reminders (the REMIND action), suggesting orders (the ORDER action), prompting for data (the OBSERVE action), inferring diagnoses (the DX action), and saving data to an external file (the SAVE action). Though actions are typically embedded within conditional statements, they may also occur by themselves.

To summarize the features discussed thus far, consider the following CARE statement:

IF LAST "DIGITALIS MEDS" WAS > 0 AND LAST "POTASSIUM" WAS < 3.0 THEN ALERT This patient had a potassium level of '{VALUE} on {DATETIME}.'
 'Because this patient is on a digitalis preparation, he/she may be at risk'
 'for significant cardiac arrhythmias. Consider adjusting potassium level'
 'to >3.5 meq/L.'

This statement consists of a composite conditional clause connected by the AND operator. The first subclause determines whether or not the patient is currently taking a digitalis preparation. If this evaluates to false, the remaining statement is not executed. If true, the second subclause is evaluated to determine if the last potassium value was below a critical threshold. If this also evaluates to true, the entire conditional clause is true and the ALERT action executes, generating an electronic alert message for the clinician's immediate attention. The keywords enclosed in braces {} within the alert message are replaceable parameters that evaluate at runtime to the respective components of the active entity list, in this case containing information about the last potassium result (since this was the last list constructed). Terms enclosed within double quotes are entity names as they appear in the Entity Dictionary, or may refer to named entity lists previously created with the DEFINE statement as described below.

A number of additional constructs exists within the CARE syntax. The DEFINE statement allows the user to associate a symbolic name with an entity list created by a conditional clause. It is of the form:

DEFINE "<list name>" **AS** <conditional clause>

Defining lists in this fashion has two distinct advantages. First, it can improve the readability of protocols by assigning a descriptive name to what a list contains. Second, it can improve the efficiency of protocols where a single list may be referenced multiple times.

Consider the following example of a list definition:

DEFINE "Recent hyponatremia" AS LAST "SODIUM" WAS ON_ AFTER 3 MONTHS AGO & <130

Future references to the entity list name "Recent hyponatremia" will access the sodium result meeting the associated criteria or, if the criteria are not met, the null list.

CARE supports block structuring and flow control of protocols with the BEGIN BLOCK, END BLOCK, and EXIT statements. These allow the collection of related subunits within a protocol into discrete sections. Blocks so constructed may be exited conditionally using the EXIT command. In addition, blocks may be nested within blocks, further enhancing protocol flow control.

Consider the following example:

BEGIN BLOCK PNEUMOVAX
  IF "AGE" WAS <65 AND NO "DX'S"
  [= "DIABETES MELLITUS, "HIV" v "COPD"] EXIST THEN EXIT
  IF LAST "PNEUMOVAX" WAS AFTER 10 YEARS AGO THEN EXIT ELSE ORDER "Pneumovax"
END BLOCK PNEUMOVAX

Here, if any of the criteria for pneumococcal vaccination are not met, the enclosing block is conditionally exited. In the case of nested blocks, if a block name does not follow the EXIT statement, the immediately enclosing block is exited; otherwise, the named block is exited.

To enhance flow control within CARE, a looping construct similar to traditional FOR/NEXT loop was introduced. This construct permits looping through each entity in an entity list, thereby simplifying the task of processing multivalued lists. The following example, using nested FOR/NEXT loops, illustrates this feature:

DEFINE "LOW BS" AS "GLUCOSE TESTS" [<= 50] EXIST
DEFINE "DIS" AS "ENCTR SITE" [= "DISCH"] EXISTS
FOR EACH "LOW BS"
  FOR EACH "DIS"
    DEFINE "ADM" AS LAST "ENCTR SITE" [BEFORE "DIS" & = "ADMIT"] EXISTS
    IF "LOW BS" WAS ON_AFTER "ADM" & ON_BEFORE "DIS"
    THEN SAVE AS "LOBS"
    AND EXIT LOOP

NEXT "DIS"
NEXT "LOW BS"

Here we identify all episodes of hypoglycemia (as defined by a serum glucose of less than or equal to 50) recorded during an inpatient stay (as defined by the encoded values of "ADMIT" and "DISCH" under the "ENCTR SITE" entity). The outer loop selects glucose results one by one from a named entity list while the inner loop determines whether the date of the result falls within the time window of a hospital stay and copies it to an output file if it does. This feature eliminates the need to enumerate all hospital stays by successive conditional statements and apply each of these to the list of glucose results. Such queries were lengthy and cumbersome and, because it was difficult to anticipate the maximum number of admissions during the period of interest, users ran the risk of missing results by not including a sufficient number of conditional statements.

CARE Actions

CARE supports a number of action statements that may be invoked alone or as part of a conditional statement. Generally speaking, CARE actions serve to communicate with the external environment in some way or to alter the execution flow of protocols. The variety of actions makes CARE a powerful and flexible decision support tool.

Actions may be divided into three categories: query, informational, and flow control. *Query actions* are SAVE and OUTPUT; functioning similarly, both of these actions permit the transfer of any component of an entity list to a DHCP-compatible output file. Data exported in this manner can be manipulated by database tools in DHCP or exported to other database formats using an export utility. The syntactic form for these actions is:

SAVE | OUTPUT [<entity list>] [{<component>}] AS <identifier>

If no *entity list* is specified, the active entity list is used. The component specifier determines which component of the list is to be exported (e.g., the VALUE or DATE components). By default, the VALUE component is exported. The *identifier* is user defined and is used to label the data in the output file. SAVE and OUTPUT differ in that, if output has already been saved under the given identifier for the current patient, SAVE will overwrite the preexisting data while OUTPUT will append to the data.

Informational actions include ALERT, ASK, DISPLAY, OBSERVE, ORDER, REMIND, SEND, DX, RX, and PAGE. Each of these actions delivers information via a specific vehicle. Several share syntactic requirements.

ALERT, SEND, and PAGE all deliver informational messages to one or more users, usually a provider responsible for the patient. Their syntax is:

ALERT | SEND | PAGE ["<recipients>"] "<message>"

The ALERT action uses the DHCP Alert Processor to issue alerts to the designated recipients while SEND uses the DHCP e-mail system, MailMan, to deliver the message. Alerts are more intrusive than e-mail in that the user is repeatedly reminded that there are pending alerts until the alerts are viewed. Once viewed, alerts may be deleted or saved as mail messages. The PAGE action delivers its message by paging the recipients and displaying the message. This is most useful in conjunction with alphanumeric pagers, but has potential applications with strictly numeric pagers as well. Because DHCP does not support automated paging, a generic paging engine was developed to provide this capability.

The *message* argument to each of these actions may consist of a combination of free text, entity list references, entity component selectors, and embedded M code. References to specific entity lists are enclosed in double quotes within the message. Such references result in no output, but merely designate the active entity list. Component selectors are enclosed in braces and result in the substitution of the corresponding data in the message. Native M code may also be embedded by delimiting with vertical bars. The code is evaluated at runtime, and the result is displayed in its place. To illustrate message syntax, consider the following message:

"The last creatinine was 'CR LAST' {VALUE} on {DATE}."

"CR LAST" refers to an entity list defined earlier and sets this as the active list. {VALUE} would be replaced with the value component of the list and {DATE} with the date component. The resulting message might look like this:

The last creatinine was 2.5 on 25-Apr-93.

The optional *recipients* argument permits directing the message to someone other than the default provider. This consists of a semicolon-delimited list of user names, user identifiers, or mail groups.

The DISPLAY, ORDER, REMIND, and RX actions also share the same syntax:

DISPLAY | ORDER | REMIND | RX "<message>"

DISPLAY directs the message to the current output device (usually the user's terminal) and is most useful in interactive applications. ORDER presents the message in the orders section of a patient's encounter form. REMIND displays the message on a reminder form that is provided to clinicians during patient encounters. This can be of use in administering health maintenance reminders. The RX action permits displaying messages on the prescription profile form and can be used to suggest therapies.

DX and OBSERVE interact with the patient's encounter form. Both expect a reference to an Entity Dictionary entry as shown:

DX | OBSERVE "<entity name>"

DX automatically appends the named *entity* to the patient's problem list as it appears on the encounter form. This is useful for displaying inferred diagnoses (such as diabetes when a patient is on insulin, but has no explicit diagnosis of diabetes recorded). OBSERVE appends the entity to the observation template on the encounter form. This template permits manual recording of specific results for capture using manual data entry. This is useful, for example, in capturing the results of occult blood testing performed during clinic encounters.

Like DISPLAY, the ASK action is provided primarily for use in an interactive environment. The syntax for this action is:

ASK {<component>} WITH "<message>"

ASK requests input from the user using the *message* as the prompt. The input format is determined by the *component specifier*. This action permits a protocol to request information interactively from the user.

Flow control actions include EXIT, CONTINUE, CALL, and TASK. The EXIT action can be used to prematurely exit a block or loop and has been described previously. The CONTINUE action terminates the conditional statement in which it is embedded with execution resuming with the immediately following statement. The CALL action takes a protocol name as its single argument. The specified protocol is invoked and upon completion returns control to the parent protocol. This enables protocol authors to write modular protocols that perform common tasks such as computing a patient's age and to reference them from other protocols.

Similar to CALL, the TASK option submits the associated protocol to execute asynchronously in the background. As an option, the protocol may be designated to run at some future time. This feature enables a protocol to be self-perpetuating and permits implementation of timed monitors. Extensions to the original CARE syntax, both TASK and CALL enhance its modularity and applicability.

Protocol Integrated Development Environment

One of our design objectives was to create a full-featured protocol authoring environment, modeled after those prevalent among many computer language implementations, to facilitate creating and compiling protocols. The result was the Protocol Integrated Development Environment (PRIDE), which integrated text processing, online help, protocol compiler, and protocol authoring tools.

Although DHCP offered integrated line- and screen-oriented text processors, they had limited capabilities and proved difficult to adapt to our specialized needs. Our objective was to develop advanced text-processing capabilities that were

- interoperable with DHCP;
- highly modular, permitting selection of only those features needed for a given application;
- configurable at the user and the application levels;
- readily extendible with new features;
- constructed with a well-defined application program interface.

The text processor we developed and named the Extensible Editor exemplifies our approach to software development (Martin, 1994a). As a result of our specifications, the text processor is useful as a protocol authoring tool and in a wide variety of other domains. In addition to general purpose text processing for any DHCP application, the Extensible Editor serves as a read-only text browser, hypertext help navigator, hypertext help editor, and a software development tool.

In the CDSS environment, the Extensible Editor provides the protocol author with essential text processing functionalities, including multiple text buffers, text block commands (text selection, cut, copy, paste, etc.), text search/replace, user-defined macros, line wrap, horizontal and vertical scrolling, and text import/export capabilities.

The Extensible Editor provides the features of an integrated development environment with extensions that include:

- the ability to look up entity names using an Entity Dictionary browsing tool;
- context-sensitive hypertext help explaining syntactic requirements and operation of CARE statements;
- global hypertext help addressing more general subjects like compiling and executing protocols; and
- integrated compilation and compile-time error tracking.

By combining tools previously separate, we made protocol generation more streamlined and efficient. By approaching development from a generalistic perspective, we created a tool that is reusable across a broad range of applications, significantly reducing development needs in other areas.

Figure 15.1 shows a sample screen of the Extensible Editor as it is used in protocol authoring. Here the compiler has just been invoked and has reflected back compilation errors that it has detected. With a single command, the user can move to each line containing an error, and the compiler-generated error message appears at the top of the edit window. This demonstrates how the functionality of the editor can be extended while maintaining the appearance of total integration.

CARE Language Compiler

As originally implemented at the Regenstrief Institute, CARE source code was compiled into an intermediate pseudocode that was then executed by a

```
CARE Editor    Version 1.3                                        ALLOPURINOL
Error: Unrecognized term DEFNE
==[*MAIN*]==[ INSERT ]===[ FWD ]=================================================
BEGIN BLOCK ALLOPURINOL

DEFINE "DATE" AS EX: TODAY

DEFINE "ALLOPURINOL USE" AS LAST "ALLOPURINOL" [BEFORE "DATE"] WAS GT 0

IF NO "ALLOPURINOL USE" EXISTS
THEN EXIT ALLOPURINOL

▌DEFNE "LAST URIC ACID" AS LAST "URIC TESTS" [BEFORE "DATE"] EXISTS

DEFINE "THIAZIDE USE" AS LAST OF_EACH MEMBERS OF "THIAZIDES" [BEFORE "DATE"]
       WAS GT 0

IF "THIAZIDE USE"
THEN REMIND 'Concurrent use of allopurinol and thiazide diuretics'
     'can lead to increased levels of oxipurinol, increasing the risk'
     'for adverse reactions.'

! Recommendations for monitoring of uric acid levels in patients on allopurinol
L=======T=======T=======T=======T=======T=======T=======T=======T=======R
```

FIGURE 15.1. The extensible editor.

pseudocode interpreter. M, itself an interpreted language, presents the possibility of a more direct approach to converting CARE protocols to executable form. Taking advantage of the inherent ability of M to generate and in turn execute M code, we developed a CARE language compiler that produces executable M code as output. The result is code that can be directly invoked without an intermediate interpreter.

The CARE compiler uses a single-pass, table-driven, state-transition algorithm to compile CARE source code. It utilizes a modified Reverse Polish Notation (RPN) processor to handle complex expressions and operator precedence. The first step in the compilation process is the parsing of a CARE statement. During this process, CARE statements, which may span multiple lines, are broken down into discrete tokens that are placed in a queue. The compiler then examines each token in sequence and processes it according to the rules embedded in associated tables.

Two tables control the compilation process: the State Transition Table and the Operator Precedence Table. The State Transition Table defines of a series of discrete context states. Each context state defines which tokens are valid in that state. Associated with each token is a next state and M code to be executed upon encountering the token. This code can be used to generate compiled code, manipulate the RPN stacks or token queue, or perform special-purpose processing. For each CARE statement, the compiler always begins in state 1 and must end in state 0. Thus, state 1 will define all valid tokens that may begin a statement. As each token is processed, the compiler transitions to the next state associated with the token and continues processing with the next token in the queue. If a token is encountered that is not valid for the current state, the compiler generates a syntax error message and aborts compilation of the current statement. If the compiler enters state 0 before reaching the end of the token queue or reaches the end of the token queue in a state other than 0, it also signals an error.

Token matching occurs in up to three passes. First, the compiler looks for an exact match for the token. If one is not found, it then looks for any token patterns defined for the current state. These patterns allow specification of a general form for the token without explicit enumeration of every possible form. (Such enumeration may be impossible as, for example, in the case of entity list names, which are named at the user's discretion.) Token patterns always begin with a question mark character and are followed by standard M pattern matching syntax. Patterns define which characters or character groups are valid at each position within a string. The general form for a single specification is:

<minimum matches> · <maximum matches> <character type>

Character type may be a quoted string literal or a character group specifier such as N for numeric or A for alphabetic. For example, ?.N.1".".N specifies two integer values of any length (including zero) optionally separated by a decimal point. Note that when the *minimum matches* parameter is not specified, it defaults to zero. The *maximum matches* parameter defaults to infinity. Thus, the preceding example would match 5, 3.75, and 0.68 as well as the null string and a single period.

If a token fails to match a pattern, the compiler checks for a special entry labeled <ANY>. When this entry is present, it will match any token. If these three lookup attempts all fail, the compiler signals a syntax error.

To illustrate this algorithm, consider the State Transition Table entries shown in Table 15.1 and the following CARE statement:

IF LAST "HEMOGLOBIN" WAS > 8.5 THEN EXIT

TABLE 15.1. State transition table.

Row	State	Token	Next State	Executable Code
1	1	DEFINE	100	D PRO^RGCARC3( ),MDF^RGCARC3("SLCT")
2	1	IF	2	D IF^RGCARC5,MDF^RGCARC3("SLCT")
3	2	FIRST	3	D OPR^RGCARC3( )
4	2	LAST	3	D OPR^RGCARC3( )
5	3	?1"""""".E1""""""	4	D QUOTE^RGCARC3( )
6	4	WAS	5	
7	5	>	6	D OPR^RGCARC3( )
8	6	?N.1".".N	7	D CONST^RGCARC3( )
9	7	THEN	8	D THEN^RGCARC5
10	8	EXIT	9	D EXIT^RGCARC4
11	9	<END>	0	
12	9	AND	1	D AND^RGCARC6

The compiler parses the statement into nine tokens (including the end-of-statement token, denoted by <END>, which is always appended by the compiler). Then starting with the first token (IF) and state 1, it finds a match in row 2. After executing the associated code (the significance of which will not be addressed here), the compiler transitions to state 2 and advances the queue pointer to the next token, LAST. The compiler locates the token LAST in state 2 (row 4), executes the associated code, and enters state 3, advancing to the next token, "HEMOGLOBIN." An explicit lookup of this token fails, but a pattern-based lookup succeeds at row 5 (which matches anything enclosed in paired double quotes; note that the multiple quotes here are a syntactic requirement of M when specifying the quote character as a literal). This sequence repeats for each of the remaining tokens. Upon encountering the end-of-statement token in state 9, the compiler finds a match (row 11), and enters state 0, the only valid state upon completion.

The Operator Precedence Table controls the sequence of code generation based on the relative precedence of the operators contained therein. Each entry consists of an operator token, a precedence level, the operator's associativity (i.e., left or right), the number of operands, and executable M code. The compiler uses a modified RPN algorithm to process operators and their operands. This algorithm defines two stacks, an operator and an operand stack. The compiler manipulates these stacks by explicit calls from the executable code within the State Transition Table. For example, the M code in row 7 of Table 15.1 calls a subroutine called OPR^RGCARC3. This subroutine pushes the associated operator (in this case the > operator) onto the operator stack. Similarly, in row 8 the subroutine CONST^RGCARC3 pushes a numeric constant onto the operand stack.

When an operator is pushed onto the operator stack, the compiler checks to see if the last entry on the stack has an equal (in the instance of left associativity) or higher (for both left and right associativity) precedence than the operator about to be added. If it does, the last entry is popped from the stack along with the appropriate number of arguments from the operand stack and the executable code associated with the popped operator is invoked (which typically outputs compiled code). A reference to the result of the operation is then pushed onto the operand stack. This sequence is repeated until the operator stack is empty or the top entry does not meet the criteria to be popped from the stack. When either occurs, the new operator is added to the operator stack and processing continues.

The traditional RPN algorithm results in left associativity of operators. That is, sequential operators of equal precedence are executed from left to right. Thus, the sequence $1 - 2 + 3$, where the operators $-$ and $+$ are of equal precedence, is evaluated as if it were written $(1 - 2) + 3$. However, many operators in CARE are right associative, with the evaluation sequence now being reversed. For example, the partial CARE statement

IF AVERAGE OF LAST 10 "HEMOGLOBINS" WAS...

contains two operators of equal precedence, LAST and AVERAGE. However, the LAST operation must be performed before the AVERAGE operation can be performed. These operators are therefore right associative. To support this, operators can be designated as right associative. In doing so, operators of equal precedence are not removed from the stack and are eventually processed in reverse order as they are later popped from the stack.

The compiler converts CARE source code into native M code. The compiler first outputs this compiled code to a temporary holding area. Because M routines are limited in length (which varies by implementation but is usually on the order of a few thousand bytes), the compiler must keep track of code length and may have to distribute code generated from a single protocol across several routines. This means that the compiler must keep track of all line label references and resolve branches to entry points outside the current routine. Because all code generation done by the compiler is limited to a single subroutine, this is relatively easy to do. This subroutine examines each line of code for a label reference as it is generated. If it encounters a label that is not yet known to the compiler (i.e., a forward reference), it adds the label and the referencing line to a list of unresolved references. As the compiler generates line labels, it examines this table for unresolved references to each newly defined label. If the branch must cross routine boundaries, the original label reference is extended to include the routine name so the M interpreter will know where to find it at runtime. At the end of compilation, if no errors were encountered, the compiler then outputs all generated code to the appropriate routines. The compiler uses an eight-character routine naming convention consisting of a two-byte header that is site determined, a four-byte representation of the protocol's internal entry number in the protocol file, and a two-byte routine sequence number used to distinguish multiple routines generated from a single protocol. The two numeric components are represented in base 62, which makes use of the 10 numeric digits and the 52 upper- and lower-case alphabetic characters that constitute the valid character set for naming M routines. Thus, the routine ZX000a02 consists of the site-selected two-byte header ZX, the four-byte internal protocol identifier 000a (which is 36 in base 10), and the routine sequence number 02 (which indicates it is the third routine in the sequence, the first being 0).

Compiled CARE protocols consist primarily of parameterized subroutine calls that reference a runtime library of M routines. These library routines perform such tasks as loading and manipulating entity lists and performing CARE actions. Entity lists are referenced by internal integer identifiers assigned at compile time. Examples of a CARE protocol and its compiled code may be found in Figures 15.2 and 15.3, respectively.

```
BEGIN BLOCK ALLOPURINOL

DEFINE "ALLOPURINOL USE" AS LAST "ALLOPURINOL" WAS GT 0

IF NO "ALLOPURINOL USE" EXISTS THEN EXIT ALLOPURINOL

DEFINE "LAST URIC ACID" AS LAST "URIC TESTS" EXISTS

IF NO "LAST URIC ACID" EXISTS
THEN REMIND
'This patient is currently on allopurinol, yet there is no uric acid level on'
'file.  Uric acid levels should be monitored at least every 6 months in patients'
'on allopurinol and more often during dosage titration.  The proper allopurinol'
'dose for gout is that which achieves a serum uric acid level of less than 7.0.'
AND ORDER 'Uric acid'
AND EXIT ALLOPURINOL

IF "LAST URIC ACID" WAS GT HIGH_NORMAL
THEN REMIND
'This patient''s last uric acid value ({VAL}) on {DATE} was high.  The dose'
'of allopurinol should be titrated to achieve a uric acid level of less than'
'7.0 in patients with a history of gout.  Once the target value is attained,'
'the uric acid level should be monitored at least every 6 months.'
AND EXIT ALLOPURINOL

IF "LAST URIC ACID" WAS BEFORE 6 MONTHS AGO
THEN REMIND
'Patients on allopurinol should have uric acid levels monitored at least every'
'6 months.  This patient''s last uric acid level was {VALUE} on {DATE}.'
AND ORDER 'Uric acid'
AND EXIT ALLOPURINOL

END BLOCK ALLOPURINOL
```

FIGURE 15.2. Sample CARE protocol.

Debugging Tools

To facilitate compiler design, the compiler supports a special debugging mode. When activated, this mode causes the compiler to generate a trace of all steps in the compilation process. This capability has proven invaluable in debugging complex code. It can also differentiate among bugs that originate from coding errors in the compiler itself, incorrect entries in the tables that drive the compilation process, and errors in the runtime library routines. In addition, M language development environments provide special code-level debugging tools that can be used to debug protocols during their execution, by setting break points and tracing through the compiled code to monitor the state of entity lists and the operation of library functions to determine where errors have occurred.

Because there are no end-user tools for debugging protocols, authors of protocols may resort to inserting DISPLAY actions in strategic locations and running the protocol interactively against a small cohort of patients to determine what is happening along the way. We are presently developing a source-level debugger integrated with the CDSS development environment to permit more sophisticated debugging by the end user.

```
          ZX000E00(RUN,DUZ,RST,SLT,DFN,PRV,DS,SC,DAT,PID)
                  ; CARE Version 1.3
                  ; Protocol ALLOPURINOL
                  ; 20-Nov-94 21:18
                  S $ZT="ZT^RGCAREEX"
                  D ^RGCAREIN
   NX             D ^RGCARENX
                  G:'DFN ^RGCAREEX
   SUB            ; BEGIN BLOCK ALLOPURINOL
                  D ^RGCARELL(1,1,505,DFN)
                  D ^RGCARELS(1,4,1,-1,1,0)
                  D ^RGCARELA(5,1,"0^^^^0")
                  D ^RGCARELC(4,6,1,5,5)
                  D ^RGCARELZ(4,6)
                  S TPL=4
                  D ^RGCARELS(4,10,1,-1)
                  D ^RGCARELN(11,10)
                  G:'$$^RGCARELT E1
                  G B1
   E1             D ^RGCARELL(13,1,3967,DFN)
                  D ^RGCARELS(13,16,1,-1,1,0)
                  S TPL=16
                  D ^RGCARELS(16,20,1,-1)
                  D ^RGCARELN(21,20)
                  G:'$$^RGCARELT E2
                  D ^RGCAREAR(SC,DAT,DFN,1)
                  D ^RGCAREAO(SC,DAT,DFN,2)
                  G B1
   E2             S TPL=16
                  D ^RGCARETP(16,25,2)
                  D ^RGCARELC(16,26,1,25,5)
                  D ^RGCARELS(26,27,1,-1)
                  G:'$$^RGCARELT E3
                  D ^RGCAREAR(SC,DAT,DFN,3)
                  G B1
   E3             S TPL=16
                  D ^RGCARETD(30,0,-180)
                  D ^RGCARELC(16,31,1,30,51)
                  D ^RGCARELS(31,32,1,-1)
                  G:'$$^RGCARELT E4
                  D ^RGCAREAR(SC,DAT,DFN,4)
                  D ^RGCAREAO(SC,DAT,DFN,5)
                  G B1
   E4
   B1             ; END BLOCK ALLOPURINOL
   B0             ; END BLOCK <MAIN>
                  G:RST'<0 NX
```

FIGURE 15.3. Compiler-generated code.

Invoking Protocols

Protocols can be invoked in a variety of ways. In each case, the protocol is bound to a process known as the initiator. It is the initiator that actually controls protocol execution. Initiators are triggered by certain system events, such as a patient encounter, the storage of a laboratory result, or the request of the user. Because compiled protocols are M routines, the list of possible initiators extends to virtually any DHCP application. However, for the typical end user, there are five formal methods for invoking protocols, as described below.

One of the great strengths of CDSS is its ability to perform complex queries of the clinical repository, producing patient cohorts with selected characteristics for further study. The initiator for CARE protocols applied in this manner is the CARE Query Manager. Using the Query Manager, the user associates (binds) a protocol with a named output file that will contain data produced by the SAVE or OUTPUT actions. The software also needs to know the input patient cohort against which it is to run. The input cohort can be the entire clinical repository, a cohort produced by another query, or individual patients selected by the user. The query manager then permits the user to submit the query to run as a background task or interactively in the foreground. Queries that have aborted due to user request or other events (e.g., system crashes) can be restarted where they left off using a special restart option.

A patient encounter, scheduled or unscheduled, can trigger one or more protocols. This is accomplished by binding protocols to the report generation engine of the Flexible Outpatient Report Management System (FORMS) (Martin, 1994b). A subsystem of RMRS, FORMS is used to generate all encounter-related forms and reports, including those that can be targets of decision-support information, namely the encounter form, prescription profile, and reminders form. Protocols are bound to specific encounter sites by entering them into a control file that also specifies which reports are to be generated for which encounter sites. The report generation engine, which may be invoked in batch mode (for scheduled patients) or interactively (for unscheduled patients), first executes each of the associated protocols before generating the reports. In this manner, protocols can be made clinic-specific by associating them only with encounter sites where they logically apply. For example, a reminder to perform a prostate exam might be appropriate in a primary-care clinic, but not in an ophthalmology clinic where protocol requirements would be quite different.

Yet another powerful use of protocols represents one of the many extensions we have added to the original CDSS. By binding protocols to the background task responsible for storing new data in the clinical repository, protocols can be invoked when results for selected entities are stored. This is extremely useful in the creation of automated clinical alerts for the rapid identification of critical laboratory values, potential drug-drug interactions, and other urgent events. We have accomplished this binding by adding a field to the Entity Dictionary to support bound protocols. For example, a protocol associated with the serum potassium entity issues an alert upon storage of a serum potassium result if the level is below a certain threshold and the patient is taking a digitalis preparation.

The fourth formally supported method of triggering a protocol takes advantage of a feature in DHCP's VA FileMan database manager whereby M code can be associated with any database field. Such code is automati-

```
CARE Protocol Status Display                              Version 1.3

CARE query identifier: POTASSIUM          CARE set = POTASSIUM PROTOCOL

Status of query POTASSIUM PROTOCOL on 18-Nov-94 16:47:
Care Set:                 POTASSIUM PROTOCOL
Submitted by:             MARTIN, DOUGLAS K., M.D.
Task identifier:          2236507
Execution status:         completed
Last started:             18-Nov-94 15:46
Last ended:               18-Nov-94 15:46
Records processed:        2752
Alerts generated:         85
```

FIGURE 15.4. Protocol status utility.

cally executed when the field's value changes. Because compiled protocols are M routines, any protocol may be invoked in this way. This is done simply by making an appropriate entry in the data dictionary definition for the field of interest. This can be useful for triggering certain actions when a database field is changed, such as issuing a page or alert to another party.

The final method commonly used to trigger a protocol has been discussed earlier. Through the use of the CALL or TASK actions in the CARE language, one protocol (the initiator) invokes another. Here, the binding actually occurs at the CARE source code level.

The CARE Protocol Status utility permits the user to monitor the progress of his or her protocol. The utility displays a variety of information including the protocol's execution status (pending, running, aborted, completed), the number of patient records processed, and the number of selected types of actions performed. The utility also permits the user to abort the protocol if desired. The appearance of a typical monitored protocol is illustrated in Figure 15.4.

We continue to explore ways to bind protocols to important system events. For example, there is great interest in triggering protocols during clinician ordering of laboratory tests and medications. We hope to attach protocols to selected orders in DHCP's Order Entry package, providing realtime, interactive decision support during this process. Given that this is a prime window in which to influence clinician decision making, this capability has great promise. We anticipate further expanding CDSS's awareness of and interaction with its surroundings by establishing associations with other DHCP subsystems in the future.

Promoting Modularity

One drawback to CARE protocol management in the past has been its lack of modularity. All reminder protocols were combined in a single source file. As the number of protocols increased over time, growing to over 1500 lines of source code, this practice became unwieldy. Modifying one protocol presented the risk of having unforeseen interactions with other protocols. Even isolating a given protocol was difficult. To circumvent these problems, we borrowed the approach used by implementors of the Arden Syntax (Hripcsak, Clayton, Pryor, Haug, Wigertz, & Van der Lei, 1990) using the paradigm of the Medical Language Module (MLM) (Hripcsak, Johnson, & Clayton, 1993; Pryor & Hripcsak, 1993). The Arden Syntax is an industry standard decision support language largely derived from existing languages, including CARE. Taking a highly modular approach to protocol representation, the Arden Syntax defines protocols as mutually independent units called MLMs. A single MLM, for example, might embody a clinical reminder for influenza vaccination while another might represent a critical value alert for potassium. In this way, sites may pick and choose which protocols to implement and where. The advantages in terms of protocol maintenance are obvious.

We have adopted an approach to protocol representation very much like that of the MLMs of the Arden Syntax. Each protocol typically represents a single (or sometimes two or more very closely related) clinical guideline. How, when, and where an individual protocol is activated depends on the initiator to which it is bound. This modular implementation is consistent with our design philosophy and far more flexible than earlier monolithic approaches.

Application and Impact

The CDSS has profoundly affected the dynamics of health care at the Indianapolis VAMC. This is not surprising, given two decades of informatics research documenting the efficacy of decision support systems in a wide range of patient-care settings. Applicable across the spectrum of clinical, research, and management needs, the CDSS continues to gain acceptance and to grow.

Presently we have over 30 active clinical protocols in place. Early protocols were written by a physician, the author of this chapter; more recent protocols have been written by a registered nurse who serves as clinical coordinator and by a certified laboratory technician, our laboratory information manager. Because we emphasize primary care at our facility, most of our protocols target primary-care providers, as shown in Table 15.2. Some protocols were requested by a medical service, others by committees engaged in quality-assurance activities. The value of CDSS in performing database queries was underscored by the 77 ad hoc queries performed within the last year, most of which originated with clinical researchers wishing to identify cohorts of study candidates. Several of these researchers now author their own queries.

To manage and prioritize the growing number of requests for protocols, we established a multidisciplinary review committee, which is charged with

TABLE 15.2. Protocol use at the Indianapolis VAMC.

Category	Number	Examples
Queries	77	atrial fibrillation following coronary bypass
Preventive care	10	occult blood testing, cholesterol screening
Critical values	9	increasing creatinine, low hemoglobin
Drug interactions	4	allopurinol-thiazides, warfarin-antibiotics
Informational	5	new protime assay, formulary change
Drug monitoring	2	theophylline, allopurinol
Study protocols	2	histoplasmosis prophylaxis

ensuring that protocols are consistent with the mission and policies of our medical center. In evaluating a newly requested protocol, the committee

- considers the likelihood that it will achieve its desired effect;
- confirms that an evaluative mechanism is in place to validate *a priori* assumptions; and
- explores the possibility of secondary, perhaps untoward, effects of a requested protocol. For example, what effect will the protocol have on resource utilization? (Our pharmacy completely depleted its stock of pneumococcal vaccine within 2 weeks of implementing a protocol reminding primary-care clinicians to administer these in eligible patients.)

In addition to reviewing new requests, the committee continually reexamines existing protocols for consistency with evolving healthcare practices. We are hopeful that this committee, still in its formative stages, will provide a much needed consensus-building approach to the implementation of healthcare policy as it is embodied within the protocols of the CDSS.

We continue to find new uses for the CDSS. The case reports that follow serve to illustrate the versatility and utility of the CDSS in addressing a variety of clinical, research, and management needs and some of the pragmatic considerations surrounding the actual implementation of protocols.

The Impact of Protocols

Case Report #1: Improving Preventive Measures

Despite guidelines recommending administration of tetanus toxoid every 10 years, this intervention is typically triggered by the occurrence of an event placing the patient at risk (e.g., penetrating trauma). To improve physician compliance with this recommendation, we implemented a decision support protocol to issue a reminder whenever an eligible patient presented for care at our primary-care clinics.

Because we estimated that most of our patients had not had the booster within the 10-year interval, we designed the protocol to ease the initial burden of immunizing large numbers of patients. The protocol included a workload distribution algorithm using the patient's day of birth within the calendar year to determine whether the reminder would be issued for a patient otherwise eligible. With the average visit interval in our primary-care clinics about 4 months, patients initially passed over were almost certain to present themselves once again within several months, thus ensuring that the majority of our population would be vaccinated within the first year. This pseudorandom selection of eligible patients to receive reminders also allowed us to study the impact of the protocol. We performed a CARE query to extract data on visits occurring over the final 3 months of 1994. If a patient had multiple visits during this interval, we examined only the initial visit. Of the 4613 patients visiting our primary-care clinics during that time, 4174 (90.5%) had no documented tetanus booster within the previous decade. (This is certain to be an underestimate of the true immunization rate as our ability to document such events outside our clinic setting is limited.) Providers received reminders to order tetanus immunization in 2539 (60.8%) of the eligible patients. The workload distribution algorithm suppressed reminders for the remaining 1635 who served as the control group. Of those patients for whom reminders were issued, 265 (10.4%) received a tetanus booster on the day of the index visit compared to only six (0.4%, $p < 0.0001$, Chi-square) in the control group.

This 26-fold increase in immunization rates was not an unexpected effect of computerized reminders. Projected impacts on workload and resource consumption must clearly be taken into account when planning the implementation of such protocols. Policies that mandate certain protocols be followed by default further amplify their effect. While compliance with the tetanus vaccination protocol was strictly at the discretion of the provider, we have "default-action" policies in place for influenza and pneumococcal pneumonia vaccination. In the presence of such a policy, unless the provider actively cancels the computer-generated order, the patient refuses the intervention, or the nurse determines the intervention to be inappropriate based on explicit guidelines, the intervention is carried out. While default-action protocols are not appropriate in all situations, they can be remarkably effective in selected cases, suggesting that nursing staff are even more susceptible to such interventions than are physicians.

Case Report #2: Assisting in Inpatient Care

Our quality assurance team identified several episodes of documented hypoglycemia occurring in inpatients receiving no hypoglycemia-inducing medications and having no established diagnosis of diabetes mellitus. While most of these cases could be explained by other factors (e.g., hypoglycemia seen in advanced liver disease), the lack of timeliness in identifying these events as they occurred hampered the team's ability to investigate them effectively. In response to this need, we designed and implemented a CARE protocol triggered by the storage of a serum glucose measurement in the clinical database. The protocol first determined if the glucose level was below a threshold of 60 mg/dL and if the patient was currently an inpatient. If the occurrence did not meet both criteria, the protocol terminated. Otherwise, the protocol then determined if the patient had a documented diagnosis of diabetes mellitus and if the patient had been taking a hypoglycemia-inducing medication such as insulin or an oral hypoglycemic agent. The protocol then relayed this information along with the patient's name and ward location as a realtime alert to members of the investigatory team. In this manner, the team could quickly identify each occurrence, locate the affected patient, and complete their assessment. After about 2 months, we terminated the protocol after it was determined that all cases could be explained on clinical grounds.

This case report illustrates the value of the CDSS in providing realtime decision support; such support can target not only the actual care provider, but other members of the healthcare team as well.

Case Report #3: Performing Outcome Research

A local health services researcher was interested in determining the efficacy of a specialty clinic in reducing the morbidity and mortality of patients with advanced congestive heart failure. To identify a comparable control group, we performed a CARE query to construct a patient cohort matched on multiple attributes including patient demographics and disease severity (as objectively measured by echocardiographic parameters). All of the required data existed within the clinical repository. In the absence of the CDSS, the researcher would have had to resort to extensive and costly chart review to accomplish this task.

Manual chart review is perhaps one of the costliest aspects of outcomes research. By providing not only a rich and highly optimized clinical database but also sophisticated query tools, we can dramatically reduce the amount of effort required to conduct such research. We believe these secondary benefits more than justify the resources required to maintain systems such as ours.

Case Report #4: Recruiting for Clinical Trials

Multiple study protocols for HIV-infected patients exist at our institution. Eligibility criteria vary by protocol, and physicians are often unaware of these criteria or even of the existence of these ongoing studies. To improve patient recruitment while preserving confidentiality, we implemented physician-directed reminders that identify potential study candidates by criteria available within the database (e.g., T-cell subsets, current drug therapy, or previous infections). These reminders provide the primary-care physician with details about the study and identify a contact person responsible for patient enrollment. The primary-care physician can then be knowledgeable about ongoing clinical trials and discuss their benefits and risks directly with the patient. Because patients are presumably more likely to place trust in a physician with whom they have had an extended therapeutic relationship, rather than a study investigator they have never met, this could be expected to augment recruitment efforts.

Lessons Learned

Recasting RMRS and CDSS within a different computing environment presented a rare opportunity to build upon the expertise of two decades of development, combining the best features of successful legacy systems with the inherent strengths of the new environment provided by DHCP.

Establishing specific development guidelines prior to writing a single line of code enabled us to weigh quickly the myriad of design options in the context of explicit priorities. By focusing on the production of flexible, generic, modular, reusable and extendible solutions, we have created a system that will continue to evolve along with our needs and will undoubtedly be used in ways we do not now envision. By creating interoperable subsystems, we have been able to integrate multifaceted functionalities to streamline the processes involved in creating and maintaining protocols.

Our ability to recreate the functionality of RMRS and especially CDSS in a short development cycle demonstrates the suitability of the M programming language as a rapid prototyping tool and an implementation environment. And because the M language specification is administered by the American National Standards Institute (ANSI), M-based applications in compliance with this standard will run unaltered on a variety of hardware platforms. Since our initial implementation, we have migrated RMRS and CDSS from VAX minicomputers to Intel 80486-based microcomputers and finally to DEC Alpha microcomputers. By following the lead established by DHCP and enforced by VA programming standards, all references to platform-specific extensions to the M language are isolated by invoking them through internally standardized subroutine calls. Special installation procedures load the extensions appropriate to the underlying hardware platform. This ensures a smooth migration pathway even where complete hardware independence is not possible.

These development approaches build upon a singular design philosophy: seek generic solutions for specific needs. This approach requires that we step back from the specification of a given need and view it in a more global context. Fulfillment of a single need can result in the fulfillment of a large number of closely related needs. This approach translates to less development time and less maintenance overhead. At the Indianapolis VAMC, it has paid great dividends by enabling us to move rapidly forward despite limited resources. We expect to reap the benefits of this approach for a long time to come.

The success of the Indianapolis model has inspired VA's national strategy, known as the CIRN. Clinical Information Resource Network extends the concept of a centralized clinical data repository to support VA's recently adopted network healthcare model by providing mechanisms for the movement and unification of patient data across all facilities involved in the care of a given patient. The result is an integrated view of each patient's history that belies its distributed origins.

Acknowledgments. The author thanks Dr. Clement McDonald for his support and expertise during the course of the project and Mr. M. Randall Cox and Mr. Phillip Salmon for their technical input and support. The author's work is supported by a career development award (CDS 91-306) from Health Services Research and Development of the Department of Veterans Affairs.

References

Department of Veterans Affairs. 1995. *Decentralized hospital computer program.* Monograph. Washington DC: Veterans Health Administration.

Hripcsak G, Clayton PD, Pryor TA, Haug P, Wigertz OB, & Van der lei J. 1990. The Arden syntax for medical logic modules. *Proceedings of the 14th Annual Symposium on Computer Applications in Medical Care*, pp. 200–204.

Hripcsak G, Johnson SB, & Clayton PD. 1993. Desperately seeking data: Knowledge base-database links. *Proceedings of the 17th Annual Symposium on Computer Applications in Medical Care*, pp. 639–643.

Martin DK. 1992. Making the connection: The VA-Regenstrief project. *MD Computing, 9(2),* 91–96.

Martin DK. 1994a (March 22). *Extensible editor technical manual. Version 1.4.* Indianapolis, IN: VAMC Department of Medical Informatics.

Martin DK. 1994b (November 11). *Flexible outpatient report management system (FORMS).* Version 1.1. Indianapolis, IN: VAMC Department of Medical Informatics.

McDonald CJ. 1976a. Protocol-based computer reminders, the quality of care and the nonperfectability of man. *New England Journal of Medicine, 295,* 1351–1355.

McDonald CJ. 1976b. Use of a computer to detect and respond to clinical events: Its effect on clinical behavior. *Annals of Internal Medicine, 84,* 162–167.

McDonald CJ. 1981. *Action-oriented decisions in ambulatory medicine.* Chicago: Yearbook Medical Publishers.

McDonald CJ, Blevins L, Chamness D, & Haas J. 1989 (August). *CARE language user's guide.* Indianapolis, IN: Regenstrief Institute for Health Care.

McDonald CJ, Blevins L, Tierney WM, & Martin DK. 1988. The Regenstrief medical records. *MD Computing, 5,* 34–47.

McDonald CJ, Hui SL, Smith DM, Tierney WM, Cohen SJ, Weinberger M, & McCabe GP. 1984. Reminders to physicians from an introspective computer medical record: A two-year randomized trial. *Annals of Internal Medicine, 100,* 130–138.

McDonald CJ, Murray R, Jeris D, Bhargava B, Seeger J, & Blevins L. 1977. A computer-based record and clinical monitoring system for ambulatory care. *American Journal of Public Health, 67(3)*, 240–245.

McDonald CJ, Tierney WM, Overhage JM, Martin DK, & Wilson GA. 1992. The Regenstrief medical record system: 20 years of experience in hospitals, clinics, and neighborhood health centers. *MD Computing, 9*, 206–217.

McDonald CJ, Wilson GA, & McCabe GP. 1980. Physician response to computer reminders. *Journal of the American Medical Association, 244*, 1579–1580.

Pryor TA. 1988. The HELP medical record system. *MD Computing, 5*, 22–33.

Pryor TA, & Hripcsak G. 1993. Sharing MLMs: An experiment between Columbia-Presbyterian and LDS Hospital. *Proceedings of the Annual Symposium on Computer Applications in Medical Care, 17*, 399–403.

Stead WW, & Hammond WE. 1988. Computer-based medical records: The centerpiece of TMR. *MD Computing, 5*, 48–62.

Tierney WM, Hui SL, & McDonald CJ. 1986. Delayed feedback of physician performance versus immediate reminders to perform preventive care: Effects on physician compliance. *Medical Care, 24*, 659–666.

Whiting-O'Keefe QE, Whiting A, & Henke J. 1988. The STOR clinical information system. *MD Computing, 5*, 8–21.

Section 4
Clinical and Support Applications

Chapter 16
Meeting Clinical Needs in Ambulatory Care
   John G. Demakis   231

Chapter 17
Surgical Systems
   Shukri F. Khuri   240

Chapter 18
Nursing Use of Systems
   *Bobbie D. Vance, Joan Gilleran-Strom, Margaret Ross Kraft,
   Barbara Lang, and Mary Mead*   253

Chapter 19
Developing Clinical Computer Systems: Applications in Cardiology
   Ross D. Fletcher and Christopher McManus   275

Chapter 20
Anesthesiology Systems
   *Franklin L. Scamman, Holly M. Forcier,
   and Matthew Manilow*   293

Chapter 21
The Library Network: Contributions to the VA's Integrated
Information System
   Wendy N. Carter and Christiane J. Jones   308

Chapter 22
Using Data for Quality Assessment and Improvement
   Galen L. Barbour   330

Chapter 23
Developing and Implementing the Problem List
   Michael J. Lincoln   349

16
Meeting Clinical Needs in Ambulatory Care

JOHN G. DEMAKIS

The Ambulatory Care Program

The Department of Veterans Affairs (VA) was formally established as an agency of the federal government in 1930. From the onset, VA was primarily an inpatient institution. Outpatient care was only allowed for one visit preadmission and one visit postdischarge. Primary care and continuity of care were not considered part of the VA's mission. Indeed, a tour of most the VA medical centers built before 1980 would convince anyone that ambulatory care was of minor significance.

In 1973, however, the Omnibus Health Care Act was passed in which VA was authorized for the first time to care for patients in the ambulatory care who had not been hospitalized. This formally established the VA's Ambulatory Care (AC) Program. The actual wording of the bill allowed patients to be sent directly to ambulatory care "to obviate the need of hospitalization." However, it was still required that these patients be recertified every 12 months as continuing to need AC "to obviate the need of hospitalization."

Although the VA's AC program began to grow, it was still small and there was no separate funding for AC. The most dramatic change in the AC program occurred in 1984 when VA adopted the methodology used by Medicare. This methodology was adopted to conform to the private sector, which in 1983 had begun to use diagnosis-related groupings (DRGs) for Medicare payment. There was the expectation that this would reduce length of stay and decrease costs in the VA. However, VA went far beyond Medicare practices. To decrease costs even further, VA Central Office staff decided it was time to shift VA care away from the hospitals into the outpatient areas. The VA model, therefore, included AC for the first time in a prospective payment system. The new AC payment system was quite generous and encouraged facilities to enroll more patients and to perform more procedures in the clinics.

The results were dramatic. The number of veterans and number of visits surged in VA clinics, as shown in Table 16.1. The level of success gave rise

TABLE 16.1. The Department of Veterans Affairs workload: inpatient versus outpatient.

Year	Inpatient average daily census	Average length of stay	Inpatient total patients treated	Outpatient visits
1980	67,475	23.6	1,047,573	18,204,000
1981	65,739	22.9	1,047,533	18,165,000
1982	64,261	22.5	1,044,315	18,202,000
1983	64,347	21.8	1,078,919	18,754,000
1984	62,440	21	1,085,885	18,836,000
1985	58,637	19.6	1,094,257	19,831,000
1986	56,257	18.4	1,114,217	20,437,000
1987	53,865	17.6	1,113,948	21,890,000
1988	52,111	16.9	1,125,318	23,534,000
1989	49,040	16.8	1,062,573	22,854,000
1990	46,728	16.6	1,028,584	22,856,000
1991	44,073	16.3	984,217	23,389,000
1992	42,756	16.2	966,980	24,195,000
1993	41,663	16	952,124	24,406,000
1994	39,941	15.5	940,043	25,442,000
1995	37,016	14.8	910,133	27,565,000

to the fear that the AC program would bankrupt VA. There was also widespread belief that people in the field were gaming the system. Allegations were made that some hospital directors were sending their staffs out to the local shopping malls to enroll healthy veterans! By the early 1990s, it was clear that this type of outpatient DRG was not the answer to funding AC. Nonetheless, it had succeeded in putting more emphasis on outpatient care and refocusing VA priorities. Hospitals built in the late 1980s and early 1990s now had large, state-of-the-art AC centers.

By 1992 the AC DRG system was suspended, and new methods were explored. This coincided with the change in the administration in Washington. A new president was elected, a new Secretary of Veterans Affairs was appointed, and soon thereafter a new Under Secretary for Health was chosen. Dr. Kenneth Kizer had never worked in the VA before. He had been Director of the State of California's Department of Public Health. Dr. Kizer started as Under Secretary in October of 1994, determined to do things differently. Through a series of meetings and input from many people in the field, a new vision was developed by late 1995. A new mission was created which would transform VA from a primarily inpatient institution to one that was mainly outpatient care. Heavy emphasis was to be placed on primary care and continuity of care. This was a dramatic break with the VA's original mission but one that had been slowly evolving since 1984.

Once this decision was made, implementation brought up many problems, including space (the VA was built for inpatient care), personnel (most physician staff at our tertiary-care facilities were subspecialist), etc. One of the most interesting and challenging problems was the AC database.

Existing AC Database

The VA's main database is the Decentralized Hospital Computer Program (DHCP). The DHCP is a Massachusetts General Hospital Utility Multi-Programming System (MUMPS) based series of packages developed over the years to serve clinical and administrative needs. It has traditionally been geared to the inpatient setting. Its main clinical packages include Pharmacy, Laboratory, and Radiology. Administrative packages include patient demographic data and discharge data. The discharge database includes International Classification of Diseases, 9th ed., Clinical Modification (ICD-9-CM) codes and Current Procedural Terminology (CPT) codes of diagnoses treated and procedures performed in the hospital. The discharge data from all patients discharged from a VA hospital are maintained at the VA's Data Center in Austin, Texas. The "rolled up" data from all VA hospitals constitute the Patient Treatment File (PTF). The PTF is an extraordinarily rich database and is used for trend analysis and research.

On the outpatient side, many of the same databases existed. The Pharmacy, Laboratory, Radiology, and Demographic packages also included outpatient data. However, there was one important exception. There was no diagnostic information because none was collected at the time of each visit. As a result, there was no way of knowing the medical problems of the patients treated in the outpatient setting at each facility. Worse yet, if a patient came into one of the VA emergency rooms or was seen without a chart, there was no way of knowing what his/her problems were. Although problem lists were mandated in the VA for many years, in the clinics they were kept sporadically and were not computerized.

In the early 1990s, a new software package was released that gained immediate popularity with the outpatient clinical staffs. This was the Health Summary (HS). Although very simple in design, it made an immediate impact on patient care in the clinics. The HS allowed the staff to view the last year's worth of data that had accumulated on any patient, including laboratory results, x-ray results, medications taken, hospitalizations, discharge diagnoses, and clinic visits. These were all presented in an easy-to-read format. Again, one of the few links missing in the HS was the outpatient diagnoses.

Outpatient data from all VA facilities are also send to the Austin Data Center and "rolled up" into the Outpatient Clinic File (OPC). This also is a helpful database for trend analysis, although limited because of the lack of diagnostic information.

When the VA's new mission was developed with its emphasis on primary care and continuity of care, the lack of an outpatient diagnostic database was a glaring exception. To bill, do trend analysis, develop quality measures, take performance measures, etc., some type of problem/diagnostic database would be a necessity. Dr. Kizer, therefore, mandated that by the end of 1996 every patient encounter in the VA outpatient clinics must include a minimum data set that corresponds to that collected in the private sector. This must include a diagnosis and/or purpose of visit and procedures performed at each encounter.

Existing Packages

Demographic Package

Often known as the Medical Administration Package, this includes very specific administrative data, such as years of active duty, branch of the military, etc. It has many features of importance to clinicians, including name, address, age, birth date, number of clinics in which the patient is enrolled, dates of clinic appointments, past clinic appointments, hospital admissions and discharges, and discharge diagnoses.

Pharmacy Package

Includes names of prescribed drugs, dosages, how many refills remaining, when patient picked up drugs, and names of providers for each prescription.

Laboratory Package

Lists all laboratory test ordered and the results.

Radiology

Lists all radiologic tests ordered, when completed, and the diagnostic reading.

Health Summary

Since all of the packages mentioned are stand-alone packages, the clinician must enter and exit each package get information from each of the specific packages. This can be time-consuming, especially in the clinic setting. The HS allows the clinician to type in the name of the patient only once, and all the above packages will be arrayed sequentially. This has simplified the procedure and decreased the time necessary to access information. This package was received with widespread enthusiasm from clinicians.

Ambulatory-Care Databases Newly Released or in Development

Problem List

One of the main weaknesses of the present AC data systems was the lack of any diagnostic information. This had been identified by a survey of VA clinicians as the highest priority for correction. As a result, work had been progressing for several years to develop a problem list package in the VA. Dr. Kizer's insistence on diagnostic information gave impetus to the work already in progress. In 1995, a Problem List (PL) package was pilot tested and released to the field. One of the hallmarks of this new package was a clinical lexicon that allowed clinicians to choose terms familiar to them. Each term was mapped to an ICD-9-CM code, which would allow billing.

The new package was a direct-entry package. The clinician could type the diagnosis directly into the computer, or the clerk would enter in the information from the chart or some type of encounter form that had been filled out by the clinician.

Although the Problem List has much to commend it, there are limitations on its use. Not all VA facilities have enough terminals to allow ready access to the clinicians. Many clinicians are not used to typing or computers and some are intimidated by the new systems. Direct entry by clinicians is also time-consuming because regular chart notes are still required.

Clerical entry, which requires extra clerical time, is also a problem because many VA sites have still not enhanced their clerical staffs in the clinics.

Progress Notes

The Progress Note package was released to the field in late 1995. It allows clinicians to type their patient progress notes directly into the computer. The package is not yet widely used, but it is anticipated that, as direct entry becomes more feasible, more and more clinicians will use this option.

Reminders

Presently being tested, the Clinical Reminders package is scheduled to be released to the field in late 1996. The package will focus on preventive medicine and patient education reminders; it has the potential to include many other of types of standards of care.

Patient Care Encounter

The Patient Care Encounter (PCE) is a dramatic breakthrough. This will be the first VA software package that is patient centered. All information on a patient for each encounter will be entered in one patient-centered package instead of a series of separate packages, such as demographic, pharmacy, laboratory, and so on. The PCE will be the repository for the Problem List, Progress Note, Reminders, and Patient Education. The PCE is scheduled to be released in the summer of 1996.

History and Physical

An automated history and physical package was identified as a high priority by the clinician survey. The VA is presently evaluating several proprietary packages to see if one can be adopted for the VA.

Minimum Data Set

In late 1995, the VA adopted the Uniform Ambulatory Medical Care Minimum Data Set (MDS) as set forth by the National Committee on Vital and Health Statistics (Hanken & Waters, 1994, pp. 70–71). By the end of 1996, all outpatient encounters in the VA should be collecting these data. Although some of these data are now being collected by the existing VA databases, much will be new. The MDS will include information on the following:

- *Patient Identification*: Most items are now being collected.
- *Provider Identification*: No data are now being collected.
- *Encounter Data*: Although some of these data are being collected, no data are collected on diagnoses, reasons for visit, procedures, preventive medicine, or patient education.

Data Collection

The emphasis on collecting more data in the outpatient clinics has highlighted the issue of how to collect the additional data. The existing methods of direct entry by clerical or clinical staff are labor intensive and have met with widespread reluctance.

In preparation for the institution of the AC Minimum Data Set by the end of 1996, VA has explored many different methods of data collection. In late 1994, a decision was made to develop an optical scanning methodology that would serve as the primary choice of data collection in the VA outpatient clinics. This decision was based primarily on cost consideration. Optical scanning was considered accurate, easy to use, and, most importantly, reasonably priced. It was also felt that this technology would be easy to develop and readily accepted by clinicians.

Optical Scanning

There are two types of optical scanning devices: the image scanner and the mark sense scanner. The mark sense scanner scans each sheet to pick up the black round marks. This type of scanning is well known to test takers because it is often used for multiple-choice tests. The image scanners take a picture of the encounter form and then reads any marks on the form. The advantage of the image scanning is that it allows for character recognition. Thus, alphanumerical values such as blood pressure, temperature, etc. can be captured.

Image Scanning

The decision was made to institute image scanning as the method of choice in the short term so that data collection could begin in all VA sites by the end of 1996. The Automated Information Collection System (AICS), which uses image scanning, was developed at the VA's Albany Information Resource Management Field Office (IRMFO). It will allow clinic-specific encounter forms to be developed and printed out at each facility. The encounter forms will have precoded the most common diagnoses, procedures, preventive measures, and patient education in the clinic and facility. Clinicians will also be able to write in specific diagnoses that are not found on the form. The forms can then be rapidly scanned into the computer, and the various components will be stored in the PCE. Information that is handwritten must be typed in by the clerical staff. Information stored in PCE can be extracted through a variety of methods for billing purposes, trend analysis, quality assurance, and so on. Problem lists will be automatically generated from the information in the PCE.

Mark Sense Scanning

Although the AICS has not yet been released, two different prototypes have been used in the VA for the last 2 years. A large cooperative trial has been ongoing for 2 years, which uses preprinted mark sense scannable encounter forms in 12 large VA General Medical Clinics. The experience has been generally favorable. There is a high degree of completion (95 to 100%), rapid scanning (up to 1000 forms an hour can be scanned accurately), and a minimum of problems. About 10% of all forms completed have handwritten diagnoses that require manual entry by a clerk. One problem with this system is the lack of flexibility. The forms are printed up in batches of 10,000; hence, changes cannot be easily made.

Testing has also been done of a proprietary mark sense scannable encounter form developed by Scanning Concept Incorporated (SCI). This form can be easily configured at each site and for each clinic. Presently 26 VA sites are testing out this system. It differs significantly from the 12-site research project in that the forms are not preprinted but are printed out at each clinic site. Clinicians can easily make changes in the forms. In this way, it is closer to the AICS system. There have been problems in printing out the forms so that they can be scanned accurately. Any small deviation in how the forms are printed can throw off the accuracy of the scanning. The added flexibility can thus create problems. For such a system to work well, a careful validation procedure is required to ensure that the scanning is accurate.

Assessment of Scanning Technology

Strengths

Scanning technology is relatively simple and well tested, having been in use for many years. It is highly accurate, with error rates of less than 1%. It takes very little time of a clinician to complete a form, usually 30 seconds or less. Scanners can scan 1000 forms an hour. The scanning technologies are also the cheapest of the existing methodologies. This is an important consideration because the VA operates about 235 outpatient clinics nationally and will probably be opening more in the next decade.

Weaknesses

Despite its strengths, scanning technology has several inherent problems. First, it does not allow clinicians to use their own words in documenting a problem. Rather they are forced to use a diagnosis printed on the form. Second, it does not allow annotation of the problem. For encounter forms that are printed on site, care must be taken to make sure that the printers are carefully aligned or the mark sense scanning will be inaccurate. (This is not a problem with the image scanning.)

Other Data Collection Methods

There are, of course, many other forms of data collection available. Besides the scanning technologies, these include mobile pen based, grid pad, voice recognition, and workstations (direct entry by clinicians). Each has its strong points and weak points. Each has its advocates and its detractors. For a system as large and as varied as VA, there is probably no one correct technology. Clinics vary in size, number and type of providers, complexity of treatment, and number and type of support staff. In a system so diverse, there are opportunities for several different technologies. An important task is to decide which technology is best for any given situation. This will require a careful evaluation. In 1995, a contract was signed with investigators at the University of Alabama at Birmingham to evaluate several of the more readily available data collection methods including image scanners, mark sense scanners, mobile pen based, and workstations (clinician direct entry). This will be a rigorous evaluation at several sites involving many clinicians. Outcome measures will include data completeness, degree of data accuracy, and availability of clinically relevant information. It is anticipated that as other forms of data collection are developed and perfected that they will also be evaluated for VA use. Although VA has decided to go with the scanning technologies in the short term, there will be ample opportunities for other technologies to be used.

Opportunities and Challenges

In the next few years, as VA moves into the primary-care and continuity-of-care arena, one of the main tasks for VA administrators is to make the delivery of care in the clinics as efficient and effective as possible. Adoption of the MDS will put new strains on the existing clinic system. Resources will have to be allocated or reallocated to the clinics.

Databases presently in use or soon to be released to the clinics have the opportunity to significantly affect how care is delivered and the quality of that care. These databases must be carefully evaluated to assure maximum efficiency. The data-entry methods must also be rigorously evaluated. Costs are important; so are accuracy, speed, acceptance by the clinicians, etc.

As these databases become widely available, the question is how the clinician will access them. Will the information be printed out for the clinician (as is often done for the HS today at many VA sites) or will the clinician be expected to bring up the data on the terminal? If clinicians will be expected to access the databases directly and do direct entry themselves, how will this impact patient care? Will it allow more rapid and accurate decision making or will it slow the process and subvert the patient–physician interaction?

There are many new tools now being developed that will allow a clinician to interact more rapidly with new and existing databases. The Text Integration Utility (TIU) will make it easier for clinicians to type directly into the databases. The Graphical User Interface (GUI) will allow clinicians to move rapidly between databases. Decision support systems (DSS) will aid clinicians in choosing the correct diagnoses or intervention, etc.

Each of these new systems must be evaluated rigorously to measure the impact on the provider and on patient care. The ultimate test must be this: Does it improve patient care? Clearly, VA is developing an excellent outpatient database. With the adoption of the MDS and the many new databases and systems now being tested, VA should be well situated to deliver excellent primary care and to ensure continuity of care.

Reference

Hanken M, & Waters TA. 1994. *Glossary of health care terms*. Chicago: American Health Information Management Association.

17
Surgical Systems

SHUKRI F. KHURI

Introduction

The Department of Veterans Affairs' (VA) Decentralized Hospital Computer Program (DHCP) is one of the richest repositories of clinical information on surgical patients in the world. Full utilization of various DHCP modules allows for the electronic acquisition and storage of more than 80% of the total clinical record of the surgical patient. With the realization of ongoing developmental and the Hybrid Open System Technology (HOST) initiatives, the electronic acquisition of the totality of the surgical patient's clinical record should become a reality within 2 to 3 years. This chapter describes the electronic information currently available on the surgical patient in the various DHCP modules, how this information is utilized daily in patient management and quality improvement, and the ongoing efforts toward the realization of the complete electronic surgical record.

Surgery Module

The Surgery Module is designed to be used by surgeons, surgical residents, anesthetists, operating room nurses, and other surgical staff. This module integrates booking surgical cases and tracking clinical patient data to provide a variety of administrative and clinical reports. The Main Menu of the Surgery Module, displayed in Table 17.1, provides a bird's-eye view of the range of functionalities that this comprehensive software provides. Access to the various sections of the Surgery Module is determined by the provider's DHCP access and verification codes. Only the Chief of the Surgical Service has complete access to the full surgery menu.

Surgery providers enter the data directly into terminals placed at various work sites, including each operating room. The flow of data in and out of the Surgery Module is shown in Figure 17.1. Typically, the surgeon or the surgical resident sees the patient in the outpatient clinic, and an operation

TABLE 17.1. Main menu of the surgery module in DHCP.

W	Maintain Surgery Waiting List...
R	Request Operations...
LR	List Operation Requests
S	Schedule Operations...
LS	List Scheduled Operations
O	Operation Menu...
A	Anesthesia Menu...
PO	Perioperative Occurrences Menu...
NON	Non-O.R. Procedures...
C	Comments
SR	Surgery Reports...
L	Laboratory Interim Report
CH	Chief of Surgery Menu...
D	Surgeons' Dictation Menu...
M	Surgery Package Management Menu...
RISK	Surgery Risk Assessment Menu...

is planned. They enter key patient information into the electronic waiting list, which also contains demographic information, including the patient's and the referring physician's addresses and telephone numbers. This latter information is automatically imported from the Medical Administration Service's module in DHCP. When the time comes to request an operation for this patient, the surgeon or the surgical resident accesses the Request Operations submenu, checks the availability of space in the operating room, and then requests an operation for a certain date. Prompted by the requesting process, the provider inputs additional clinical information required by the anesthesia and operating room staff. The Head Nurse in the operating room pulls from the computer all the requests for the next day and prepares the operating room schedule. In the preparation of the schedule, additional patient information is entered, including the names of all the providers (surgeons, house staff, nurses, and anesthesiologists) who will partake in the operation. Once the schedule is finalized, it is disseminated by printing it simultaneously on a large number of predefined DHCP printers, located throughout the hospital and on all patient wards.

The intraoperative information is entered, as it is generated chronologically, into terminals placed in each operating room, mostly by the operating room nurses and the anesthesiology staff. This information, including the Current Procedural Terminology (CPT)-4 coding, is verified by the surgeon on a separate verification screen, immediately after the termination of the operation. The surgeon's dictation of the operation record is transcribed into personal computers and uploaded into DHCP, merging the transcription with intraoperative patient information and producing the Operation Report that becomes part of the patient's permanent electronic record.

Figures 17.2 and 17.3 show the operative information data fields that are filled by the time the patient leaves the operating room. These data provide

242   Clinical and Support Applications

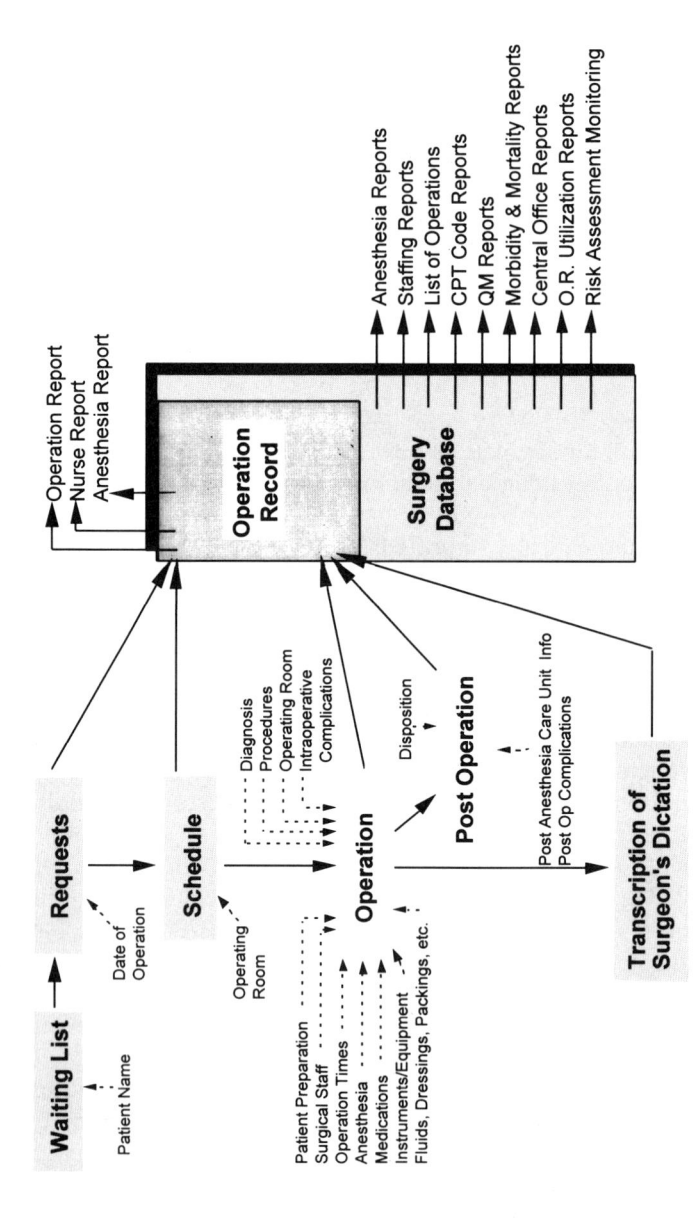

FIGURE 17.1. Schema of the flow of data in the surgery module in DHCP.

```
** STARTUP **    CASE #41282                    PAGE 1 OF 2

1    DATE OF OPERATION:       APR 11, 1996 AT 07:30
2    PRINCIPAL PRE-OP DIAGNOSIS: CAD
3    OTHER PREOP DIAGNOSIS: (MULTIPLE)
4    OPERATING ROOM:          OR3
5    SURGERY SPECIALTY:       CARDIAC SURGERY
6    MAJOR/MINOR:             MAJOR
7    REQ POSTOP CARE:         SICU
8    CASE SCHEDULE TYPE:      URGENT
9    REQ ANESTHESIA TECHNIQUE: GENERAL
10   PATIENT EDUCATION/ASSESSMENT:
11   CANCEL DATE:
12   CANCEL REASON:
13   CANCELLATION AVOIDABLE:
14   DELAY CAUSE:             (MULTIPLE)
15   VALID ID/CONSENT CONFIRMED BY: O'CONNOR,JOYCE

** STARTUP **    CASE #41282                    PAGE 2 OF 2

1    ASA CLASS:               4E-LIFE THREAT-EMERG.
2    PREOP MOOD:              ANXIOUS
3    PREOP CONSCIOUS:         ALERT-ORIENTED
4    PREOP SKIN INTEG:        SCAR
5    TRANS TO OR BY:          ICU BED
6    PREOP SHAVE BY:          WESLEY,ARTHUR L.,JR.
7    SKIN PREPPED BY (1):     BUTLER,CHARLES E
8    SKIN PREPPED BY (2):
9    SKIN PREP AGENTS:        IODOPHOR SCRUB SPONGE
10   SECOND SKIN PREP AGENT:  POVIDONE IODINE
11   SURGERY POSITION:        (MULTIPLE)(DATA)
12   RESTR & POSITION AIDS:   (MULTIPLE)(DATA)
13   ELECTROGROUND POSITION:  RIGHT BUTTOCK
14   ELECTROGROUND POSITION (2): LEFT BUTTOCK
15   IN/OUT-PATIENT STATUS:   INPATIENT

** OPERATION **   CASE #41282                   PAGE 1 OF 3

1    TIME PAT IN HOLD AREA:   APR 11, 1996 AT 07:30
3    TIME PAT IN OR:          APR 11, 1996 AT 07:30
4    ANES CARE START TIME:    APR 11, 1996 AT 07:30
5    TIME OPERATION BEGAN:    APR 11, 1996 AT 09:00
6    SPECIMENS:               (WORD PROCESSING)(DATA)
7    CULTURES:                (WORD PROCESSING)
8    THERMAL UNIT:            (MULTIPLE)
9    ELECTROCAUTERY UNIT:     #5510 & #5511
10   ESU COAG RANGE:          30-50
11   ESU CUTTING RANGE:       30-40
12   TIME TOURNIQUET APPLIED: (MULTIPLE)
13   PROSTHESIS INSTALLED:    (MULTIPLE)(DATA)
14   REPLACEMENT FLUID TYPE:  (MULTIPLE)(DATA)
15   IRRIGATION:              (MULTIPLE)(DATA)

** OPERATION **   CASE #41282                   PAGE 2 OF 3

1    MEDICATIONS:             (MULTIPLE)
2    SPONGE COUNT CORRECT (Y/N): YES
3    SHARPS COUNT CORRECT (Y/N): YES
4    INSTRUMENT COUNT CORRECT (Y/N): NOT APPLICABLE
5    SPONGE, SHARPS, & INST COUNTER: AUGUSTA,JEAN
6    COUNT VERIFIER:          O'CONNOR,JOYCE
7    SEQUENTIAL COMPRESSION DEVICE: NO
8    LASER TYPE:              N/A
9    NURSING CARE COMMENTS:   (WORD PROCESSING)(DATA)
10   PRINCIPAL PRE-OP DIAGNOSIS: CAD
11   PRIN DIAGNOSIS CODE:
12   PRINCIPAL PROCEDURE:     CABG X2 WITH CRYOVEIN TO LAD
13   PRINCIPAL PROCEDURE CODE: 33511
14   OTHER PROCEDURES:        (MULTIPLE)(DATA)
15   INDICATIONS FOR OPERATIONS: (WORD PROCESSING)(DATA)

** OPERATION **   CASE #41282                   PAGE 3 OF 3

1    BRIEF CLIN HISTORY:      (WORD PROCESSING)
2    OPERATIVE FINDINGS:      (WORD PROCESSING)
```

FIGURE 17.2. Screens of operation information in the surgery module.

244   Clinical and Support Applications

```
** ANESTHESIA INFO **    CASE #41282                    PAGE 1 OF 2
1    ANESTHESIOLOGIST SUPVR: POLANZAK,LEE
2    ANES SUPERVISE CODE:    2. STAFF ASSISTED BY RESIDENT OR C.R.N.A
3    PRINC ANESTHETIST:      POLANZAK,LEE
4    RELIEF ANESTHETIST:     ROSENKAIMER,SIGURD
5    ASST ANESTHETIST:       MACNEILL,J
6    ANES CARE START TIME:   APR 11, 1996 AT 07:30
7    INDUCTION COMPLETE:     APR 11, 1996 AT 07:55
8    ANES CARE END TIME:     APR 11, 1996 AT 16:40
9    ASA CLASS:              4E-LIFE THREAT-EMERG.
10   BLOOD LOSS (ML):        1000
11   MIN INTRAOP TEMPERATURE (C): 34
12   FINAL ANESTHESIA TEMP (C): 36.1
13   TOTAL URINE OUTPUT (ML): 350
14   OP DISPOSITION:         SICU
15   POSTOP ANES NOTE:
```

```
** ANESTHESIA INFO **    CASE #41282                    PAGE 2 OF 2
1    ORAL-PHARYNGEAL SCORE: CLASS 3
2    MANDIBULAR SPACE:       90
3    REPLACEMENT FLUID TYPE: (MULTIPLE)(DATA)
4    MEDICATIONS:            (MULTIPLE)
5    MONITORS:               (MULTIPLE)
6    GENERAL COMMENTS:       (WORD PROCESSING)(DATA)
7    THERMAL UNIT:           (MULTIPLE)
8    ANESTHESIA TECHNIQUE:   (MULTIPLE)(DATA)
```

```
** POST OPERATION **    CASE #41282                     PAGE 1 OF 2
1    DRESSING:               4X8'S, PAPER TAPE AND ACES
2    PACKING:
3    TUBES AND DRAINS:       #16 FOLEY, #36 & 2 #32 R/A CHEST,
4    BLOOD LOSS (ML):        1000
5    TOTAL URINE OUTPUT (ML): 350
6    GASTRIC OUTPUT:
7    WOUND CLASSIFICATION:   CLEAN
8    WOUND POINT CLASSIFICATION: 2
9    POSTOP MOOD:            RELAXED
10   POSTOP CONSCIOUS:       ANESTHETIZED
11   POSTOP SKIN INTEG:      SUTURED INCISION
12   TIME OPERATION ENDS:    APR 11, 1996 AT 16:00
13   ANES CARE END TIME:     APR 11, 1996 AT 16:40
14   TIME PAT OUT OR:        APR 11, 1996 AT 16:20
15   OP DISPOSITION:         SICU
```

```
** POST OPERATION **    CASE #41282                     PAGE 2 OF 2
1    DISCHARGED VIA:         ICU BED
2    PRINCIPAL POST-OP DIAG: CAD
3    PRIN DIAGNOSIS CODE:
4    OTHER POSTOP DIAGS:     (MULTIPLE)
5    PRINCIPAL PROCEDURE:    CABG X2 WITH CRYOVEIN TO LAD
6    PRINCIPAL PROCEDURE CODE: 33511
7    OTHER PROCEDURES:       (MULTIPLE)(DATA)
8    ATTENDING CODE:         1. ATTENDING IN O.R.
```

FIGURE 17.3. Anesthesia and postoperative information screens in the surgery module.

the information contained on top of the Surgeon's Operative Report, and form the basis of the Operating Room Nurse's Report. Postoperatively, information from the recovery room is entered into the Surgery Module. Nonoperating room procedures performed in-hospital postoperatively are also captured. Morbidities and mortality encountered during the first 30 postoperative days are captured as part of the Risk Assessment Module (see below).

DHCP Output and Management Reports

Aside from the surgeon's dictation of the operative record, the nurse's intraoperative report, and the anesthesiologist's report, the Surgery Module generates more than 30 customizable reports that provide a cumulative assessment of the productivity and activities of the Surgical Service. It also generates a quarterly report of the productivity and outcomes of the Surgical Service that is automatically transmitted to the Director of Surgery's office in Washington. The generation of these reports, which are listed in Table 17.2, is restricted to authorized personnel and is determined by the user's DHCP access code.

The Chief of Surgery's access code allows him/her full access to this report-generation capability, including a special set of reports that were designed exclusively for the use of the Chief. An examination of the reports listed in Table 17.2 should impart to the reader the value of this automated data acquisition system in monitoring and improving the administrative management of the Surgical Service and the quality of care of the surgical patient. For example, the OR Utilization Report prints utilization information for a selected date range for all operating rooms or for a single operating room. The report displays the percent utilization, the number of cases, the total operation time and the time worked outside normal hours for each operating rooms individually and all operating rooms collectively. The percent utilization is derived by dividing the total operation time for all operations (including total time patients were in the operating room, plus the cleanup time allowed for each case, plus 1-hour allowance for start-up and shutdown daily for each OR that had at least one case) by the total functioning time as defined in the site-specific Surgery Utilization File. Another example, the Report of Delayed Operations, will display all cases that have been delayed within a specified date range, sorting by the surgical subspecialty and including both the delay cause and the delay time. The report will also list, in a cumulative manner, reasons for delays and the number of postoperative occurrences (morbidity and mortality) for delayed operations.

Knowledge of and access to FileMan, the Massachusetts General Hospital Utility Multi-Programming System (MUMPS) database utility used in DHCP, allows the local facilities to develop their own specific reports to

TABLE 17.2. Management reports generated from within the surgery module in DHCP.

Management reports (for selected ranges of dates)

Activity Reports
  Schedule of Operations
  Report of Surgical Procedures (by CPT codes; cumulative)
  List of Operations (Text and information; chronological; by Surgical Specialty or Priority)
  Report of Surgical Priorities
  List of Undictated Operations
  Report of Daily Operating Room Activity

CPT Code Reports
  Cumulative Report of CPT Codes
  Report of CPT Coding Accuracy
  List Cases Missing CPT Codes

Staffing Reports
  Attending Surgeon Reports
  Surgeon Staffing Report
  Surgical Nurse Staffing Report
  Scrub Nurse Staffing Report
  Circulating Nurse Staffing Report

Anesthesia Reports
  Anesthesia AMIS
  List of Anesthetic Procedures
  Anesthesia Provider Report
  Principal Anesthetist Report
  Anesthesiologist Supervisor Report

Chief of Surgery Management Reports
  Morbidity & Mortality Reports
  M&M Verification Report
  Comparison of Pre- and Post-op Diagnosis
  Delay and Cancellation Reports
  List of Unverified Surgery Cases
  Report of Returns to Surgery
  Report of Daily Operating Room Activity
  List of Undictated Operations
  Report of Cases Without Specimens
  Report of Unscheduled ICU Admissions
  Operating Room Utilization Report
  Wound Classification Report
  Quarterly Report—Surgical Specialty

monitor the activities of the Surgical Service, and to improve efficiency and productivity. For example, because the Surgery Module keeps a running average of the operation time for each CPT code in each facility, a report was generated in one facility that compared the operation time of each procedure to the average time for its CPT code, flagging those procedures in which the operation time exceeded twice the average time. As part of the quality management process at that facility, the flagged operations were reviewed and specific actions taken accordingly.

The Health Summary Module in DHCP is also a powerful tool for generating patient reports encompassing data from the Surgery Module as well as data from all other modules within DHCP. In one facility, a special Preadmission Testing Health Summary was designed that generated a report detailed enough to allow for the elimination of the paper record at the Surgery Preadmission Testing Clinic. As the new Decision Support System (DSS) software is integrated into DHCP at various VA medical centers, it will provide an additional tool for generating useful managerial reports, particularly in relation to the cost accounting of surgical procedures.

Risk Assessment and Monitoring of the Quality of Surgical Care

As discussed above, using a combination of standard reports generated by the surgery module, specific facility-related reports generated by the use of Fileman, the Health Summary Module, and DSS, numerous VA medical centers have been able to monitor and improve the quality of their surgical care.

However, one of the most innovative approaches to the global assessment of the quality of surgical care in the VA has been the institution, in 1991, of the National VA Surgical Risk Study and its ultimate evolution into the National Surgical Quality Improvement Program (Khuri, Daley, Henderson et al., 1995; Khuri, Daley, Henderson et al., submitted; Daley, Khuri, & Henderson, submitted). Neither the study nor the program would have been possible without the development of a specific software package within the Surgery Module, which was entitled the Risk Assessment Module. This module provides medical centers with a mechanism to track information relating to surgical risk and 30-day outcome (morbidity, mortality, and length of stay). At each of VA's 128 medical centers performing surgery, a Surgical Clinical Nurse Reviewer, under the direction of the Chief of Surgery, collects risk and outcome on all patients undergoing major surgery requiring general, spinal, or epidural anesthesia, and enters them into the Risk Assessment Module. Completed noncardiac assessments are electronically transmitted to the Hines Center for Cooperative Studies in Health Services (CCSHS), while cardiac assessments are transmitted to the Denver Cardiac Coordinating Center for data analysis (Figure 17.4).

Using state-of-the-art statistical methodology, these two coordinating centers generate risk-adjusted morbidity and mortality rates for each of the hospitals performing major surgery in the VA. Risk-adjusted outcomes for each hospital and for each of the major surgical subspecialties in it are expressed as observed/expected (O/E) ratios. The O/E ratio is a risk-adjustment tool that allows for the assessment of interhospital variation in outcomes and that identifies those institutions with morbidity and mortality rates that are significantly above or below the rates expected on the basis of the patients' preoperative risk. An O/E ratio at a specific institution that is

248   Clinical and Support Applications

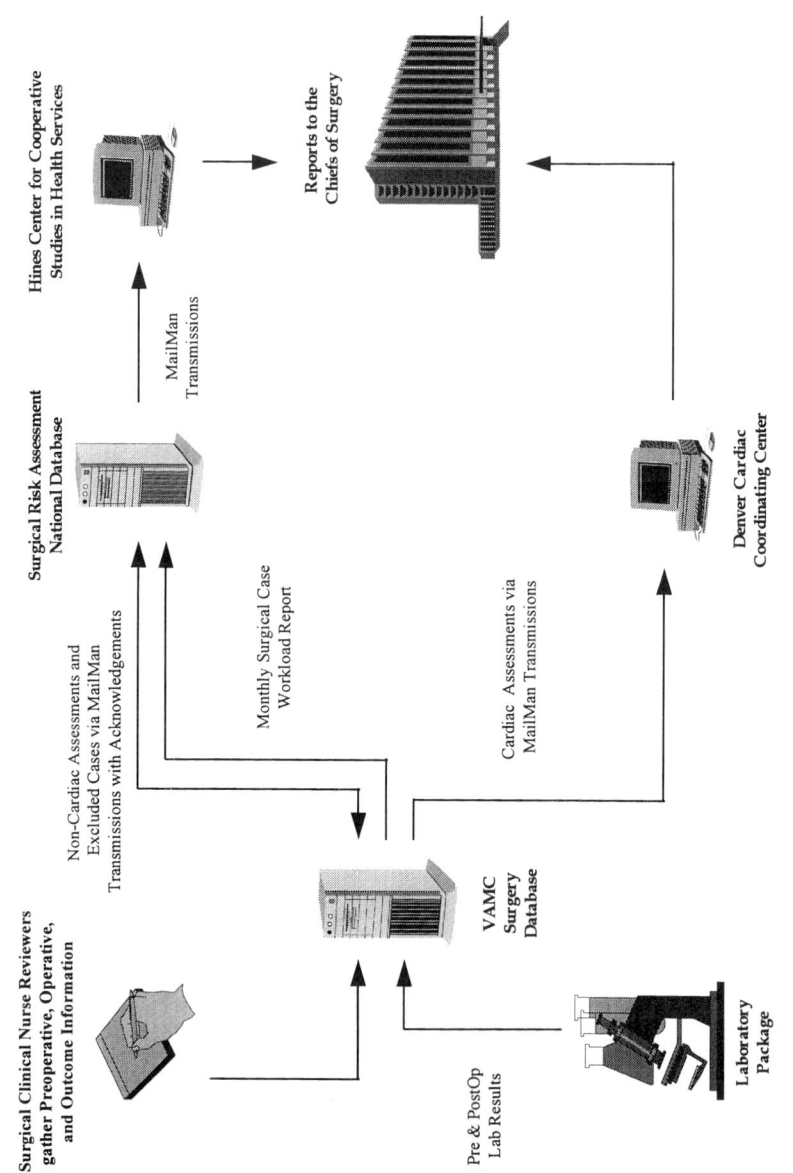

FIGURE 17.4. Data flow in the surgery risk assessment module.

significantly higher than 1.0 indicates a mortality rate higher than that accounted for by the severity of the patient population's preoperative risk and suggests suboptimal processes and structures of care. A low O/E ratio, on the other hand, indicates superior quality of care (Khuri, Daley, Henderson et al., submitted; Daley, Khuri, Henderson et al., submitted; Hammermeister, Johnson, Marshall, & Grover, 1994). Managerial reports are prepared at the coordinating centers to provide all the Chiefs of Surgery in the VA with their own risk-adjusted data, compared to the VA national averages.

The following are key features of the Risk Assessment Module:

- Provides for entry of noncardiac surgery assessment information including preoperative information, laboratory test results, operation information, and intraoperative and postoperative occurrences.
- Provides for entry of cardiac surgery assessment information including clinical information, cardiac catheterization and angiographic data, operative risk summary data, cardiac procedures requiring cardiopulmonary bypass, and intraoperative and postoperative occurrences.
- Creates a Surgery Risk Assessment report on each patient assessed.
- Transmits completed Surgery Risk Assessments to Hines and Denver.
- Lists Surgery Risk Assessments by categories including complete, incomplete, and transmitted assessments, as well as lists of major surgical cases and all surgical cases.
- Generates a monthly Surgical Case Workload Report.
- Prints follow-up letters to patients 30 days after a procedure.

The Risk Assessment Module has enabled the VA to provide the surgical community, for the first time, with risk-adjustment models that allow for the comparative assessment of the quality of surgical care among various surgical facilities. It also made it possible for the VA to take the lead in using risk-adjusted outcomes as performance measures.

Toward the Complete Electronic Medical Record

In its new Strategic Information System Plan (SISP), Veterans Health Administration (VHA) has identified the pursuit of an automated electronic record, which will totally replace the patient's paper record, as one of its priority goals. The Surgery Module, and all the other clinical modules and functionalities currently operative within DHCP, provide enough clinical information about the surgical patient to make the achievement of this goal possible within a relatively short period of time. At this day and age, however, no stand-alone informatics system can provide the totality of automated information necessary for the institution of a complete paperless record for the surgical patient. The DHCP, as rich a repository of clinical information as it is, lacks vital information that is normally captured by

250   Clinical and Support Applications

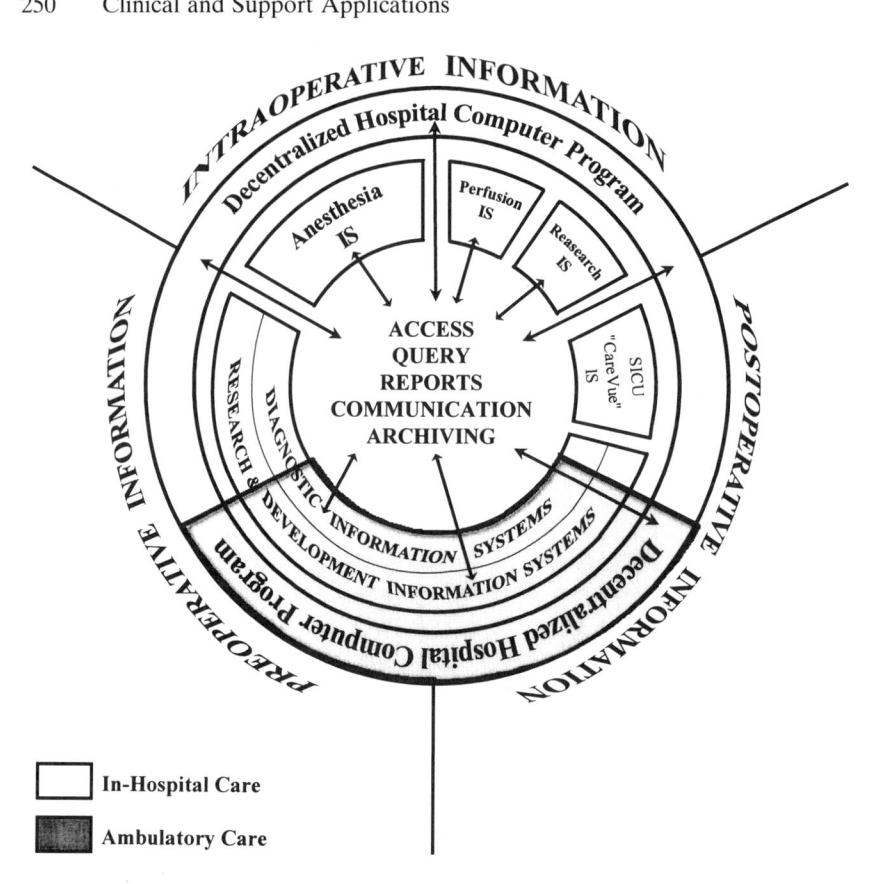

FIGURE 17.5. Conceptual framework for the integration of disparate databases that contain clinical information about the surgical patient. Integration of clinical information residing inside and outside DHCP, within a framework that allows for access, query, report-generation, communication and archiving, is imperative for the realization of a totally automated surgical patient record. IS = Information System; SICU = Surgical Intensive Care Unit.

other automated information systems, such as the commercial systems that capture online hemodynamic and other digitally acquired data.

As depicted in the conceptual schema in Figure 17.5, data acquisition systems other than DHCP provide information on the surgical patient through the pre-, intra-, and postoperative course, in and out of the hospital. Anesthesia information systems are now available that automatically capture the data from the various monitors, anesthesia machines, and infusion pumps, and store them digitally allowing the creation of a dynamic, fully automated, anesthesia record. In open-heart surgery, important information related to the conduct of extracorporeal perfusion is also captured and archived digitally. Numerous systems have become available that totally automate the postoperative intensive care data acquisition and archiving, most notably the Careview Clinical Information System developed by the Hewlett Packard company. The DHCP has developed the functionality of capturing digitized images, but there are still a number of diagnostic digital imaging systems that contain vital patient information that reside outside DHCP. Large-scale clinical research databases also contain important patient information not normally captured by DHCP.

To enhance the functionality of DHCP with an eye toward the implementation of the paperless record, VHA has established mechanisms for integrating clinical information obtained through commercially available information systems with the various clinical modules in DHCP. One such mechanism is the Hybrid Open System Technology (HOST) initiative, which encourages and funds such integration efforts at a number of VA medical centers nationwide. Two ongoing HOST initiatives are of particular relevance to surgery, and their fate will be an important determinant of whether the totally paperless surgical patient record will become a reality. The first is the integration of two commercial anesthesia information systems with the Surgery Module in DHCP. This effort is currently ongoing at VA medical centers in New York City and San Diego. The second HOST initiative is the application of third-party integration tools to DHCP, allowing the generic integration with DHCP of all patient information residing in any disparate database within the hospital environment. As a proof of concept of this initiative, Hewlett Packard's Careview System and the Nuclear Medicine computer information system at the West Roxbury VA Medical Center will be integrated with DHCP in a DEC Pathworks network using a Windows NT workstation. As depicted in Figure 17.5, what is needed for the totally electronic surgical record, is an information system with a user-friendly graphic interface that allows for (1) the simultaneous and transparent access and manipulation of patient data residing inside and outside DHCP; (2) the ability to conduct simultaneous online queries and to generate cumulative reports on all automated patient data, irrespective of the type of data repository; (3) digital communication of patient data to other providers, facilities, and national VA patient databases; (4) automatic archiving of selected data that are deemed to be important for the permanent electronic record.

Surgical systems in the VA are well on their way toward achieving these promising functionalities.

Acknowledgments. The author acknowledges the help, dedication, and support of the numerous individuals who, over the years, have worked hard on developing, testing, and implementing the Surgery Module and the Risk Assessment Module in DHCP; in particular, the help and support of the ISC staff in Birmingham under the leadership of Roy Swatzell, Mike

Montalli, Alan Monowsky, and Steve Musgrove. The help of Lynne Santangelo and Craig Miller at the West Roxbury VA Medical Center is also acknowledged.

References

Daley J, Khuri SF, Henderson W, et al. (submitted). Comparative assessment of thirty day morbidity following major surgery: Results of the national Veterans Affairs surgical risk study. *New England Journal of Medicine.*

Hammermeister KE, Johnson R, Marshall G, & Grover FL. 1994. Continuous assessment and improvement in quality of care; a model from the department of Veterans Affairs cardiac surgery. *Annals of Surgery, 219,* 281–290.

Khuri SF, Daley J, Henderson W, et al. 1995. The national Veterans Affairs surgical risk study: Risk adjustment for the comparative assessment of the quality of surgical care. *Journal of the American College of Surgery, 180,* 519–609.

Khuri SF, Daley J, Henderson W, et al. (submitted). Comparative assessment of thirty day mortality following major surgery: Results of the national Veterans Affairs surgical risk study. *New England Journal of Medicine.*

18
Nursing Use of Systems

BOBBIE D. VANCE, JOAN GILLERAN-STROM, MARGARET ROSS KRAFT, BARBARA LANG, AND MARY E. MEAD

Introduction

Above all else, a computer system designed for nursing must support the delivery of patient care. Doing so requires that the system address clinical, administrative, education, research, and quality improvement needs in nursing. In today's networked environment, a nursing system must also interface with other healthcare systems that are relevant to nursing.

Automated nursing information systems have been recommended as labor-saving devices that conserve nursing time for direct care and increase the efficiency of management practices. Generally, efficiencies have been realized by nursing through automating patient acuity tracking, budgeting, and productivity monitoring. When standards become an integral aspect of the nursing information system, more effective evaluations can be conducted on the quality of care delivered.

Whatever the benefits, nurses have not consistently responded positively to the introduction of computers in the nursing environment. The attitudes of nurses toward computers in the work setting have been extensively researched (Murphy, Maynard, & Morgan, 1994; Newton, 1995; Stockton & Verbey, 1995). Nonetheless, the proliferation of computers in health care has challenged nurses to increase their computer knowledge and skills to maintain their effectiveness.

The application of computers to nursing can be described as the combining of computer and information sciences with nursing science (Walker & Walker, 1995). These linkages facilitate the processing and management of nursing data for the purposes of patient-care delivery and administrative decision making.

In developing their nursing information system, the Veterans Health Administration (VHA) adopted a model that stressed involving nurses to define strategies that yielded the greatest benefits to patients and nurses alike. Nurses collaborated with computer programmers to write all the specifications for the nursing software. Nurses with expertise in clinical, administrative, educational, and research practice areas joined to form a special interest users group (SIUG) and assumed accountability for nursing software design.

As development proceeded, the SIUG addressed issues associated with the nationwide implementation of nursing software as part of the Veterans Health Administration (VHA's) Decentralized Hospital Computer Program (DHCP). The SIUG acknowledged the critical role that training must play in the introduction and use of nursing information systems. This emphasis on training and the level of user involvement at the development stage have been credited with the high degree of user acceptance the nursing software has enjoyed.

The nursing system replaced pen and paper (i.e., manual systems) with computerized systems for major aspects of nursing administration and clinical documentation. The use of DHCP diminished inefficiencies and redundancies that erode patient care, and increased the accuracy and timeliness of nursing administrative reports. Information collection, processing, retrieval, and reporting have been enhanced. Computer outputs have been tailored to reflect VHA reporting requirements for local, regional, and national reports.

Nursing System Modules

The nursing software package is one of many programs included in DHCP. The modular design of the nursing software package allows for the computerization of clinical, administrative, research, education, and quality-improvement data. Designed to complement and supplement other DHCP packages, the nursing software package interfaces with the entire integrated hospital information system that is DHCP.

The original nursing SIUG, later called an expert panel (EP), agreed that the nursing software framework must be holistic, include bio-psycho-social elements, and use a taxonomic structure capable of addressing functional patterns, nursing diagnoses, and body systems. The SIUG further agreed that the nursing system should support nurses as they work to deliver care and to keep current with initiatives in their profession.

As a result, the data collection system was designed to include elements of the Nursing Minimum Data Set (NMDS) (Werley & Lang, 1988). In addition, specifications were developed using the North American Nursing Diagnosis Association (NANDA) diagnostic taxonomy. The resulting design for the nursing software package within DHCP included provisions to:

- plan and document nursing care provided to veteran patients;
- support nursing administration in the management of nursing service;
- provide nursing quality management data to managers, clinicians, educators, and researchers
    to aggregate data for nursing research and
    to support education efforts for staff and patients.

The VHA nurses practice in a variety of settings, including clinics, home health, and acute and long-term care. Depending on a clinician's area of practice, data may be required from a number of departments in a variety of formats: images, graphs, diagnostic results, dietary evaluations, medication regimens, and consults. The VHA nursing software can be used across all practice settings to support decision making and delivery of quality patient care. As end users, VHA nurses were interested in applications and output, not background hardware and software. Their goal was to develop nursing software that was economical and capable of supporting the broad scope of nursing practice.

Clinical Nursing Package

The philosophical basis for the clinical nursing software development was closely aligned with policy statement issued by the American Nursing Association (ANA), which defines nursing as "the diagnoses and treatment of human responses to actual or potential health problems" (ANA, 1980). Nursing diagnoses comprise the database for patients' health problems, while the care plan format reflects the nursing process. In other words, design of the module was guided by the profession's definition of nursing and the methods of processing information to guide the provision of direct patient care. Having information available in formats that increase usability can indeed enhance the clinical decision-making process.

Nursing Care Planning

The first clinical nursing software application developed was to support and document nursing care planning. Known as the patient-care plan, this application allows for care planning by medical diagnoses or by all approved NANDA nursing diagnoses. It can be individualized by the user according to the patient's needs. Features allow for the addition of user-specific or editing of specific care plans. The program has been described as quick and easy to use, because nurses creating the care plan need only select the most appropriate choices from a series of cascading screens. After identifying the patient, the nurse creates the plan by nursing diagnosis or medical diagnosis. Screens cascade through a taxonomic structure of functional patterns and resulting nursing diagnoses, or body systems, medical diagnoses, and resulting nursing diagnoses.

Once the nursing diagnoses are established, the user defines measurable goals and sets an evaluation date for each nursing diagnosis or patient problem as well as a target date for each goal. Once goals and dates are defined, the user selects the appropriate interventions. The individualized-care plan can then be printed for inclusion in the permanent paper medical record. Because each change made to the patient-care plan is documented, the patient-care plan contains a complete record of the patient in chronological order, by defined problem.

As the patient progresses during the course of treatment, the plan of care is easily edited, adding problems, goals, and/or interventions through the use of an option designed specifically for this purpose. The status of the patient's problem can remain active or be designated as resolved, suspended (due to severity of another problem), or unresolved at discharge.

Once released to the field, the care plan software was used by different sites in different ways. Some sites used the software exactly as released, producing individualized nursing care plans. Other sites used the customizing features of the software to build in their own site specific standards of practice, effectively giving life to standards, which previously all too often remained as reference documents in binders at the nursing station. Still other sites created care plans that included clinical diagnoses or procedures, such as stroke or aneurysm repair. These streamlined plans were found to be useful and timesaving; more than 70 of them were included in a subsequent release of the software (Figure 18.1).

As defined by the nursing user, patient-care problems serve to establish some basic elements in the patient's clinical database. Just as laboratory data elements can be retrieved in many ways, elements described in the patient-care plan can be put to a multiplicity of uses. One output of the care plan software is a patient assignment worksheet. Available in two different formats, it includes patient problems and interventions; it can also include information entered into the patient record via the Patient Information Management System (PIMS) and the allergy-tracking software package. This tool helps staff nurses familiarize themselves with patients assigned to them; it is also used for orientation of new staff and for training of student nurses (Figure 18.2).

As care planning evolved, the Joint Commission on the Accreditation of Healthcare Organizations (JCAHO) developed standards to encourage the creation of care plans that were multidisiplinary and/or interdisciplinary. Within the VHA, some sites were able to use customizing features available in the original care plan software to produce such plans, which have received favorable reviews by the JCAHO. Today the formal care planning process continues to grow and change.

Clinical Pathways

What began as a single discipline nursing-care plan is now serving as a conceptual foundation for creation of clinical pathway software. All over the country, nurses are working as case managers and serving on committees to create facility-specific clinical pathways and practice guidelines.

Work has begun within the VHA on computerizing this process. A subgroup of the expert panel (with membership representing nursing, medicine, social work, dietetics, and the chaplain service) is developing

PATIENT PLAN OF CARE - Patient Print

4-11-96
STARK, ANTHONY L. Diagnosis: Pancreatitis

Page 1

Date/Stat/Unit	Patient Problems	Goals/Expected Outcomes	Nursing Interventions/Orders	Date/Stat/Unit
4-4-96 8M 4-8-96 E	Discharge Planning (Health Maintenance, Alteration in)	Identifies S/S of infection Verbalizes minimal discomfort or absence of pain Verbalizes post discharge instructions from patient teaching	Assess deficits/capabilities to determine D/C needs including: activity restrictions / dietary restriction such as excess cholesterol / lipids / follow up care / medications to be taken Reinforce importance of adhering to treatment regime	4-9-96 T
4-4-96 8M 4-6-96 E	Fluid Volume Deficit	Maintains fluid / electrolyte balance Balanced I/O, urine output WNL / Moist mucous membranes / Skin turgor normal	Mouth care frequency / Assess, monitor & record: I&O, site & character frequency / skin turgor & oral mucous membranes frequency	4-9-96 T
4-4-96 8M 4-7-96 E	Nutrition, Alteration in : (Less than required)	Maintains nutritional intake to meet metabolic requirements / Absence of negative nitrogen balance indicators / Stable weight	Resume / advance diet as indicated	4-9-96 T
4-4-96 8M 4-5-96	Pain, acute	Verbalizes effect of pain relief interventions	Assess pain (location, duration) q2-4H Administer pharmacological agents as ordered/per protocol; monitor and document effects / Teach pain control interventions	4-6-96 T

Diagnosis: Pancreatitis   Physician: HOGUE, MARILYN
Reactions: E-prob eval dt   U-prob unresolved at discharge   S-prob susp
            T-goal target dt   DC-order/goal discontinued   M-goal met
            @-entered in error   R-prob resolved/order reinstated
STARK,ANTHONY L. 589-63-2111   64SICU   PROT M

FIGURE 18.1. Patient plan of care.

specifications for clinical pathway software that will support the development and use of clinical pathways. When completed, the exported software will not contain the actual clinical content for a pathway, as was done with the nursing-care plan software. Instead, the software will enable a site to create its own pathways to meet its unique needs and community standards of practice.

```
                    PATIENT CARE ASSIGNMENT/WORKSHEET
SEP 17, 1989                                                    PAGE 1
WARD: 3 AM
TOUR: _____
STAFF:_____
```

```
RM/BED: 301/1      BEDSECTION: MED       ADM: AUG 16, 1989 @ 09:00
NAME: PARKER, PETER P.      SSN: 931-84-8807    PHYSICIAN: STEELE, V.
CATEGORY: 1        FACTORS: (279)
ADMITTING DIAGNOSIS: COPD
REACTIONS:
PATIENT PROBLEMS                        NURSING INTERVENTIONS
Infection Potential (Specific to Respiratory System)   I&O q 24 HRS
Injury Potential                        assess effects of bronchodiliators
Mobility, Impaired Physical             assess knowledge base
Self-Care Deficit Specify               out of bed q DAILY STARTING
Skin Integrity, Impairment of (Potential)   8-17-89
                                        assess for signs of fatigue and
                                        weakness
                                        determine ability to learn &
                                        implement plan
                                        incentive spirometer q 2 HR
                                        oxygen numeric value% per flow rate
                                        teach prevention of infection
```

ADL	SAFETY	BATH	DIET	BP	TPR	WT	TREAT MENTS	I/O	OTHER

FIGURE 18.2. Sample assignment sheet.

The software will allow customization of a patient pathway, including selection of patient-specific outcomes with associated timeframes to complete and automatically generate necessary orders and treatments. Closely interfaced with other DHCP packages (e.g., physician order entry, patient problem list, treatment record, progress notes), the software will take full advantage of work now underway to develop clinical decision support. Specifications call for the software to run in inpatient and outpatient settings, making it appropriate for use in either ambulatory or primary care.

Each facility will be able to define its own paths and timelines for achievement of outcomes. Variance tracking will be defined as relatively complex or simple by each site, and can be tracked as critical or not critical. Critical variances are those that result in a delay in discharge or cause a significant delay in patient progress. In a clinical pathway, critical variances will generate a patient progress note. Both forms of variance may be tracked and retrieved in aggregate or individual data for process analysis system improvement and patient documentation.

The software will allow users to merge more than one pathway and customize the resulting pathway according to their prescriptions. Clinicians will be able to define the anticipated length of time for completion of the pathway. Shown average timeframes for outcomes as default values, the user will be able to edit these values to create appropriate and achievable goals for individual patients.

Reports generated by the software will include standard facility approved pathways for given diseases or patient problems. These reports will be available to clinicians online or in printed form to help in the delivery of patient care. A dietitian or social worker, for example, will be able to search all pathways to identify quickly who needs what type of care.

The clinical pathway software will be heavily integrated with DHCP programs now in place or under development. Definition of a clinical pathway will automatically generate necessary physician orders. Interventions or orders classified as treatments will automatically appear on treatment records; medications will automatically appear on the medication administration record; assessments made during the course of treatment will automatically be entered into the assessment database.

Vitals/Measurements

A second project assigned to the nursing SIUG was the development of software to record vital signs. This exercise was relatively straightforward. Vital signs were initially limited to those routinely taken by nursing personnel, that is, temperature, pulse, respiration, blood pressure, height, and weight. The software consisted of a number of files. Some contained the actual vitals data; others established which physiological parameters would be tracked, provided qualitative descriptors of these parameters, and defined abnormal values.

Subsequent versions of the software have expanded parameters to include different types of measurements, including tympanic temperature, hemodynamic monitoring values, ventilator modes, and postural positions. Today an individual site can add parameters to the files and enter descriptors for these parameters.

Given that some of these parameters were frequently measured by non-nursing staff, such as dietitians and respiratory therapists, the vitals/measurements package became the first truly interdisciplinary package for clinical use. It was decided that this package should become separate from the nursing software, and as such it was exported to a field in a separate General Medical Record (GMR) name space. Other patient-oriented computer packages that are nondiscipline specific, such as the Health Summary and Progress Notes, are also stored in this GMR name space.

Vital data entry is simple. Once the user establishes the date and time of the measurement and parameters measured, data may be entered for a single patient, an entire patient unit, or a group of patients, based on their room and bed number. It is possible to enter data by single parameter or by commonly grouped parameters, such as temperature, pulse, and respira-

260   Clinical and Support Applications

tion. The actual data are entered in a string of characters, such as 100.2R-60A-34C. In this example, the temperature, 100.2°, is followed by an indicator that this is a rectal temperature. The "60 A" indicates apical pulse rate of 60 beats per minute, and the "34 C" indicates this is the respiratory rate controlled by a ventilator. If no qualifying letters are entered, the system assumes an oral temperature, a radial pulse, and a spontaneous respiration. After data entry, the data are displayed to the user for confir-

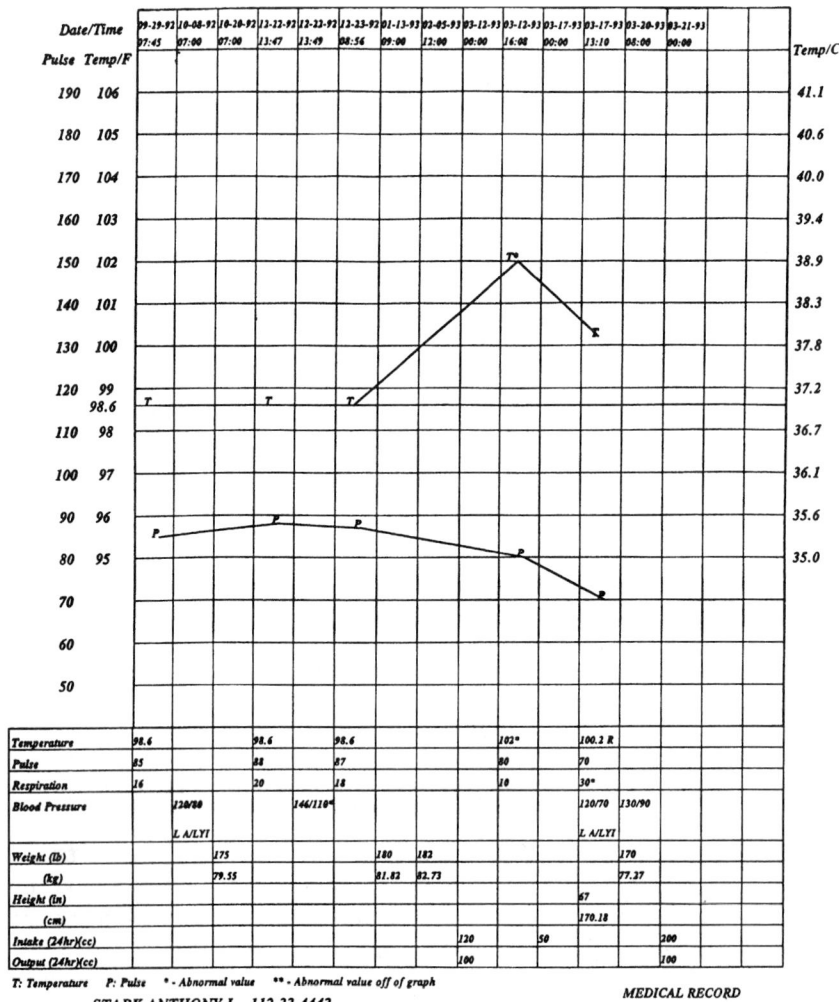

FIGURE 18.3. Vitals flowsheet.

mation before actually being stored. Conversion features can display data entered in fahrenheit and centigrade, centimeters and inches, and pounds and kilograms.

Vitals/measurement data are available for retrieval in various formats (Figure 18.3). These data may be retrieved by patient or location, and by tasks or cumulative measurements. They can be displayed in a list or graph format, on a terminal screen or paper copy. Hard copies generated are intended for inclusion in the permanent paper medical record. There are also package interfaces that greatly enhance use of this software. The vitals/measurement package interfaces with the health summary package and the unit dose package. A popular example of this data integration is the incorporation of vitals measurement data with dietetics, PIMS, and allergy tracking in the end-of-shift report (Figure 18.4).

Changes have been made in the software in response to feedback from users in the field. For example, dietitians requested that the combination of height and weight be available; this was added to a later release of the package.

Intake and Output

The Intake and Output (I&O) Module is designed to store all patient I&O information associated with an episode of care. This application is not service specific and currently interfaces with PIMS, nursing, and pharmacy applications. Its basic features allow for documentation of patient intake (oral fluids, tube feedings, intravenous fluids (IF), irrigations, and other types of intake as defined by the facility) and patient output (urine, nasogastric secretions, emesis, drainage, liquid feces, and other types of output as defined by the facility). The quick route documents totals, while the detailed option requests the entry of specific types of fluid, as well as quantity.

A major component of the I&O application is the intravenous (IV) therapy function, which contains seven protocols:

Start IV;
Solution: replace/discontinue/convert;
Replace same solution;
Discontinue line;
Site care and maintenance;
Adding additional solutions;
Restarting discontinued IV.

The IV Module interfaces with the IF Module of the pharmacy software where nurses can select the appropriate ordered IV medications and admixtures without additional typing. Once the IV function is implemented, the clinician may print needed I&O data. These data are available in several forms, such as summary information for each type of category of I&O by

```
SEP 29,1995@08:43   END-OF-SHIFT REPORT WARD: MICU                       TOUR: 7:00-15:00            PAGE:1
ROOM-BED/NAME (SSN)/         |ADMITTING      |PT |LATEST VITALS
DIET/ALLERGIES               |DIAGNOSIS/DATE |CAT|TPR/BP/WT             |PATIENT PROBLEMS
-----------------------------|---------------|---|----------------------|------------------------------------
500-3                        |MALNUTRITION   | 3 |99.8(ORAL)(37.7 C)    |Coping, Ineffective Family
STARK,ANTHONY L.  (4443)     |01/03/93@10    |   |80(RADIAL)            |Health Maintenance, Alteration in
Tube Feeding:                |69 yrs. MALE   |   |10                    |@08:00 DEXTROSE 5% admix 1000 mls started Site: LEFT LOWER
  SUSTACAL, Full             |               |   |200(ACTUAL)(90.91 Kg) |   ARM  IV cath: NEEDLE 1 1/4 IN-16  Rate: 120 ml/hr
  strength, 2000             |               |   |                      |         Discontinued on MAY 27,1993@13:27 Reason: ORDER CHANGED
  KCAL/per day               |               |   |                      |@08:00 LEFT LOWER ARM: NO REDNESS/PAIN/SWELLING, tubing
Allergy: PENICILLIN G        |               |   |                      |   changed, dressing changed
  200,000U TAB, SULFA        |               |   |                      |@08:30 FAT EMULSION 10% piggy  500 mls added  Site: LEFT
  DRUGS                      |               |   |                      |   LOWER ARM
Adv rxn: ACETAMINOPHEN       |               |   |                      |         Discontinued on MAY 27,1993@10:00 Reason: INFUSED
  SUPPOSITORY 650MG          |               |   |                      |@08:30 LEFT LOWER ARM: tubing changed
-----------------------------|---------------|---|----------------------|------------------------------------
```

FIGURE 18.4. Sample end-of-shift report.

patient, day, and shift. If a nurse, dietitian, or physician needs to review detailed information about a patient's I&O, the 24-hour itemized shift report is available.

An IV flowsheet provides detailed information about therapy, including start and discontinuation dates/times, solutions, lines, ports, needles, site condition, and dressing changes. This information can also be displayed on the nursing end of shift report. Nurses have acknowledged the utility of the IV flowsheets, and IV therapy teams utilizing this software have given positive feedback.

Patient Assessment

The driving force behind patient assessment software, nurses joined with other clinicians, including physicians, respiratory therapists, social workers, and dietitians, to develop the application. The task group also drew representatives from several DHCP expert panels (e.g., ambulatory care).

Current components of the patient assessment package include the following: vital signs, activity, hygiene, nutrition, IV monitoring, I&O, treatments, patient education, neurological, psychological/behavioral, circulatory/cardiac, respiratory, gastrointestinal, genitourinary, musculoskeletal, pain, skin, dressings, irrigations, tubes/drains, sleep–wake cycle, and patient problems.

Functionally, the software documents the initial patient assessment, daily shift assessments, focused assessments, and transfer and discharge assessments. Screens provide a logical path of progression through the assessment process. At any time, the user may break out of an established pathway and select any component of the assessment.

The newest version of the patient assessment application offers a graphical user interface (GUI), developed by programmers using tools provided by Borland's Delphi software.

Treatment Record

The patient's treatment record documents ward- and clinic-based therapies or treatments associated with the patient assessment. Upon viewing a list of clinical services that provide treatments, the user can call up a list of all treatments associated with a specific service on a selected patient, documenting the date and time specific treatments were done. Predetermined patient assessment screens appear in an ordering scheme until the clinician finishes the additional documentation associated with the treatment. For point-and-click selections, the software provides a listing of treatments offered by the various clinical services.

The latest software is designed to link the Assessment Module with clinical pathways and the treatment record and with all other DHCP applications. Constructed with a seamless interface, the software supports the delivery of care and generates a component of the electronic patient record.

Nursing Administration Software

The computer has become a decision support tool for the nurse executive and management staff. As administrative modules within the VHA nursing package have come online, nursing service administrators and nurse managers have ready access to an array of information: personnel database, payroll and other fiscal data, patient census, acuity data, education, research, quality improvement, staffing, and productivity reports.

These administrative nursing modules create databases that provide valuable information to support managment decisions and facilitate more effective use of resources. Standardization of these modules across the national VA system allows comparisons of cost-effectiveness between VHA nursing services. Administrative reports are available on a realtime basis, allowing the unit nurse manager to access unit data, clinical chiefs to access section data, and top-level nurse administrators to access total service data.

Personnel Database

The nursing service personnel database contains employee demographics, educational preparation, work experience, licensure, certification, credentialing, and payroll information. The organization of this database allows for computerized searches to identify nurses with specific educational and/or experiential qualifications, certification status reports, advanced practice credentials, and evaluation due dates. Reports can present either individual or aggregate data. A position control function in the database allows tracking of positions (past, current, and future) and vacancies (current and projected).

The DHCP nursing software also contains an interface that can link the personnel database with commercial staff schedulers for those facilities that have automated the scheduling process. Various VA medical center nursing services have been involved in testing and using a variety of commercial schedulers, such as ANSOS, RES-Q, M-DEX, InCharge, ACUSTAF, PROMEDEX, and Scheduling Sculptor. No one commercial scheduler has been identified as meeting all scheduling needs or fulfilling all VHA nursing expectations for such a package.

An employee may review his/her own database on a read-only basis. This capability has contributed to employee acceptance of the system and increased confidence in the accuracy of the information. Editing of the personnel data is restricted, and the privacy of employee information is protected by defined levels of access security.

Patient Classification

The patient classification system was the first nursing software released. This software allows online access to the census/acuity/staffing data on each unit on a shift-by-shift basis. Patients are classified at least once every 24 hours and when significant acuity changes occur. The classification system is used to support staffing guidelines and resource allocation within and between units on a shift-by-shift basis. The classification system distinguishes medical, surgical, intensive-care, intermediate-care, nursing home, psychiatry, substance abuse, and spinal cord-injured patients. Review of longitudinal classification reports allows identification of acuity trending by unit and clinical sections for budgetary purposes. Each VHA nursing service maintains interrater reliability for the classification system.

Manhours

The Manhour Module of the nursing package records actual on-duty staffing hours by staff level and distinguishes between the direct- and indirect-care hours of nursing staff. Indirect-care hours are further broken down to identify hours for orientation, education, and administrative duties. This database allows clear identification of available direct-care hours and scheduled hours on a tour-by-tour basis. Data can be entered prospectively to allow review of projected staffing/workload/productivity. Daily staffing changes may be entered to keep the manhour data current.

Workload Statistics

The manhour and patient classification package data can be accessed to produce the daily workload report. This report compares the staffing figures recommended within the classification system with the actual on-duty staffing by level of staff and generates productivity reports available by tour of duty, 24-hour period, and week, month, or quarter. The capability to generate productivity reports is an option widely used by nurse managers and they have an increased understanding of the relationship between productivity and resource utilization. These computerized reports have also proved a useful tool for review of nurse staffing during JCAHO survey visits.

Fiscal Information

Computerization of the timekeeping and payroll processes online data has provided nursing administration with online realtime cost information for hours worked, earned time off, leave usage, overtime, holidays, and differentials for weekend and off-tour hours. This capability has been valuable to the nurse executive in budget management. Increased accuracy and efficiency have been noted in the timekeeping process. The integration of fiscal data with patient census/acuity data can produce reports on the cost/hours per patient day. All employees can access their personal annual leave and sick leave balances, review their timecards, and request leave through the computer. This feature has significantly reduced the volume of paper related to personnel management processes.

Recruitment

Software specifications have been developed for a recruitment/retention database that allows required application information to be put online either by the applicant or by support personnel. Progress through the recruitment process can be tracked and data of those candidates selected for hire can be transferred into the active nursing personnel database.

Communication

The widespread acceptance and use of e-mail has greatly improved communications throughout the VHA. Nurses make extensive use of the local e-mail system, Mailman, and the national e-mail system, Forum. The national system has, indeed, become a forum for information exchange. Currently more than 30,000 VHA users are registered in the national system. Groups working on particular tasks are able to access all members and readily receive input and review work in process. E-mail is the quickest way to reach the greatest number of facility employees on a timely basis. E-mail is also widely used by local, regional, and national workgroups to create and distribute minutes.

Some facilities have chosen to put nursing and hospital policies and procedures online, making information readily available to clinicians without maintaining policy and procedure books in individual units. Revisions are easily entered through a word processing application; a print option can generate needed hard copy of any policy or procedure.

All the modules that contribute to more accurate objective data contribute to better administrative practices and increased effectiveness and efficiencies in resource management.

Nursing Research

The computerization of nursing clinical, administrative, and educational data has resulted in the development of databases capable of supporting a variety of nursing research endeavors. Computerization of the nursing process in particular yields data for large numbers of patients and facilitates data retrieval and analysis. Reports can be generated that rank order both nursing diagnoses and nursing interventions as identified in the patient's plan of care. In sum, it is now possible to begin to study the correlation between nursing practice and patient outcomes (Kraft & Lang, 1994).

When an interdisciplinary clinical group developed the software specifications for the Problem List Module of DHCP, nursing persuaded clinical colleagues to support the inclusion of the NANDA, Omaha, and Saba taxonomies (Pohlmann, 1995). Future database studies will allow researchers to identify which diagnostic labels VHA nurses find most useful in inpatient and outpatient settings.

Nursing Education

Designed by nurse educators as a generic package to be used by all clinical and administrative services, the education tracking software provides online documentation of employee training. Requests can be entered and processed online; completion of training courses can be noted. Mandatory education requirements can be tracked and monitored for level of compliance. Options included in the program are

Type program or class name;
Supplier/presenter;
Type of media used in the presentation;
Location of the student/attendee;
Purpose of the training;
Category of program/class;
Type of training (mandatory, inservice, and continuing education);
Cost per student;
Accreditation/continuing education units.

Reports can be generated with information pertinent to the content and purpose of training, source of financing (employer or employee), and scheduling of training (on or off duty).

The program has been invaluable in tracking mandatory training to meet external review agency and national headquarters requirements. Reports can be generated for an entire medical center, a service or work unit, or an individual employee. For example, training required of all employees can be grouped as mandatory inservice (MI); training required of registered nurses can be grouped under RNs.

Employees are trained to use the system to register for training programs and to track their personal training records, which are stored in the fiscal files. For programs sponsored by the VHA, the software opens and closes registration for classes, limiting enrollment. Information on programs offered by a VA medical facility may be entered as continuing education (CE); a listing of noneditable options of classes and class purposes is provided, eliminating the need to type specific entries.

Quality Improvement

Nursing makes extensive use of DHCP software to support quality management and quality-improvement (QI) efforts. Computerized nursing process data can be used to support the continuous QI activities of a nursing service. The nursing software provides nurse users with an audit tool to assess the appropriateness and timeliness of care planning. The routine reads existing care plans by patient, room/bed assignment, or patient unit, and lists all patients and problems that need evaluation by date. This greatly speeds up the process of updating care plans and gives nurse managers an easy tracking mechanism to evaluate staff performance.

Complementing the nursing software, the DHCP provides a QI package with a variety of applications. These include clinical monitoring, external review, patient event tracking, and other applications which support QI in nursing.

The clinical monitoring system allows the development of user-designed monitors to capture patient data. The external review management information system tracks reports and recommendations, compiles reports, and provides service-level worksheets for each recommended action plan. The QI checklist automatically extracts information from DHCP to identify patient-care and management issues that need to be addressed on a local, regional, or national level. The Patient Incident Tracking Module initiates reports online, providing the hospital risk manager an immediate record of the incident. This realtime system shortens the time required to investigate incidents and targets locations/areas needing attention.

The Occurrence Screen Module identifies events requiring follow up in compliance with VHA policies (i.e., readmission within 20 days, return to operating room on same admission, and death within 24 hours of admission) and generates worksheets to follow these events. Patient representative software can track and trend patient compliments and complaints.

Other Clinical Modules

In 1988, the VA targeted the integration of patient data for clinical care as the development priority for DHCP. Efforts have focused on promoting program interfaces that allow one-time data entry and thus minimize redundancy. Since the inception of this integration process, nurses have been involved, and their input has been cultivated and appreciated at all levels of program development.

As coordinators of patient care, nurses need information from a variety of modules originally developed by other disciplines. They also make daily use of such packages as they deliver patient care. The paragraphs that follow give a brief overview of the modules most widely used by nurses to supplement and complement their own nursing software.

Health Summary

The most popular integrated module available to nurses and other clinicians, the Health Summary integrates clinical data from DHCP modules.

The user can customize the summary by indicating what data are needed, specify occurrence and time limits for data, etc. Health summaries can be viewed on screen, printed, and batched to run at specified times and locations. Widely used in outpatient clinics, the health summary is especially valuable when a chart is not immediately available. Specific patient data can be requested and transmitted from one VA to another using the Patient Data Transfer (PDX) software or the Network Health Exchange. Nursing staff on inpatient units use this software to access brief summaries of patient data, either by patient or by location.

Computerized Patient Record System

An expansion of the original order entry/results reporting software, the Computerized Patient Record System (CPRS) integrates packages used during the order process. The first clinical module in the DHCP to incorporate a GUI, the package is designed for direct physcian order entry. The focus is on the patient, not the package. Orders entered through the CPRS appear on daily assignment sheets, medication administration records, and care plans. Direct order entry eliminates the need for ward clerks to transcribe and RNs to verify orders, saving time and decreasing the possibility of errors. Orders go directly to the appropriate services for acknowledgment and implementation, making the care delivery process more efficient. The CPRS also supports the design of physician- or service-specific order sets that can be used to expedite nursing-care delivery.

Adverse Reaction Tracking

This software supports data entry of allergies and adverse reactions to drugs, foods, and other causative agents. Alternately, it can document a patient as having No Known Allergies (NKA). Each entry is identified as historical or observed, and includes signs and symptoms experienced/reported by the patient, comments by the user, and verification by site selected individuals. The software can automatically generate a progress note with the electronic signature of the clinician entering the data attached, or produce a MedWatch report to meet FDA requirements. Adverse Reaction Tracking (ART) is accessed during the ordering process to alert a prescriber that a reaction has been documented; this alert occurs after the system has searched the drug ingredient and VA drug class files, as well as the national and local drug files. Numerous reports are available through the ART package, including dietetics, nursing, laboratory, health summary, and any other package that calls up ART data.

Pharmacy

Nurses have multiple daily interactions with modules in the pharmacy package, including automatic replenishment/ward stock, controlled substances, inpatient medications, and outpatient profiles. Probably most frequently used by nursing is the Unit Dose Module. This module includes a medication due list by patient, unit, bed, or room number for any time parameter; this feature is particularly helpful for odd-hour medications. The module also provides Medication Administration Records (MARs). The 24-hour MAR is used in acute care units with frequent medication order changes; 7-day and 14-day MARs are used for longer lengths of stay and more stable medication regimes. The MAR provides a list of all active medications for an individual patient; used as the reorder sheet, it has helped reduce medication errors related to incorrect order transcriptions. The module also reads from the vitals/measurements and allergy (ART) package and alerts physicians to possible reactions.

Problem List

Included as part of the Health Summary, the computerized Problem List provides the caregiver with a current and historical view of a patient's healthcare problems across clinical specialties. Nursing diagnoses are incorporated in the problem list, providing linkages with the nursing package.

Progress Notes

The Progress Notes module uses a boilerplate format to facilitate efficient and legible documentation of patients' progress. Flowsheet emulation and free-text entries offer flexibility in documenting outcomes of care. Online Progress Notes eliminate the dependence on paper charts and make patient information widely available on a timely basis.

Consult/Request Tracking

This software allows users to speed up the process of consult request by allowing electronic request and responses. This application facilitates nurse-to-nurse consultations and allows tracking of consultation workloads.

Mental Health

This software provides a miniclinical record that includes a patient profile, physical examination, psychological test results (a test patients can take at computer terminals), diagnostic results, past medical history, and seclusion and restraint episodes. Nurses utilize this software to document daily patient case management.

Surgery

This package supports scheduling of operating room cases, data input by operating and recovery room nurses, downloading of orders, and extracting of data needed for risk management. A risk assessment used for every patient undergoing surgery is a primary section of the module. Nurses input surgery notes and document the progress of procedures using this module.

Laboratory

Nurses utilize this software on a daily basis to obtain immediate access to laboratory results. This has been especially facilitative in preoperative patient care. Other features used by nurses include transfusion reactions and unique patient traits. The package is interfaced with CPRS; reports can be accessed through the health summary.

Dietetics

The dietetics software module documents diet orders, supplemental and tube feedings, nutritional assessments, and patient preferences; this information is vital to nurses providing daily patient care. Information from dietetics can be inserted into patient assignment tools and end-of-shift reports.

Social Work

This package includes case management, clinical assessment, and community resource information. Social work reports are vital to the nurse when implementing the discharge process.

Training

Introducing computers in nursing has been equated to introducing change, and resistance to change is a characteristic of human nature. As more data are gathered and made accessible for critique, some nurses fear a loss of control over their practice. Some also fear a possible distancing from patients if time must be used to input data into computers. Training for nurse users is one method of counteracting these negative attitudes, and has been part of the software development process from the onset (Sweeny & Post, 1994).

Today the nursing expert panel has a training subgroup; approaches to training nurses serving as software package coordinators have included standardized curricula and locally developed educational materials. A preceptor program for nurse coordinators encompasses such topics as data management, report generation, nursing modules functionality, PC networking, and moving data between DHCP and a windows environment. Training for end users at local sites addresses the learning needs of users, emphasizing the rationale for new technology and potential positive outcomes for patients and care providers.

Comprehensive training manuals have been completed for each of the nursing modules. Other training materials include video productions and quick reference guides. All training materials have an evaluative component.

Future Focus

Point-of-Care Approach

Because patient data are key resources required by clinicians to provide quality patient care, a commitment must be made to the implementation of a point-of-care approach to facilitate data capture into the clinical database. Several studies (Lang, Gallien-Patterson, & Chou, 1994; Miller, Mayher, & Wild, 1995) in VA medical centers have found that the point-of-care approach gives clinicians greater access to patient information and thus supports improved clinical decision making, as well as documentation that contains fewer omissions. The point-of-care approach has also been found to be labor-saving based on the reduced time required to document patient care.

Research on the point-of-care approach was conducted at two VA medical centers. At the Chicago-Westside VA, 10 measures were collected for each of 5 events, that is, vital signs, intake and output, patient admission, new care plan development, and care plan revisions. Findings revealed that each time an activity was performed, there was a decrease in the amount of time required to take and record vital signs (1 minute), measure and record intake and output (1.5 minutes), admit a patient (5 minutes), develop a patient-care plan (1.5 minutes), and update a care plan (3.5 minutes). Overall, savings of 13.5 minutes/patient potential time savings were realized on a single tour of duty when all patient activities were performed with a point-of-care system. Extrapolating from these data, taking vital signs on all patients on a 28-bed unit on a single tour of duty would result in savings of 42 minutes; if five of the patients were on intake and output, at least 7.5 additional minutes could be saved, making an overall savings of approximately 50 minutes.

The second study, at the Marlin VA Medical Center, focused on documentation of vital signs. Measures were taken of 200 daily occurences of vital sign documentation. Results indicated that the point-of-care approach saved 1.44 minutes per occurrence, producing a savings of 288 minutes or 4.8 hours per day.

Together, these two studies suggest that implementing point-of-care documentation hospitalwide should have significant positive results on the quality of care provided to veteran patients. A point-of-care system should also have significant impact on the cost savings related to clinical full-time equivalent employees and the associated operating costs of multiple hospital services.

Other Challenges

There are still many hurdles to overcome in reaching the VHA's goals for nursing systems. Without a nursing taxonomy and clinical lexicon, nursing information is not transportable to other databases, reducing the richness afforded by this technology. It is only a matter of time before healthcare reforms demand the data that can come only from computerization.

Nursing must be ready and flexible. Nursing must continue to be involved at all levels of the evolutionary process of healthcare information systems. Only then will nursing be able to track each patient across the entire continuum of care from homecare to intensive care. Only then will nursing be able to evaluate and improve interventions and outcomes, and address issues that lie behind the cost of care.

Like it or not, nurses have been on the frontlines of computerization for many years, making systems work as they struggled to coordinate the care process. They have knowledge born of experience. Above all others, nurses must be involved in the design of systems that support integrated healthcare delivery networks.

References

American Nurses Association. 1980. *ANA social policy statement*. Kansas City, MO: Author.

Kraft M, & Lang B. 1994. Spinal cord injury nursing: Database analysis, a pilot study. In S Grobe & E Pluyter-Wenting (eds): *Nursing informatics: An international overview for nursing in a technological era*. Amsterdam: Elsevier, pp. 134–138.

Lang B, Gallien-Patterson Q, & Chou JY. 1994. Point of care: A move toward a paperless process. In S Grobe & E Pluyter-Wenting (eds): *Nursing informatics: An international overview for nursing in a technological era*. Amsterdam: Elsevier, pp. 263–267.

Miller ED, Mayher PA, & Wild G. 1995 (Winter). Evaluation of a wireless LAN point-of-care nursing computer system. *Journal of Military Nursing Research*, 24–27.

Murphy AC, Maynard M, & Morgan G. 1994. Pre-test and post-test attitudes of nursing personnel toward a patient care information system. *Computers in Nursing*, *12(5)*, 239–244.

Newton LC. 1995. A study of nurses' attitudes and quality of documents in computer care planning. *Nursing Standards*, *9(38)*, 35–39.

Pohlmann B. 1995. A computerized problem list: the VA experience. In N Lang (ed): Emerging data systems. Washington, DC: American Nurses Association.

Stockton AH, & Verbey MP. 1995. A psychometric examination of the Stonge-Brodx nurses' attitudes toward computers questionnaire. *Computers in Nursing*, *13(3)*, 109–113.

Sweeney ME, & Post HD. 1994. The collaborative approach to the development of a nursing information system environment in a medical center setting. In S Grobe

& E Pluyter-Wenting (eds): *Nursing informatics: An international overview for nursing in a technological era.* Amsterdam: Elsevier, pp. 207–211.

Walker PH, & Walker JM. 1995 (February). Informatics for nurse managers: Integrating clinical expertise, business applications, and technology. *Seminars for Nurses, 2(2),* 63–71.

Werley H, & Lang N. 1988. *Identification of the nursing minimum data set.* New York: Springer Publishing.

19
Developing Clinical Computer Systems: Applications in Cardiology

Ross D. Fletcher and Christopher McManus

Early Activities

Clinicians in the Veterans Affairs (VA) health system have been afforded a valuable experience by participating in the development of clinical computer systems. They have participated in the development of a user-directed, innovative, and increasingly more sophisticated, more comprehensive computer technology. Over a multidecade span, VA clinician users have never failed to move ahead just because the tools were not yet available for a perfect application. The available tools were always used to their maximum capability, thereby stimulating new capacity. As new capabilities became available, the developed applications readily assumed new functionality and improved user friendliness.

The initial VA experiences were with well-defined applications. A pioneer of computer development at the VA Medical Center in Washington, DC was Dr. Hubert Pipberger. From the late 1950s into the early 1980s, Pipberger used a suite of rooms on the fourth floor to conceive and develop a system for analyzing vectorcardiograms and electrocardiograms (ECGs). His collaboration with the noted statistician Jerome Cornfield produced one of the first automated systems in the world capable of digitizing and accurately interpreting ECGs.

Such individual applications gradually changed to broad database applications used by all clinicians. By 1979, the VA had developed many computer tools, and these were running at the Washington VA Medical Center (VAMC). These proto-efforts have been affectionately called "the underground railroad." The process was simple. The tools to organize one's department or study existed on a PDP 11-34 loaded with the VA FileMan database program developed by George Timson. The system literally resided in the basement but was made available to any interested clinician with a terminal. The managers' policy was to allow users with a clinical or research need the platform to develop applications to organize their data and solve their needs. Several programmers were available, but they usually needed only to introduce the clinician to the tools and then give the clinician free rein. All applications resided in the VA FileMan database, which at that time far outstripped other databases, especially DBase and those bound by slow personal computers (PCs) with limited memory. The VA FileMan had no limitations in file size, number of fields or number of entries, and intrinsically the MUMPS system did not require reserving space for future data, as did many of the more cumbersome programs. Full error checking of entries, computed fields, and easy look up by typing just a few initial letters were available.

At the same time that engineering, security, and other groups developed their prototype applications, clinicians in cardiology and pulmonary medicine developed files for pulmonary function tests, exercise testing protocols, and arrhythmia data. Many of these original applications were vital for collecting data in formal systematic research studies. The versatile search, sort, and print capacities of the original VA FileMan were more than sufficient to meet the needs of our departments. The search tools, at that time, permitted complete Boolean logic with automatic lookup of available files. Simple statistics were an option in all print templates. In addition, fields from the files were easy to download to statistical programs.

Some of our research assistants became adept at creating printouts with statistics on the fly. These printouts included fields automatically computed, for example by subtracting two input fields, and thereby computing a total range, mean, standard deviation, and standard error of the differences. The standard error of the difference, of course, is the basis of the commonly used Student's *t*-test and the associated probability values. At times these adaptations required a limited but easily acquired use of MUMPS language such as the *$Select* functions. As with other databases, the programming statements could be embedded in any field, search routine, or print template.

Cardiac Pacemaker Surveillance

In 1982 the VA was challenged by Congress to produce evidence that the VA was adequately following veterans with implanted cardiac pacemakers. Several VAMCs had already instituted regional follow-up centers and had substantial expertise in the pacemaker field. Ten of these centers met and decided to create a nationwide follow-up system centered at the VAMCs in Washington, DC and San Francisco. At the Washington VAMC, Dan Maloney developed within days a basic VA FileMan pacemaker follow-up database, on the basement PDP 11-34, which was chosen over all existing databases. This program had the additional advantage of being totally compatible with the developing hospital system. Data elements, their formal definitions, and a common registration form were soon devised collaboratively to serve the Eastern and Western United States. Initially, Dr. Gertz at the San Francisco VAMC developed the Western region's parallel database program in Fortran. After 1988, however, both centers operated in a MUMPS environment, with a single programmer analyst maintaining identical file structures under VA FileMan.

Pacemakers and other cardiac devices are a classic example of a vital, complex medical need where patient care is enhanced by substantial computer assistance for safety and efficiency. Cardiac pacemakers and implantable defibrillators are adding years of problem-free, active life for hundreds of thousands of Americans. The problem is to determine when a given patient needs to have his device replaced. No formula exists as an answer, because battery depletion and other causes of pacemaker failure are extremely variable. The only efficient way to determine when the cardiac device is ready for replacement is to have patients enrolled in a periodic follow-up program by telephone, supplemented by less frequent on-site clinic checks.

Telephone-Based Monitoring

The Cardiac Pacemaker Surveillance Centers in Washington, DC and San Francisco were established in May of 1982 as the two regional centers for telephone-based monitoring of veterans with implanted cardiac pacemakers. They have become one of the largest and most innovative computer pacemaker surveillance databases in the world. The Eastern Pacemaker Surveillance Center (ECPSC) coverage extends through the Eastern and Central US time zones. The Western Pacemaker Surveillance Center, under Dr. Edmund Keung, covers the Mountain and Pacific time zones. Collectively, the two sites actively monitor the pacemaker performance in over 11,000 patients nationwide, with an average follow-up interval of 1 to 3 months between telephone checks. These patients receive their primary health care from 172 VA medical centers. Figure 19.1 shows the growth in ECPSC caseload. This growth has not been due to mandatory participation, but instead to patient and physician satisfaction in the program. The ECPSC has also become the National VA Pacemaker Registry and is the VA national center for testing all pacemakers explanted at VA medical centers. The National VA Pacemaker Registry includes clinical histories on 43,000 present and former pacemaker patients.

Each VAMC where cardiology patients are treated has an integrated hospital information system running a common database package (VA Kernel), which includes a Pacemaker Procedure Module. This module was developed in light of the ECPSC experience and directly incorporates hundreds of data field definitions and tables of standard codes originally developed at the ECPSC. Figure 19.2 is a simplified view of the ECPSC system's major relational files and their sizes. Integrating and taming such a welter of diverse information is a task that clearly requires automation, and tests a database program's mettle. A 60% increase in the ECPSC caseload over 10 years, as depicted in Figure 19.1, was accomplished without addi-

278   Clinical and Support Applications

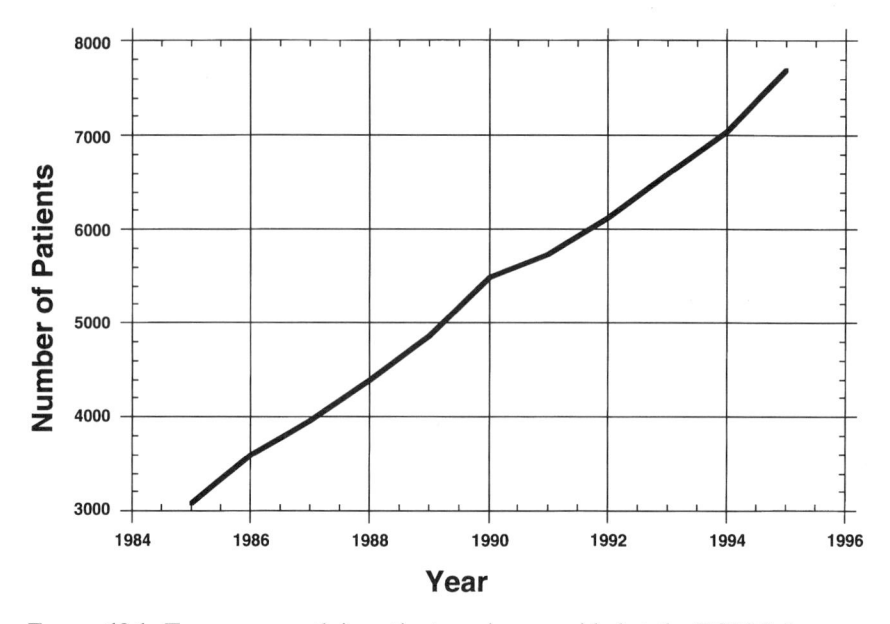

FIGURE 19.1. Ten-year growth in patient services provided at the ECPSC. Increasingly comprehensive automation has allowed 150% increase in patient load with static staff size.

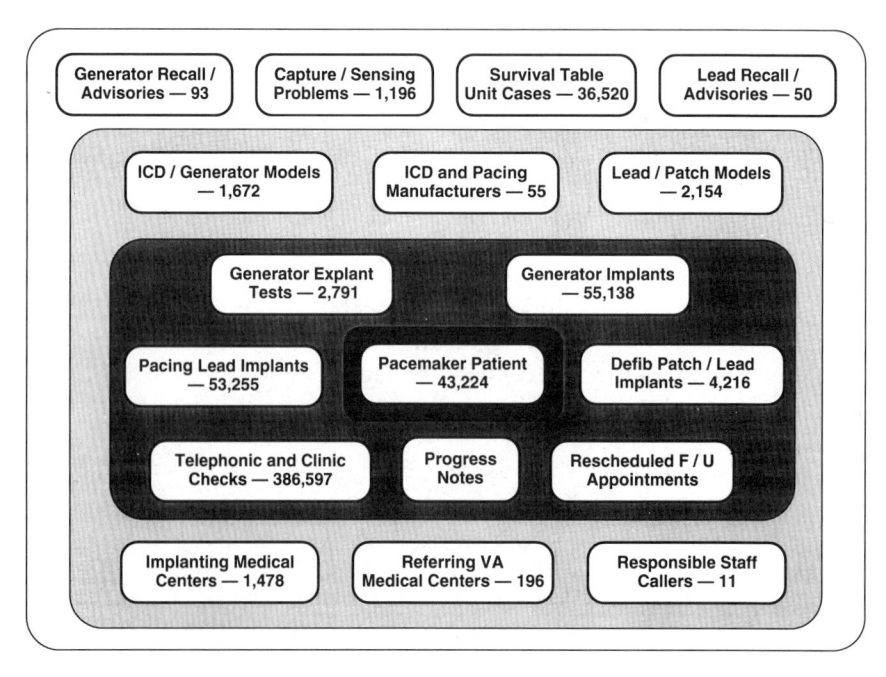

FIGURE 19.2. Principal relational files comprising the ECPSC database.

tional ECPSC staff. This was done by progressively improving the efficiency, while checking efficacy, using well-planned automation and an increasingly "paperless" record system. Without automation, the increase in workload would have required a comparable increase in staff.

Electronic integration vastly simplifies the process of registering patients for ECPSC follow-up. To have the ECPSC start monitoring a patient's device, a VAMC needs only submit a patient registration form. This registration can be accomplished totally electronically, as easily as having a clerk choose the option *Transfer a Patient Registration to National Center*. The local medical center computer then automatically sends an electronic message containing all the patient's pacemaker-related data and needed demographic data to the ECPSC. At the ECPSC, another computer routine can parse the incoming message, and add the patient data to the national database.

Once a patient is enrolled, another computer routine is invoked to generate a schedule of all the patient's telephone checks for the current or following calendar year. Schedule frequency depends primarily on the pacemaker's age and any history of problems associated with the pacemaker model or the implanted leads, as seen in our patient population by our internally generated device survival curves or as reported by the Food and Drug Administration (FDA). The schedule always increases to one or two weekly calls after any reprogramming, and also increases in frequency when the first appearances of battery depletion occur.

The backbone of the ECPSC is highly trained technicians and nurses who conduct the individual telephone checks. Every morning, each caller obtains a computer-generated list of all the calls they are scheduled to make that day, and the appointment times. Before each call, the ECPSC caller enters the patient's name into the computer; the terminal then displays various data related to the patient. This information includes the pacemaker, leads, and transmitter. Also displayed are any special reminders about problems known for specific pacemaker generator or lead models, such as an FDA warning. This is an important safety feature that computerization makes possible: even a dedicated human can nod and not notice the presence of a device under alert, but the computer is totally consistent. Patient-specific difficulties, such as deafness, are also displayed to the caller, further facilitating communication.

Next, the patient is reached by telephone, and asked to send his ECG, including pacer signal by telephone, through a supplied transmitter, to a receiver located in the Pacemaker Center. Separate pacemaker tracings are recorded first without, then with magnet application. The caller then reviews the electrocardiographic tracings, especially for evidence of appropriate pacemaker sensing and capture. The data for this particular pacemaker check is entered into the computer. A copy of the pacemaker ECG tracing and the computer-generated report are printed out and sent to the patient's physician at his referring medical center. This printout can be sent by e-mail. Automation has reduced this entire process to less than 5 minutes. The patient is never rushed, and feels free to communicate any new symptoms. The center in turn leaves the patient with a preliminary report of the pacemaker's current status. If a pacemaker problem is recognized, the caller additionally alerts the patient's own hospital immediately and, after consultation, advises the patient if hospital care is warranted. If the patient was not able to be reached on our first call, auxiliary computer menus and files are automatically invoked that capture the reason and create a new appointment. Loss of contact with patients is obviated by this computer feedback, as well as by several other electronic reporting and searching tools.

Telemedicine Prototype

This effort, beginning nationwide in the VA in 1982, is an early prototype of the efforts now being made in telemedicine. To contact a patient, review his symptoms, and look at the realtime ECG, has been a powerful way to manage a patient's illness in his home. The patients' confidence in the surveillance centers has been reflected in their frequent calls to ask questions about symptoms unrelated to the pacemaker.

One of our first administrative officers enhanced this vital element of patient confidence by asking our programmer to provide a monthly mailing list for the birth dates of all veterans over 75 years of age. A volunteer veterans group took on the project of providing birthday cards, so that each older veteran now receives on his birthday personalized congratulations from the service. What our administrative officer knew was that birthdays are kept secret after 39 but again become appreciated milestones after one reaches 75 years. The return mail from this small computer-assisted project has been heavy and heartwarming. The confidence and trust of the patient–clinician relationship have further enhanced the timeliness of calls for help from our patients when they had a genuine need.

A plethora of periodic management reports are generated for the ECPSC and medical centers associated with the ECPSC program. As mentioned above, these include reports to ensure that patients are not left uncalled because of a short-term difficulty in contacting them. Occasionally manufacturers will issue a recall or advisory on a particular lead or pacemaker model. This warning may be broad or may affect only a handful of serial numbers. In either case, the ECPSC can generate within minutes a listing of all patients affected, sorted by the associated medical center. Only electronic integration of pacemaker data allows such prompt recognition of problem units, allowing immediate change of coverage. Additionally, a powerful database computer program has allowed the ECPSC to achieve close cooperation with other federal agencies monitoring healthcare providers, notably the FDA, the General Accounting Office, and the VA

Inspector General, with whom we have shared our files of pacing and lead model histories.

While the computer system began as a relatively large PDP 11-44, the system has migrated to the ECPSC system of network-linked 486-based microcomputers, terminal servers, and Internet connections to other VA medical computing systems. These connections allow 24-hour seamless access of ECPSC data by surgeons and emergency room physicians, directly from their hospital system menus. The ECPSC system also interfaces with the local VA medical center's MUSE 5000 computer, a multisite environment for analysis, storage, and retrieval of ECGs. All electrocardiographic records back to 1982 can be retrieved from this system.

To provide true 24-hour service, the ECPSC has a toll-free 800 telephone number that patients may call any time they are experiencing a problem. During off hours, this number is answered by Intensive Care Unit personnel. Any incoming tracing can be forwarded to electrophysiologists who share 24-hour coverage. Because all ECPSC records are computerized, intensive care unit nurses can instantaneously access the latest patient information on their screens at any hour of the day or night and thereby handle after-hours calls fully informed.

The first internal cardiac pacemaker was implanted at a VA hospital in 1960, by Dr. William M. Chardack (Chardack, Gage, & Greatbatch, 1960). The 36 years since then have witnessed a radical transformation in the design of electronic cardiac pacemakers. The introduction of multiprogrammable pulse generators, dual chamber pacing, rate responsive pacing, and finally implanted defibrillators, have given rise to a vastly expanded vocabulary of pacing, reflecting an enhanced scope of pacemaker functions. For example, programmable characteristics include atrial and ventricular stimulus amplitudes, durations, and refractory periods, postventricular atrial refractory period, escape rate, atrial-ventricular interval, upper rate limit, sensitivity, hysteresis, and pacing mode. These changes in technology have significantly increased reliability, versatility, and longevity of pulse generators. Keeping track of their profusion is increasingly a task only a well-designed computer database can master.

Impending Pacemaker Malfunction

The primary task of any monitoring program remains the timely identification of impending pacemaker malfunction. Despite technological breakthroughs, cardiac pacemakers continue to have finite life expectancies for their generators, batteries, and leads. Pacemaker replacement based primarily on manufacturers' prediction of battery longevity would often occur prematurely, before full use of the battery. This would mean unnecessary surgical procedures, with increased cost, risk, and pain. A combination of monitoring by telephone and examinations in clinic is the recommended procedure to determine when the battery voltage has decreased. Close pacemaker monitoring additionally provides a valuable feedback loop: age and output parameters are observed in a constantly updated population, leading to better implant criteria for pacemaker follow-up. The ECPSC identifies generator survival curves by its internal data, and can confidently follow pacemaker models that are performing well at less frequent intervals, thereby allowing safe coverage of an expanded number of patients.

Emerging technologies have complicated the task of recognizing when pacemakers reach end of life. Manufacturers have adopted a wide variety of end-of-life parameters and criteria, sometimes several for a model, depending on programmed mode. For the callers these differences need not be remembered for each patient, because the computer brings up the specific end-of-life criteria for each patient's pacemaker at the time of the telephone call.

One invaluable tool for examining the performance of particular models of cardiac devices is a statistical method known as cumulative survival tables. First suggested by Cutler and Ederer (1958), these tables reveal the rate at which a given unit fails over time. Examples are shown in Figures 19.3 and 19.4. The marked difference in failure rates is apparent. It is not enough to know that a patient's generator will not fail for 50 months. The generator is connected to a lead, which conveys the impulse to the myocardium. Usually leads last far longer than the battery-dependent generator. On occasions the leads fail early and will determine the way a patient is cared for and followed. Figure 19.5 superimposes graphs for three lead models. One of these models can be seen to fail significantly faster than the

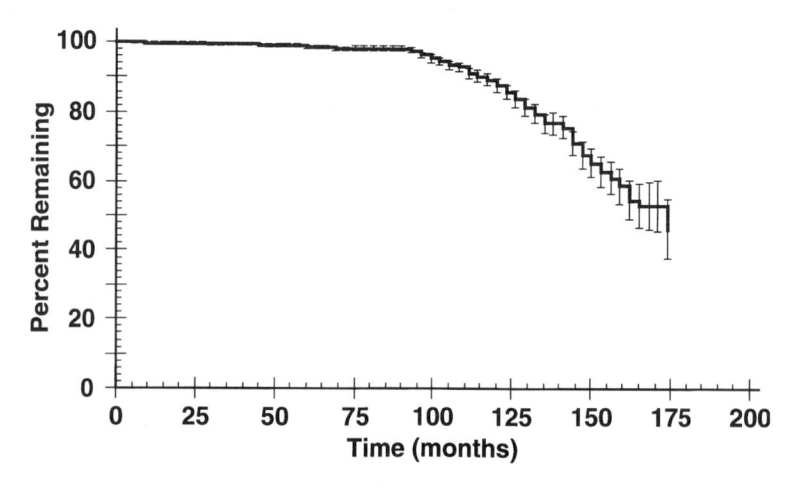

FIGURE 19.3. Cumulative survival table for a pacemaker model (medtronic 5985) implanted in ECPSC patients. Curves are derived entirely from implant histories in the ECPSC database. Also shown are standard errors of each interval, calculated by the Peto method.

19. Developing Clinical Computer Systems   283

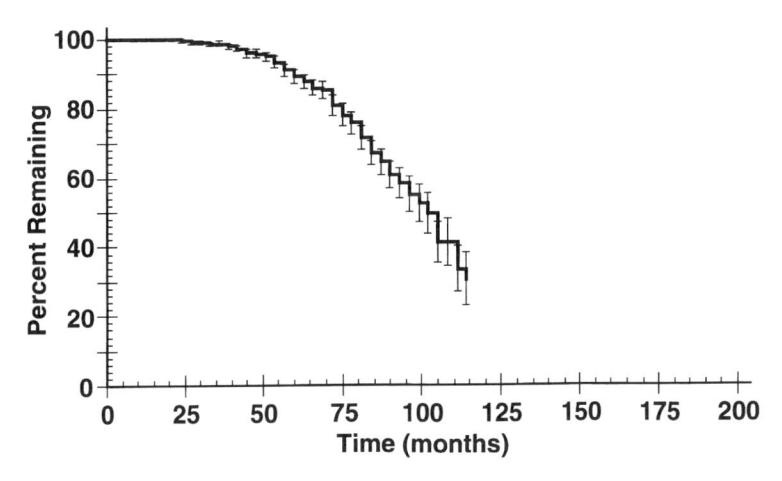

FIGURE 19.4. Cumulative survival table for a pacemaker model (medtronic 7005) implanted in ECPSC patients. Curves are derived entirely from implant histories in the ECPSC database. Also shown are standard errors of each interval, calculated by the Peto method.

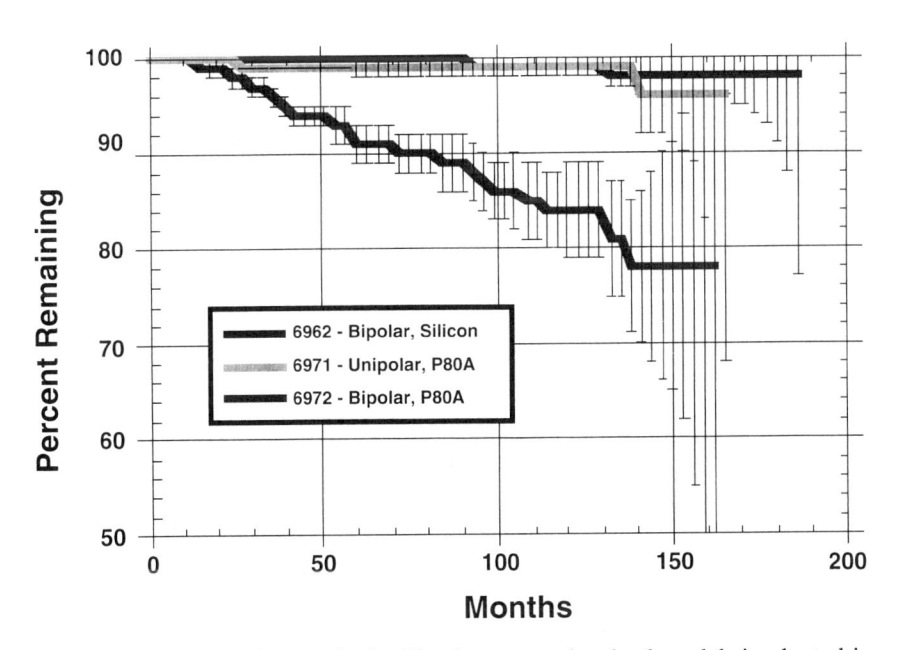

FIGURE 19.5. Cumulative survival tables for pacemaker lead models implanted in ECPSC patients. Curves are derived entirely from implant histories in the ECPSC database. Also shown are standard errors of each interval, calculated by the Peto method. Differences reveal the conjunction of bipolarity and P80A polyurethane coating as a risk factor for early failure.

others. One of the other models has a polyurethane coating, the other is bipolar; only the failing model is both bipolar and polyurethane coated. Such vivid insights, made possible only by a comprehensive computerized database, then inform future cardiologists' future selections of cardiac devices and leads.

Cumulative survival tables require at least 40 unit histories to be reliable. Currently, the ECPSC has sufficient case histories to generate table for 100 generator models, covering 72% of the ECPSC's active caseload, and 92 cardiac lead models, reflecting 75% of its active caseload. The method includes a formula for calculating confidence limits, as revised by Peto and colleagues (Peto, Pike, et al., 1977).

Survival tables such as Figures 19.3, 19.4, and 19.5 provide a basis for computer-generated follow-up schedules. If the implant duration for the patient's specific generator model falls on or after beginning of failures in the survival curve, follow-up calls are made monthly. If the patient is on the flat position of the curve, then calls are made every 3 months. A "standard guideline" for follow-up that covers both generator models would needlessly accelerate the follow-up for the generator with the longer life. The quality-assurance question that can be asked and answered by the computer database, is whether any patient lost pacing of the ventricle because his battery failed. The basic principle of a follow-up service is to allow pacemakers to remain implanted as long as possible, but replace them before the pacemaker can no longer produce enough voltage to pace the heart. Because there is always a rate change when the battery voltage decreases, it is a simple process to search the database for all generators where no capture was present, and examine it for a change in rate. In our database, loss of capture has occurred, but never with a change of rate. Thus, the method of follow-up determined battery failure before loss of pacing in all instances. Battery failure has not been responsible for any loss of capture in our service.

A particularly exciting experience has been the productive interplay between functions. The registry draws information from all VAMCs and immediately reflects any movement of a patient under ECPSC care from one medical center coverage to another. Testing of all pacemakers explanted from VA patients impacts on the other functions of the ECPSC, not only by providing critical but often unreported facts needed for patient coverage and failure analysis, but also by revealing specific generator failure patterns on an oscilloscope. Explant test results are also helpful to ECPSC callers as confirmation of their monitoring judgments that a pacemaker has exhibited end-of-life or malfunction indications. Additionally, integration of functions helps develop a total clinical outlook and enhances motivation and knowledge: registry and explant testing are done by staff who know pacemaker technology and can see the impact of their results on patient care. The combined efforts mutually enhance each other's efficiency, accuracy, and completeness of coverage and eliminate redundant efforts.

Integrated Cardiology Database

The original PDP 11-44 available for the pacemaker center spawned many other applications. Because the original Massachusetts General Hospital Utility Multi-Programming System (MUMPS) programs were intrinsically multiuser, they formed a network for our department long before personal computers and Novell networks became part of other sections. It was therefore a small step for the cardiology to enter the network scene. From an early point, the cardiology section had terminals in each office and laboratory, and used a common word processing program. From that time, expansions in the cardiology database were designed with integration of information in mind. These expansions grew to include linked files for catheterization, Holter, echocardiography, and electrophysiology procedures, as well as conversions of previous research exercise testing files to clinical exercise tests.

The section had a secret weapon in the presence of a remarkable patient and volunteer, David Lum, Ph.D. He had inoperable coronary artery disease, but responded to many of the antianginal drugs that were part of our research protocols. He aided our clinicians in setting up files for their individual areas, and rapidly became adept at producing high-level outputs with VA FileMan. Those outputs were clinically superior summaries, but ran aggravatingly slowly until the tools in VA FileMan advanced to allow compiled print templates. Speed of response is perhaps the single most important element for enhancing user friendliness. In this instance, enhanced software came to the rescue. In other instances, hardware upgrades changed unbearably slow response to a suddenly adequate and welcomed response time. In our current era of faster and faster technology, software should never need to be kept mundane to avoid a slow response in an outmoded technology. Catheterizations, exercise tests, Holters, electrophysiology (EP) tests, and echoes were all components of this integrated file structure originally developed on the ECPSC computer system.

Expert Panel Input

An important synergy was the inclusion within our walls of the VA's regional Information Service Center (ISC), headed by Dan Maloney. Under the ISC, the cardiology files were reorganized and integrated into the emerging Decentralized Hospital Computer Program (DHCP) structure. Under the ISC aegis, the files were improved and modified by a select group of cardiologists from throughout the VA system. This expert panel included the chiefs of cardiology from major teaching centers who each had developed research and clinical databases in cardiology. Small groups used the strawman database to critique and modify until a consensus was achieved. A common concern at that time was not whether the critical data elements could be identified, but whether the VA software was quick enough or user friendly enough. Screen editors were not officially available except as locally maintained (Class III) software. These were incorporated, nonetheless, to improve acceptance of the software. These changes produced the cardiology component of DHCP, which rapidly evolved into the medicine package. The mandate from the VA Central Office had always been to rapidly expand the utility of the cardiology module to the rest of Medicine. The expert panel expanded to include members from gastroenterology, hematology, endocrinology, oncology, pulmonary, renal, and rheumatology. Of necessity, the expanded expert panel lost some of its critical cardiology specifiers. Each specialty concentrated first on reporting all its procedures and then on the other vital aspects of that field.

The expert panel structure naturally progressed to the current Procedure Expert Panel, with its mission to collect and report not just medicine but all clinical procedures for the computerized clinical record. While their generalized task is vital to development of an electronic record, it is still important to keep smaller groups organized and active to maintain current accurate data and general consensus within the group of practicing physicians. The electronic record can in some instances be simple text uploads of clinician reports, but much more valuable is the inclusion of specific data elements. With specific fields such as ejection fraction, the hospital database becomes a much more robust engine for inquiry by diverse groups with a multitude of goals and needs.

The principles established by the expert panel included three levels of procedure entry and reporting. The first level was a one-line summary. The second was a brief one-page input containing all fields important for patient care, but specifically including fields that could be used for quality assurance. The brief entry and brief print templates included patient, date and time, procedure with its International Classification of Diseases, 9th ed., Clinical Modification (ICD-9-CM) code, indications, results, complications, and medical personnel. Whenever possible, the third level is a complete dataset, including all elements normally found in a procedure report. In cardiac catheterization for example, the sedatives, access, and catheters used are part of the full report; pressures measured, results of coronary arteriography, ventriculography including ejection fraction, are all part of both brief and full reports.

Clinical and Research Benefits

After adopting full automation of procedures, the performing section, the clinicians relying on the results, and the patients all gain in many ways. Because data entry into the hospital database is part of the usual task of providing reports, the section gradually builds a large complete record with very little added effort. When asked for quality-assurance reports or caseload reports, the section can frequently provide the answer immediately with no added effort. Most importantly, results are immediately available in the treatment areas. No time is wasted in calling the section that performed the procedure, answering such calls, or searching the paper record. Procedure reports have a tortuous path to the paper record that defies completeness and accuracy. Online data, however, are always available using direct inquiry or healthcare summaries when the patient arrives in the clinic or the emergency room.

One of the primary goals has been to automate, as much as possible, the data entry into DHCP. In medicine there is a profusion of computer systems with clinical data. Some of these computers are used directly to obtain or analyze information. An example is the ECG computer system. Central ECG storage, begun by such pioneers as Dr. Pipberger, has been part of our routine for years. The results should be able to be seen by DHCP automatically. The Medicine/Procedure program incorporated automatic machine reading of the Central ECG system earlier when the program existed on a PDP 11-44, and more recently and more directly when it assumed a PC platform. Earlier efforts used a Laboratory Service Instruments (LSI) interface between external systems and DHCP, but a direct connect (soon to use Health Level 7 [HL7] messaging) is now available. While technology and software have changed, tracings from 1989 to the present reside on the present system and are integrated into the DHCP database. For example, because the Pharmacy package is on the database, all patients with long corrected QT interval and antiarrhythmic drugs can be listed not only for the past (where it is no longer relevant to the patient), but in realtime, day by day, to prevent side effects including sudden death. Using the ECG data, similar searches have been made to see how often widening QRS intervals occurred with elevated cocaine levels. Getting blood pressures into the chart was seen as difficult in our cardiology clinics. The ECG technician took the blood pressure and entered it into the ECG cart. That blood pressure reading is now transferred automatically to DHCP with the rest of the ECG data and becomes available for review and comparison.

Other searches of the catheterization database strikingly illustrate the clinical importance of an integrated database covering many years of patient entries. Research colleagues Michael Franz and Marcus Zabel (1996) demonstrated that stretch channels could produce ventricular premature beats. They asked if there was a correlation between left ventricular (LV) diastolic pressures and arrhythmia. Because our catheterizations and Holter records resided in the integrated hospital database, it was easy to download all patients where both procedures occurred within a month of each other. Over the last 5 years, a significant correlation has been demonstrated between patients with greater than 17 mmHg left ventricular end diastolic pressure (LVED) and high-grade ventricular arrhythmia such as ventricular tachycardia and couplets. Complex arrhythmias did not, however, correlate significantly with ejection fraction in the study group, consisting of patients with heart failure (Zabel, Franz, & Fletcher, 1993). These data imply a possible rationale for afterload reducers such as enalpril and diuretics. These retrospective analyses become more powerful with large numbers and are valuable for creating hypotheses for prospective studies. Searches have also been generated to characterize pulmonary functions in all Agent Orange patients in our hospital. A new program for lowering cholesterol with diet and the newest cholesterol lowering drugs was established in our hospital. Immediately a programmed search selected all patients with cholesterol greater than 240 mg. The printed report included their usual follow-up clinic and the method for communicating with their physician.

The fact that all patient data appear in one integrated database, with the patient's name pointing to a single central patient file, leads to high accuracy, reliability, and avoidance of duplication. The research and clinical potential is illustrated by the number of records on the system: the Washington, DC VAMC database currently contains detailed clinical measurements and findings for more than 4500 catheterization studies, 43,000 ECGs, 6100 Holter studies, 14,000 echocardiography studies, and 6000 exercise tests.

Recent and Planned Enhancements

Entry of data has progressed from line entry to screen entry and will soon adopt graphic user interfaces (GUI). This should provide a user-friendly interface to which all users can readily adapt. In the meantime, those of us who encouraged data to be entered while the interface was not as user friendly are now blessed with a legacy of records that are readily reviewed anywhere in the hospital. When a patient from long ago enters the hospital anew or when a young investigator wishes to review large numbers of patients with the same clinical profile, the needed records are readily at hand.

Adding Images to the Database

Parallel to the developments that led to entry of text-based reports was a process to capture images to the database. This process was developed by Dr. Ruth Dayhoff and Dan Maloney at the Information Service Center in Washington, DC. While intrinsically a part of DHCP, the image or multimedia system collected the images as TARGA DOS files on a Novell network. They were linked to the specific procedure and patient by image fields, and contained subfields that tracked the address of the image on the network. Thus, whenever a patient arrived, the image of his ECG, the X ray, endoscopy, cine clips from the catheterization, and any images from echo or pathology all became visible to the clinician user at an image workstation. Reduced images, 12 per page, summarize a procedure and serve as icons for enlarged images with diagnostic detail. Revascularization is made more imperative by collected images, which reveal a 90% lesion of the left anterior descending (LAD) coronary artery supplying an anterior wall, since these images also show the artery contracting well in the systolic frame of the ventriculogram.

The screen of collected images is shown in Figure 19.6a; Figure 19.6b shows a single frame from the set, chosen, enlarged and enhanced by a single keystroke. The result of opening the lesion with angioplasty is also documented on the image chart. The beneficial result of the procedure is thus communicated rapidly and is heartening to the clinician practitioner and to the patient, who at times reviews these images with his physician. What is even more surprising is the synergy between data entry compliance and online graphic display. Sections initially reluctant to spend the time entering procedures into DHCP have become 100% compliant and fans of the system when the image/multimedia system became available. The multimedia system has recently progressed from a workstation requiring two separate monitors to a single-screen Windows-based workstation. This application has produced a major improvement in user friendliness, It provides easy controls to alter contrast and brightness of X rays, obviously a great improvement on the bright light technique. A soon-to-be-completed project will include all ECG graphics records currently residing in the Muse ECG system. All ECGs recorded over the past 7 years will then be available for viewing and for comparison with current ECGs on any image workstation in the medical center. This multimedia system illustrates the dramatic adaptiveness of the legacy of the MUMPS system to new technologies to produce superb graphic images for the computerized patient record system (CPRS). The paper record increasingly becomes an anachronism when the computerized record can encompass and display information and original views not previously available in the paper record.

Teleconsultation

Physicians place far more trust in original visual images than in summary text records. In general, the x-ray or catheterization films are sent along with the patient to ensure adequate consultation when another hospital or clinic is involved. As sharing of resources becomes a more common reality in today's environment of managed care, telemedicine will become increasingly more important. Telemedicine makes consultations with distant experts possible in a small hospital setting, avoiding unnecessary and potentially harmful transportation of the patient. The DHCP image system uses VA connectivity to make this goal a reality. Multimedia e-mail is already in place, and allows messages with associated images to be sent immediately between remote sites. This capacity creates an additional synergy: it not only allows teleconsultation in the case of a patient being considered for advanced care; it further returns the results of the advanced care to the primary-care physician, in the form of images and interpretative text.

An advanced use of VA connectivity is the current ability of an expert echocardiographer at the West Los Angeles VA Medical Center, Dr.

290   Clinical and Support Applications

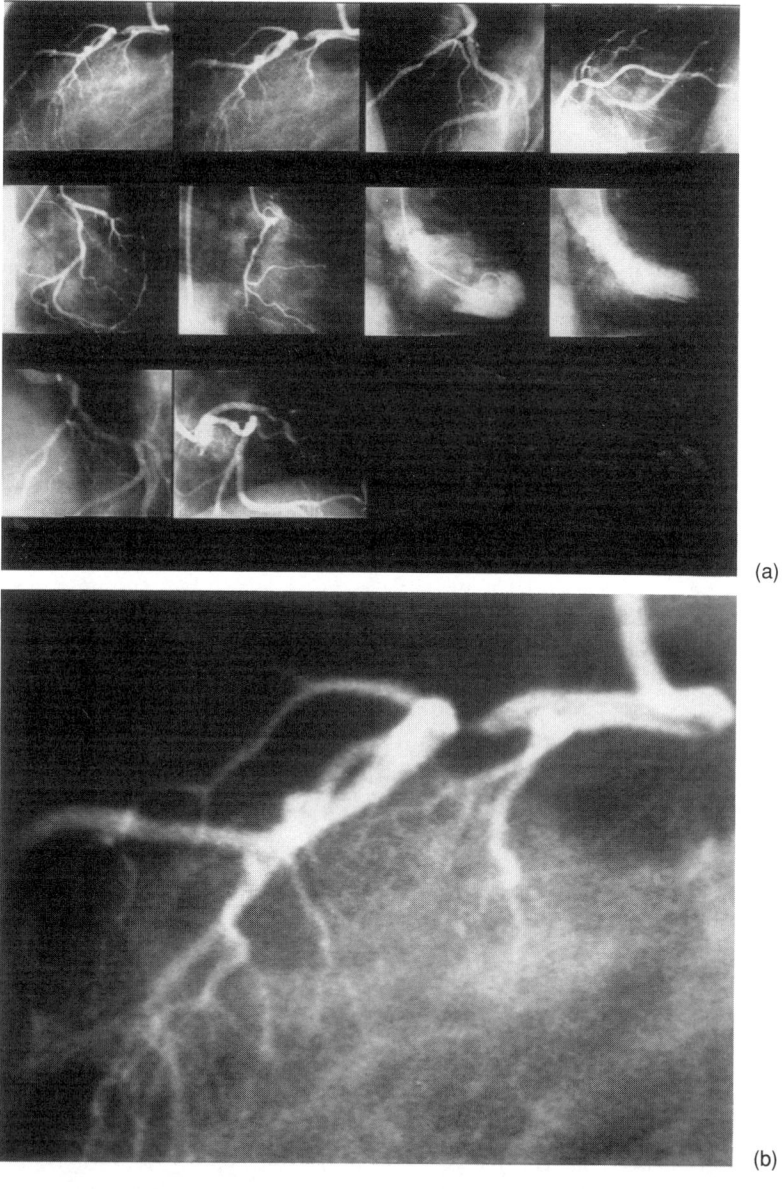

FIGURE 19.6. (a) Typical DHCP screen of cineangiographic images. Frames 1 through 6, from the patient's cineangiogram, reveal a 90% narrowing of the LAD artery. Images of the ventriculogram in diastole (frame 7) and systole (frame 8) show excellent contraction of all segments of the left ventricle. Frame 9 documents resolution of the narrowing after angioplasty. (b) An enlarged, higher resolution image of frame 2 from Figure 19.6a. This enlargement replaces Figure 19.6a on the display screen after a single keystroke is typed. This frame reveals a 90% narrowing of the LAD artery.

Maylene Wong, to overread the echocardiograms recorded in Altoona, Pennsylvania. After overreading, Dr. Wong signs on to the Altoona VAMC system and enters her interpretations. The results are immediately part of the local patient record, and can displayed on any terminal in the Altoona hospital. This application saves many thousands of dollars locally, while improving the quality of available care. Additionally, this innovative use of computer systems suggests the further synergies possible in a large, integrated healthcare system through imaginative outreaches.

The enthusiasm and willingness of clinicians to participate in development are demonstrated daily in VA medical centers, and for that reason a large part of the computerized patient record is already a reality. In April 1996, a workstation on our wards is used to enter all orders and to retrieve healthcare summaries. These healthcare summaries contain all laboratory results, a list of all medications, discharge summaries for 2 years, and medical and surgical procedures. Associated with the procedures are x-ray catheterization images, echo images, images of all endoscopies, pathology microscopy images, images of dermatological lesions, x-rays, and ECGs. If additional information is needed for individual patient care, the user can click for Internet access to obtain guideline information, or can move to a Department of Medicine network that contains access to a tower of CD-ROMs that contains tutorials in dermatology and rheumatology, synopses of lectures in primary care given in the hospital, as well as Medline and Drug description and compatibility programs. The CD-ROM tower was developed by Dr. David Nashel and John Martin from the Department of Medicine at the Washington, DC VAMC and encouraged by its Chief, Dr. James Finkelstein. The description above sounds like a future fantasy. Indeed, it is for many hospitals in other medical systems, but it exists and is being developed and deployed in the VA medical system now.

The current process will continue to grow as long as users are closely involved in development and as long as development is responsive to the clinician users' needs. The VA has been a model for developing clinical systems in this manner, and has to date progressed more rapidly than any other group toward the *paperless record* goal that many clinicians, administrators, and patients anxiously await. The process will be continuous and involve tools and advantages to patient care that none of us have anticipated. Forward motion is the essence of imaginative, effective development. Patient care has become far more efficient, timely, informed, and accurate with the current stage of development and will become even more so in the future.

References

Chardack WM, Gage AA, & Greatbatch W. 1960. A transistorized, self-contained implantable pacemaker for the long-term correction of complete heart block. *Surgery, 48,* 643–654.

Cutler SJ, & Ederer F. 1958. Maximum utilization of the life table method in analyzing survival. *Journal of Chronic Diseases, 8(6)*, 699–712.

Peto R, Pike MC, et al. 1977. Design and analysis of randomized clinical trials requiring prolonged observation of each patient. *British Cancer Journal, 35(1)*, 1–39.

Zabel M, Franz MR, & Fletcher RD. 1993. Correlation of ventricular arrhythmias on Holter monitoring to pulmonary capillary wedge pressure an ejection fraction in patients with congestive heart failure. *Circulation, 88*, I-603.

Zabel M, Sachs F, & Franz MR. 1996. Stretch-induced voltage changes in the isolated beating heart: importance of timing of stretch and implications for stretch-activated ion channels. *Cardiovascular Reseach, 333* (in press).

20
Anesthesiology Systems

Franklin L. Scamman, Holly M. Forcier, and
Matthew Manilow

Introduction

Anesthesiologists have been major players in the development of health data systems. Because of the large amount of physical science associated with anesthesia, the field has been attractive to scientists and engineers. These individuals have been on the forefront of utilizing computers to facilitate their day-to-day activities. Because of the skills necessary to organize and conduct an anesthetic, anesthesiologists are frequently positioned to be active in administrative areas. We are beginning to appreciate the contributions that the field has made to the prosperity and health of information systems within healthcare enterprises.

The first part of this chapter is organized to follow the data flow of a typical anesthesia encounter. The latter part reflects on administrative activities that are facilitated by data and information management.

Anesthesia and Automated Medical Information Systems

Preanesthesia

Current national health policy centers around increasing the amount of ambulatory care and minimizing the amount of days a patient spends in the hospital. Having a mobile patient makes continuity of care more difficult. Many healthcare providers will need to have access to the database that develops as a patient progresses through the process of having an operation. Automated systems are ideal for this.

The anesthesiologist gathers information on the patient consisting of the proposed operation, diagnosis leading to this operation, past medical history, family history, laboratory, special study results and the results of the physical examination. In conjunction with the patient, the anesthesiologist develops an anesthesia plan, discusses the benefits and risks of the planned anesthetic with the patient, and receives permission from the patient to implement the plan. The anesthesiologist then enters a note into the medical record outlining the plan and stating that the patient agreed with the plan (informed consent).

Development of Automated Preanesthesia Systems

Automating the gathering of preanesthesia information and providing results to medical personnel on an on-demand basis independent of location increases the efficiency of the process. Many individuals are in the process of developing systems to do this. Two are discussed below.

University of Chicago

Dr. Michael Roizen has studied the preanesthetic evaluation extensively from a standpoint of maximizing the amount of information needed while minimizing the number of tests and costs. He has developed a handheld computer to interview the patient. The computer has four keys and a display on the front. The keys are labeled *yes, no, I don't know*, and *go on*. The computer is programmed to step through important questions in the patient's medical history. If there is a significant reply that needs to be investigated further, the program branches until the significant data are collected and then returns to the main questions. At the conclusion of the interview, the patient's responses are analyzed. A list of significant answers is printed for the anesthesiologist along with the studies that the computer recommends be done. Because the logic leading to the recommended tests was developed from a consensus of what constitutes an appropriate medical practice, the recommendations are authoritative. Roizen has demonstrated that following the recommendations reduces the amount of tests ordered without decreasing the quality of the evaluation (Roizen et al., 1992). The computer has been interfaced with a network so that an anesthesiologist scheduled to care for the patient may obtain the data from a remote location.

University of Florida

Dr. Gordon Gibby has developed a preanesthesia evaluation system with a slightly different focus. His goal was to automate the data gathering process and reduce the amount of manual labor involved. His system interfaces with the hospital mainframe computer to gather already existing data. It also has a questionnaire, but data entry is done by the anesthesiologist during the interview. Positive findings are assigned diagnosis codes that are available to individuals in the medical records department. Having these codes available has facilitated more accurate processing of discharge data resulting in an increase in hospital income because of the ability to document the increased severity of illness. This system is networked, and the evaluation can be viewed from multiple terminals.[1]

The VA Experience

As VA continues to shift from inpatient to outpatient care, anesthesiologists will find the lessons learned from the previous examples valuable. Fortunately, VA, in its Decentralized Hospital Computer Program (DHCP), has the ability to capitalize on the wealth of data and information it contains. A package is being developed that will allow the anesthesiologist to access data on a patient from any terminal or workstation in the hospital or remotely from other VA medical centers or home. The VA's National Anesthesia Service is responsible for this development. The package will use the tools below.

Order Entry/Results Reporting (OE/RR) and Progress Notes

The VA is rapidly heading towards an all-electronic medical record. One of the major tools is Order Entry/Results Reporting (OE/RR) that has two main functions. The first, order entry, allows the anesthesiologist to enter requests for tests, procedures and consults electronically, validated by an electronic signature. The second is the retrieval of information on the patient from the databases contained in DHCP. Among these databases are laboratory, history and physical, vital signs, allergies, patient demographics, pharmacy for medications being taken, radiology and nuclear medicine reports and surgery reports. When fully implemented, OE/RR makes available almost everything there is to know about a patient and his medical care. The progress notes module allows the anesthesiologist to enter free text about a patient and is the main vehicle for replacing the handwritten progress note in the medical record. A substantial amount of standard phrases and sentences (boilerplate) is available. The combination of OE/RR and Progress Notes can make a completely electronic preanesthesia evaluation.

Data retrieval from other facilities

Many of the veterans we serve are retired and enjoy traveling. They may visit several VA Medical Centers (VAMCs) during their travels. The VA is implementing a network transfer utility that will allow an anesthesiologist to move information from one medical center to another over the VA's packet switching network. Because the information belongs to the originating medical center, the receiving medical center does not incorporate the information into its database but stores it separately. The data compromis-

---

[1] This system was demonstrated at the Hewlett-Packard display, the American Society of Anesthesiologists Annual Meeting, Atlanta, 1995.

ing the information are in the same format as the local data so the anesthesiologist can use the same tools as used locally to view the information. As VA develops the graphical interface to OE/RR, it may be possible to cut and paste information from one facility's database to another to reduce the amount of arduous data entry.

Text Integration Utility and electronic signature

When all of the preanesthesia information is available on a patient, DHCP will create a document via the text integration utility (TIU). The TIU uses a template to organize the information and present it in a form that may be printed, signed with pen and ink, and included in the medical record. Alternatively, the document may be signed electronically, stored, and retrieved as a completed document. It is recognized that the data that went into the document may change but the document, once signed, is unchangeable. If an anesthesiologist needs to view the most recent information, he may run the template in realtime. There is no problem that the realtime template contents may differ from those in the document. One must remember that the document represents a snapshot in time and remains unchangeable, the same as any piece of paper filed in the medical record.

Intraoperative

Since the time of Harvey Cushing, M.D., the great, early 1900s Boston surgeon, anesthesiologists have kept track of events and activities taking place in the operating room. Traditionally, the anesthesia record has been kept by hand using pen and ink. With the advent of computers, anesthesiologists have dreamed of having them keep track of the patient's vital signs (blood pressure, heart rate, respiratory rate, etc.) without intervention and generate a document that replaced the handwritten anesthesia record. Not until the advent of the microprocessor was this practical. Today, the majority of the physiological monitors are based on microprocessors, and automated data collection is feasible, with at least six manufacturers now offering EARs.

Early Work at Duke and Ohio State Universities

The microprocessor was introduced in the late 1970s. By 1981, anesthesia departments at Duke and Ohio State were well on the way to constructing inhouse systems that were slow, cumbersome, and not very user friendly. Only the zealots could make them work. Not highly popular, these early models taught that developing a useful tool for the anesthesiologist was not going to be easy. Industry, however, became interested in the concept and by 1990 several manufacturers had products on the market. On the order of $25,000 each, they were expensive, and their value was not proven.

Experiences with EARs Outside the VA

In some settings, EARs became popular. Dr. David Edsall purchased the Arkive system for all of his anesthetizing locations and studied this product for utility and value in several areas. He concluded that accuracy and legibility are improved. The most important feature, he claimed, was that data collection for quality management became much easier and that he was able to provide anesthesiologists with profiles of their practice. Comparison among colleagues helped increase the total level of quality. There was some concern that accurate and timely recording of data might provide damaging information if a lawsuit were filed as a result of an adverse anesthesia event. However, Edsall concluded that accurate and timely recording of data reduced the likelihood of a successful suit because the data were present and legible. Many have claimed that lawsuits are lost because of missing or illegible data. The jury is still out on the contention that EARs reduce the payout on lawsuits. There are claims that tracking usage of drugs by provider can point to those who are cost outliers. It is further claimed that education of these outliers to a more cost-effective anesthetic plan can pay for the cost of the EARs. Although this is an appealing rationale for purchasing EARs, nothing appears in the literature to substantiate this contention.

The VA Experience

It is common that EARs in a suite of operating rooms are networked together to form a system so that information can be exchanged between them and the file server. The file server is a separate computer dedicated to data archive and retrieval. The file server may also interface to workstations in the preanesthesia areas and recovery room to ensure that all perioperative data is collected. Few systems, however, interface with the main hospital information system (HIS) to retrieve patient data and to store important data and events of the patient's operative experience. Through its Hybrid Open Systems Technology (HOST) efforts, VA has encouraged the interfacing of DHCP to third-party devices to help automate data collection, analysis, and display. In this respect, the HOST program has awarded grants to interface EARs to DHCP at the Bronx and San Diego VA Medical Centers. A third initiative is underway as a demonstration project at the Philadelphia VAMC under a contract with Lockheed Information Systems.

The Bronx

The HOST anesthesiology project at the Bronx VAMC may be viewed in three phases: (1) procuring and installing the Modular Instruments Inc. Lifelong Anesthesia Data Collection System; (2) interfacing the Lifelong System to DHCP; (3) evaluating the quality of the system and the collected data, its impact on patient-care and staff activity, and assessment of cost-effectiveness.

When this project was proposed, the goals were to implement an electronic anesthesia information system to improve the accuracy and legibility of the documentation in the intraoperative record, and to allow the free interchange of information between the anesthesia system and DHCP. An additional goal was to install a system capable of allowing online access to anesthesia-related reference information for intraoperative consultation, initially through locally accessible CD-ROM based reference material and ultimately via access to the Internet. It was anticipated that the system also would be used to facilitate quality-assurance data analysis and report generation. The expectation was that with training, staff would develop expertise with the system that might reduce time spent in documentation and increase time spent in patient care.

To fulfill these goals, it was believed that a system was needed with sufficient flexibility regarding the information that was collected and shared with DHCP. The system also needed to have inherent true multitasking capabilities to support the ability to access local information and via the Internet. The OS/2-based system developed by Modular Instruments, Inc., was selected.

The system was installed and brought online relatively quickly. Initially, the software interface necessary for communication with DHCP was not completed, and the system functioned as a stand-alone recordkeeping system. The inherent flexibility of the system made training difficult, as the configuration of the system involved a significant amount of user-initiated input. Problems were encountered immediately with printing of the intraoperative record, because it requires configuration of the printer file to match the various monitors that are acquiring data. Because the monitor configuration was not identical in each operating room, multiple printer configuration files needed to be written. The operator then had to select the correct printer file to properly print the intraoperative record. This resulted in numerous errors and many reprintings of the records. The interface phase was further subdivided due to the course of the software development, which was first implemented by downloading the case data from DHCP to Lifelog prior to the start of operating room procedures and then uploading the case data from Lifelog to DHCP at the conclusion of a case.

The DHCP software used in this project was based on two packages, the HL7 package of the messaging system, developed by the Albany (NY) Information Resource Management Field Office (IRMFO), and the surgery package, developed and maintained by the Birmingham (AL) IRMFO. This created normal problems in coordination that were compounded by a reorganization of DHCP development that transferred the messaging system to the San Francisco IRMFO.

To assess the effectiveness of this system, a time and motion evaluation has been underway since the download became active. An observer is in the operating room during a fixed set of procedures chosen to provide a stable reference. Four anesthesiologist activities are observed: monitor watching recording, other patient-related activity, and idle time. The duration of these activities is recorded during three identifiable periods of the procedure: the 45-minute initial period, a 15-minute middle period, and a 15-minute end period. Data are maintained on a spreadsheet. No conclusive results are available at this time.

San Diego

ARKIVE, the first Electronic Anesthesia recordkeeping system commercially available in the United States, was derived from research performed at the University of California and the San Diego VAMC. Data exchange between the ARKIVE network and DHCP began in 1993. Information from the ARKIVE network is transferred to DHCP as follows. The anesthetic record from each surgical case is saved in a proprietary format file on the Novell file server. Paradox scripts written by the San Diego VAMC personnel extract a list of anesthetics administered from the Paradox tables into an Excel readable file on a floppy disk. Anesthesia administrative personnel copy this list into their log of anesthetics, which they maintain to keep a running total of the number and type of anesthetics administered by the anesthesia service. Totals are entered quarterly into DHCP for review.

In July 1994, funding for the development of additional data exchange between the EAR network and DHCP was awarded by the HOST program office. The intent was to expand the interface between the installed ARKIVE/EAR system with DHCP and significantly increase the use of the reports produced from the data collected. When ARKIVE filed for bankruptcy in May 1995, the HOST program office increased funding for the project by the amount estimated to replace the ARKIVE/EAR network.

Experiences with ARKIVE/EAR network have influenced selection criteria for a replacement system and plans for supporting and developing that system at the San Diego VAMC. The most significant selection criteria are anticipated maintainability, adaptability, and extensibility by the VAMC. Also important is the capacity to exchange data with DHCP, using the HL7 extension to the Surgery Package. Other criteria include user interface, data analysis capabilities, data exchange with other systems or with DHCP, using other methods, and purchase price.

When switching from ARKIVE to another EAR network, the San Diego VAMC also plans to change its network topology, from the daisy chain topology now in use to a more reliable star configuration. Plans for support began with the preinstallation staging. Equipment for the new network is being tested on site before it is put into clinical use. This will allow VAMC staff to verify the network's reliability. In addition, it will allow staff to verify that transferring the list of patients scheduled for surgery each day works with the new system, and that the list of anesthestists administered can still be routinely copied from the EAR network to a Macintosh.

Philadelphia

The Philadelphia VA Medical Center project is quite different from the above two. Congress, when it approved funding to implement DHCP throughout the VA system in the mid-1980s, wanted to ascertain that the money was being spent efficiently and effectively. It set aside five VAMCs to be non-DHCP facilities that would use commercial software as the core of their hospital information systems (HIS). When, after almost 10 years of effort, these four centers still did not have an HIS up and running that compared favorably to DHCP, the comparison project was abandoned and DHCP was installed at all five. However, the VA still wanted to maintain a test site for major installations of non-VA HIS systems; Philadelphia remained a test site. A contract was awarded to Lockheed Information Systems to be the prime contractor for interfacing non-VA hardware and software to DHCP. By May 1996, the EARs had just arrived at Philadelphia, and the network(s) were just being established.

Postoperative

The responsibility of the anesthesiologist to the patient does not cease once the patient leaves the operating room. The medical supervision of the recovery room is frequently provided by the anesthesia care team. The patient may have complications occurring postoperatively that need to be tracked and analyzed. Long-term outcomes are important to improve the delivery of patient care. Automation of these functions is underway.

Recovery Room

The process of bringing the patient back to a fully functioning status is a continuum from being anesthetized in the operating room to being nearly normal by the time he leaves the recovery room. The need for monitoring physiological variables and recording events is no less in breadth but is less in intensity as the patient recovers. The monitoring equipment in a recovery room may be no different from that in the operating room. Continuing to capture data from the continuum is reasonable; some EAR manufacturers, such as Hewlett-Packard, are beginning to incorporate recovery room data into the patient file. A separate report generator creates appropriate documents for the medical record.

Patient Tracking and the Postanesthesia Visit

The patient needs to be followed once he leaves the recovery room. Regulations require that the anesthesiologist visit the patient roughly one day following his anesthesia and surgery to determine if there have been any complications and to answer any questions the patient may have about the conduct of anesthesia. Automation allows the anesthesiologist to go to any

HIS terminal or workstation and enter information obtained during the postanesthesia visit.

Data Collection for Quality Management

Data have always been the infrastructure of an effective quality management program. The need to apply statistical methods for identifying areas of improvement as intensified as the emphasis on improving quality shifted from identifying the "bad apple" to that of identifying areas of the "system" that needed improvement. Being able to track processes and outcomes over a long period of time to establish whether the system is under control or not has become vital. The wealth of data provided by EARs has allowed supervisors to monitor and make changes in the processes of anesthesia.[1] An outstanding example of this is the VA's National Surgical Quality Improvement Program discussed later in this chapter and elsewhere in this volume.

Anesthesia Administrative Activities and Medical Information Systems

Anesthesia personnel are active outside the operating room. This is driven in part by the increasing breadth of anesthesia-related activities related to the skills they have to offer and in part by the reduction, in many parts of the United States, of the operating room workload under managed care.

Scheduling for Anesthesia Clinics

More of the surgical workload is being done with the patient not being admitted to the hospital. This shift to ambulatory surgery requires that the preanesthesia evaluation become an office visit. To prevent chaos when patients arrive at the preanesthesia clinic in a haphazard manner, many clinics are scheduling patients with defined appointment times. The VA has a package in DHCP that facilitates clinic scheduling and, as a consequence, better allocation of anesthesia personnel. The same applies to the Pain Clinic patient scheduling.

Quality

As was mentioned earlier, quality management is data driven. In the past, much of the collection and analysis was done manually. There was a general appreciation that the entire process could be made more efficient, both in

---

[1] Sall DW. Personal Communication 1993.

terms of time and personnel required and in the statistical techniques available.

Joint Commission on Accreditation of Healthcare Organizations

The Joint Commission on Accreditation of Healthcare Organizations (JCAHO) is a not-for-profit organization whose purpose is to monitor the quality of process and care in the thousands of healthcare facilities throughout the United States. This organization sets standards to assess quality that it expects the institutions to meet. Until recently, it judged each institution on its own merits, and there was little if any comparison in outcomes between institutions. As part of its agenda for change, the JCAHO has initiated comparing rates of adverse outcomes between institutions.

Beta testing on anesthesia indicators

In 1989, the JCAHO solicited hospitals to participate in the testing of a program to collect what it called "anesthesia indicators" (JCAHO, 1989). These were stroke, nerve damage, heart attack, cardiac arrest, respiratory arrest, death, unexpected admission from outpatient status, and unexpected admission to intensive care. Over 100 hospitals eventually participated, including 14 VAs. The JCAHO provided a software package to analyze the International Classification of Diseases, 9th ed., Clinical Modification (ICD-9-CM) and Current Procedural Terminology (CPT)-4 diagnosis and treatment codes that were associated with a surgical procedure. The package analyzed these codes by combinations and permutations to identify patients who may have experienced one of the eight indicators. Data entry was either manual or automated. Extracting the codes from the Patient Treatment File of DHCP required a template to be written and a connection to DHCP to initiate the execution of the template and the data to be captured while appearing on the PC screen. The resulting file was contaminated by the header and footer DHCP attached as part of screen formatting; these were removed with a text-only word processor. The resulting text file was imported into the JCAHO software, which created the list of patient encounters that needed additional data input for the logic to decide whether a suspected "hit" was a true hit. Frequently, the information contained within the dictated discharge summary available online on DHCP was sufficient to complete the data needed for the logic to make a decision. When all true hits were verified, the package automatically dialed the JCAHO file server in Chicago and uploaded data on the hits so that the indicator could be validated. The total caseload was also uploaded to become the denominator.

Rates of indicator occurrence were calculated by the JCAHO for each hospital and an analysis of the data returned to each side. Because this was a beta test, the data comparing hospitals against each other were not released. The JCAHO analyzed the nationwide data to establish the validity of each indicator.

JCAHO Agenda for Change: Perioperative indicators

The JCAHO was undergoing a massive shift in policy from looking for the "bad apples" to analyzing processes and outcomes, a process it called the agenda for change. The collection of massive amounts of data on outcomes (above) was part of this process. The validity analysis on the original eight indicators resulted in five being retained to be incorporated into a mandatory data reporting process called the IMSystem (JCAHO, 1995). The five are stroke, nerve injury, heart attack, cardiac arrest, and death. Because an adverse event happens to a patient and not specifically surgery or anesthesia, these five have been renamed "perioperative indicators." The JCAHO wants all hospitals to start collecting data on all procedures in January 1997 and sending the information to their main office. Needless to say, this will be an expensive project for both the hospitals and the JCAHO. The JCAHO intends to release reports on each hospital with comparisons to nationwide data.

The VA Experience

Congress has mandated that the VA demonstrate that medical care given at its facilities is as good as delivered at non-VA facilities. The VA responded with the Surgery Risk Program, which has been expanded into the National Surgical Quality Improvement Program (NSQIP). The anesthesia input to this program is twofold. The first data element is the physical status scale developed by the American Society of Anesthesiologists (ASA). The scale, assigned by an anesthesiologist and entered into DHCP at the time of surgery, is: (1) a normal healthy patient; (2) a patient with mild systemic disease; (3) a patient with severe systemic disease; (4) a patient with severe systemic disease that is a constant threat to life; and (5) a moribund patient who is not expected to survive without the operation. The ASA scale, after the first 87,000 cases were analyzed, was the second-best predictor of morbidity and mortality. The second element is the anesthesia technique. Only patients who received general, spinal, or epidural anesthesia are included in the study. This data element is also entered at the time of surgery. The JCAHO perioperative indicators routinely are collected by the NSQIP.

Workload, Productivity, and Value

As downsizing continues as part of balancing the federal budget, VA is concentrating more on determining the value of services and comparing the value to the cost. Because personnel costs take up the greater part of the budget, it is imperative to determine the correct staffing level. The VA's National Anesthesia Service is in the planning phase of determining clinical productivity.

For years, the ASA has used a relative value guide to determine the amount of complexity and work that goes into delivering anesthesia. The Health Care Financing Administration (HCFA) accepted the ASA guide in its entirety when it started the Physician Payment Schedule in 1992. Workload consists of two parts: (1) the complexity and skill level associated with starting the anesthetic (base units); and (2) the duration of the anesthetic (time units). Base units are linked to CPT-4 codes. Current HCFA policy allows one time unit for every 15 minutes of anesthesia time. The Surgery Package in DHCP captures anesthesia time and CPT-4 codes and the framework for determining clinical anesthesia workload is in place. The missing link is the crosswalk between CPT-4 codes and base units. This crosswalk is being developed. When units per case are available, then it becomes simple to derive total workload, workload by provider, and workload by surgical specialty. Having workload by surgical specialty would facilitate more discerning allocation of operating room resources. Workload by provider when combined with cost per provider (salary) comes very close to measuring the value of the provider. Knowing which providers are valuable will allow staffing adjustments to maximize efficiency.

The above concept of calculating the value of providers does not take into account productivity from other activities such as administration, teaching, and research. Developing measures of workload and productivity in these areas will require considerable effort.

Future Enhancements

What is coming for automation and anesthesia within the near future? The answer depends a great deal on the real and perceived value of spending large sums on equipment and technology versus the gain in favorable patient outcomes and increased efficiency.

Expansion of EARs into All Operating Rooms

The VA has on the order of 1000 anesthetizing locations in 132 medical centers. Is it worthwhile to purchase an EAR for every anesthetizing location as part of implementing the all-electronic medical record? This is an expenditure on the same order of magnitude as the cost to the VHA of installing the original DHCP. What is the benefit?

Studies on Cost, Efficiency, and Effectiveness

At a meeting of the ASA 2 years ago, one EAR manufacturer claimed that his system would pay for itself in 2 years. His claim rested on the assumption that there were a significant number of disposables and drugs that when used normally would have generated a charge to the patient that the hospital could capture through billing either the patient or his insurance company. Last year, with the advent of managed care and capitated financing, this manufacturer no longer made this claim, substituting the claim that his system could pay for itself over an unspecified period of time because the system could track the use of expensive anesthesia techniques by provider. Expensive providers could be counseled to reduce their expenses, thus becoming more efficient at the delivery of anesthesia. How effective such modification of an individual's practice would be is open to question. The senior author found no studies that set forth guidelines for the value of Electronic Anesthesia Recordkeepers (EARs) in the above respects.

Longitudinal Anesthesia Records

One could argue that the value of an EAR could come in the ability to retrieve information from one anesthetic encounter that might have value in reducing morbidity or mortality in subsequent medical treatment or care. If a patient were to be anesthetized at a facility repeatedly, being able to retrieve the anesthetic record off the file server and to review it without having to find the chart could save time and effort. Again, nothing was found in the literature to substantiate this assumption.

The economic climate at the present time is to cut the cost of providing medical care without reducing quality. It has not been conclusively shown that EARs can do this. Increasing income without a parallel increase in expenses is no longer an option. Electronic Anesthesia Recordkeepers are expensive. At the present time, the senior author believes expansion of EARs into all operating rooms should be delayed until it is proven that there is substantial value to do so.

Data Exchange Between Different EARs at Different Facilities

As mentioned in the section on preanesthesia evaluation, the VA population is mobile. If EARs do become ubiquitous and every anesthetic is captured in some form of a database, it would be desirable for anesthesia providers to be able to exchange information about anesthetics. At the present time, data cannot be exchanged among the various original equipment manufacturers because some of the databases are proprietary in structure and none of the manufacturers use the same data dictionary or lexicon.

Common Data Dictionary

Four years ago, the Society for Technology in Anesthesia established a committee, under the leadership of a prominent leader in EAR utilization, to develop a data dictionary and lexicon that would define a file structure for electronic anesthesia records. This work was in conjunction with the American Society for Testing and Materials (ASTM) and the Computer-Based Patient Record Institute (CPRI). Although the committee met at least twice per year, and every member had e-mail access, the committee did not succeed in its goal. One problem was that all members were volunteers and there was not a critical mass large enough to devote sufficient energy to the project. The second problem was that the scope of the project was underestimated. For example, the committee became bogged down working with over 100 definitions of perioperative events that required date and time stamps. A third, and perhaps more formidable problem was resolving what to do when a timed event had two values, one determined by the anesthesiologist and one by the circulating nurse. Even more of a challenge was what to do when the surgeon said that both of the above were wrong and entered his own value. The committee has quietly disbanded.

The Use of the Internet and Browser Technology

One of the very exciting new features of information transfer is the graphical user interface that has developed around the Internet. One of the transfer protocols that the Internet supports is HTML. Hyper Text Markup Language has become very popular as embodied in Internet browsers such as Netscape, Microsoft Internet Explorer, and Mosaic. Browsers are now available on commercial Internet access providers including America Online, Prodigy, and CompuServe. Files can be transferred as simply as pointing at a name and clicking with a mouse. The complicated process of transferring a file and checking for errors in transfer happens transparently to the user. Many organizations, including the VA, have found that having a server that supports these browsers is of great benefit to the public. Not all servers, however, need to be available to the full Internet. The VA has many servers that are available to only VA employees, protected from the Internet by a "firewall." At the present time, the firewall will let VA employees out to full Internet access but will not allow anyone who is not authorized in to the VA resources. The VA internal "intranet" will become a valuable resource for information exchange, including anesthesia information.

Conclusion

This chapter provides an overview of how automated medical information systems have benefited anesthesia with special emphasis on the Veterans Health Administration and its Decentralized Hospital Computer System.

References

Edsall DW, Deshane P, Giles C, Dick D, Sloan B, & Farrow J. 1993. Computerized patient anesthesia records: less time and better quality than manually produced anesthesia records. *Journal of Clinical Anesthesiology, 5,* 275–283.

Joint Commission on Accredation of Healthcare Organizations. 1989. *Beta I: anesthesia indicators*. Oakbrook Terrace, IL: Author.

Joint Commission on Accredation of Healthcare Organizations. 1995. *IMSystem software manual*. Oakbrook Terrace, IL: Author.

Roizen MF, Coalson D, Hayward RS, Schmittner J, Thisted RA, Apfelbaum JL, Stocking CB, Cassel CK, Pompei P, Ford DE, et al. 1992. Can patients use an automated questionnaire to define their current health status? *Medical Care, 30*, MS74–84.

21
The Library Network: Contributions to the VA's Integrated Information System

WENDY N. CARTER AND CHRISTIANE J. JONES

Background

The Veterans Health Administration (VHA) recently completed a massive reorganization (Kizer, 1995) designed to improve the quality and efficiency of care to veterans and better accomplish its missions of education and training, research, and contingency support during war or national emergencies. Focusing on instructional technology, information systems, and corporate realignment, workgroups considered the librarian's role in providing information for clinical and management decision making and helping end users navigate through electronic databases and information networks. In response to the worksgroups' recommendation, oversight for Department of Veterans Affairs (VA)'s library program moved from the Office of Academic Affairs to an expanded Chief Information Office (CIO) in October 1995, as shown in Figure 21.1.

To appreciate how librarians contribute to VHA's efforts to improve veteran's health care and integrate its information functions, it is important to understand the components of the VA Library Network (VALNET) and the expanding role of health sciences librarians in managing knowledge-based information.

Integrated Information Management/ Improved Quality of Care

In recent years, health sciences librarianship has shifted its emphasis from physical collections to improved information access. This change was encouraged by improvements in information and communications technologies and by the vision and guidance from professional organizations such as the Joint Commission on Accreditation of Healthcare Organizations (JCAHO).

As the Joint Commission changed its focus from structures to outcomes, there has been an accompanying shift in emphasis, from departments to functions within the healthcare organization. Functions generally affect

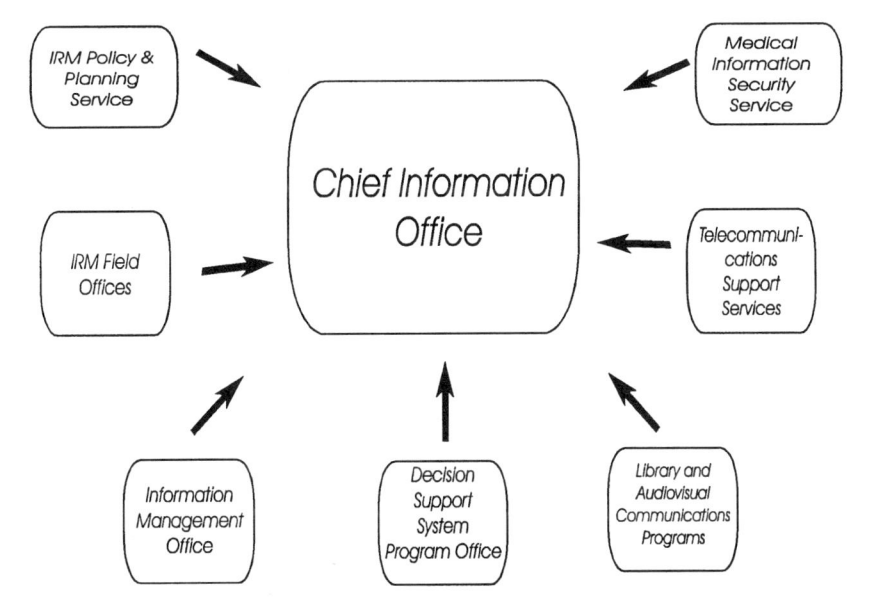

FIGURE 21.1. Office of the chief information officer.

every area of an organization and are addressed from an interdepartmental standpoint. Information management is a function of healthcare organizations that is critical to patient care and hospital management.

The management of information standards recognize the prominent role that the management of information plays in an organization's ability to function. Without the effective and efficient management of information, organizations cannot properly deliver patient care or carry out other important activities, such as performance improvement. Thus, the Joint Commission views the management of information as not only helping organizations to meet external demands, but also the critical demands arising in an organization's internal environment. Therefore, organizational leaders should view this function as an activity that needs to be planned for, just like human, materials, and financial resources management. (JCAHO, 1995, p. 10).

The information management function consists of four types of data/information:

- Patient-specific data are what have been known in the past as medical records;
- Aggregate data enable an organization to establish patterns and trends related to clinical outcomes, performance, and costs;
- Knowledge-based information is what has been traditionally referred to as "the literature"; and
- Comparative data are similar to aggregate data in that they enable an organization to establish patterns and trends related to clinical outcomes,

performance, and cost; however, these patterns and trends are about other organizations and are used for comparative purposes.

The JCAHO information management standards focus on a set of six generic activities and processes that are essential to the management of the four types of information, that is, identification of data sources, capture of data, analysis of data, interpretation of data, transformation of data into information, and transmission of data. These activities are described in the language of information science and apply to all four types of information. These six activities are typically carried out by the information enablers in the organization, that is, medical records personnel, health sciences librarians, and computer department personnel. The way that the activities are carried out differs for each type of information; however, the underlying premise for each is the same. Although these six activities are essential to the management of any of the four types of information, the information management function is not complete until the information has been used within the organization. Therefore, users are critical stakeholders in the information management function. Reaching full compliance with the new standards requires an integrated view of all types of information and collaboration among various stakeholders and enablers.

Librarians, as participants in the integrated information management function, must structure library and information programs, services, and activities to support the intent of the new JCAHO standards. In so doing, they must emphasize anticipating users' information needs and systematically linking the literature with clinical and organizational processes. Current, accurate, and convenient knowledge-based information should be available, regardless of format. Much important information can be found in audiovisuals or other nonbook materials. Systems should be in place for organizing, identifying, retrieving, and delivering knowledge-based information, and for ensuring the uniformity and completeness of this information (JCAHO, 1994).

The systems and structures used by VALNET to organize, manage, and share knowledge-based resources as a network, are further described in this chapter.

Network Structure

VALNET is a library network of national significance, which supports the VA healthcare system. The 156 network libraries and information centers are staffed by over 275 professional librarians, assisted by over 290 clerical and technical personnel. Libraries are valuable building blocks for the delivery of information and learning resources required by VA staff, students, trainees, and patients. VALNET librarians actively support and develop patient health information resources to help patients and their families become informed partners in their own health care. Because no one library can own or manage all of the information and educational resources needed to support the complex and diverse needs of VA staff and patients, it is critical that each librarian exploit the collections and skills of all other VALNET libraries, as well as those of other health sciences libraries nationally. All VALNET libraries participate in resource sharing networks, such as the National Network of Libraries of Medicine, led by the National Library of Medicine (NLM).

Although VA libraries are similar in many ways, variations exist that reflect the patient-care emphasis at each medical center. A small library may be staffed by one librarian or library technician, subscribe to 100 journal titles, own fewer than 2000 books, and provide access to NLM's MEDLINE databases. A large library in an actively affiliated medical center with significant research activity could be staffed by nine full-time staff, subscribe to over 500 journal titles, own more than 15,000 books and audiovisuals. This same library could offer a variety of services such as online database instruction, operation of a patient health education center, media support, clinical team participation, and public access to electronic library information and databases as well as the Internet (Wiesenthal, 1987).

As a network, VALNET owns approximately 800,000 books, 60,000 periodical subscriptions, and more than 100,000 audiovisuals. Many individual VALNET librarians have developed subject-specific expertise that can be shared throughout the system. Because VA staff and patients at one medical center should be able to benefit from systemwide collections and capabilities, resource sharing and networking have been major components of the library program since its inception.

During World War I, VA Library Service was established by the American Library Association as a part-time activity under the Public Health Service. In 1923, it was transferred to the Veterans Bureau, the predecessor of the VA. The library program provided recreational library service to patients, medical information to clinical staff, and administrative information to other personnel. Library operations, including selection and acquisition of books and journals, cataloging, and governance for VA libraries nationwide, were centralized at Headquarters in Washington, DC.

Responsibility and authority for configuring individual library services to support local needs became increasingly decentralized after 1940. However, core systems, enabling resource sharing, identification of library holdings, and service standards continued to be directed centrally. Currently, VA libraries operate as autonomous nodes within VALNET, local community library consortia, and within the National Network of Libraries of Medicine.

Congruent with each medical center's mission to provide quality health care to veterans, VALNET'S goals are to develop, support, and encourage the exchange of resources and expertise through shared services such as cataloging, development of consolidated listings of VA-owned books, journals, and audiovisuals, and distribution and delivery of expensive and/or program-significant materials and bibliographic databases.

Customers

VALNET librarians provide a mix of library services to a diverse clientele. The types of services that librarians provide differ based on the category of user.

Patients

The veteran patient is one of VALNET's primary customers. During 1995, more than 2.8 million veterans received care in a VA medical center (Department of Veterans Affairs, 1995b). Historically, library services to patients had been primarily recreational. In the early 1970s, the decreasing length of hospital stays and the growing interest in helping patients make informed decisions about their health care changed the focus of VA's patients libraries from recreational reading to support for patient health education. Currently, VALNET libraries offer a wide range of health education support to patients and their families, and to healthcare providers designing educational interventions to meet their needs.

The librarian identifies patient health education resources, and coordinates the selection of these materials with clinical staff to ensure appropriateness. Using these resources, patients and their families are better able to make informed decisions about their care. Databases, such as the Health Reference Center, which include basic clinical and lay-oriented journals, pamphlets, drug information, and excerpts from medical texts, and patient-oriented versions of pharmacy databases, are often available through VALNET libraries. Librarians are also searching the Internet to locate information resources and knowledge bases to enhance consumer and patient health information materials in their collection.

Management and Support Staff

Hospital managers are interested in comparing their clinical outcomes, performance, and costs with other institutions locally and nationally. They strive to continuously improve the quality of care and delivery of services, and to attract and keep a well-trained, committed work force. Knowledge-based information from management, economic, public affairs/news, and human resources literature is available through VALNET librarians. Increasingly, managers use this information to identify "best practices," compare like characteristics, and define outcome measures.

Additionally, the library collections, audiovisual equipment, and study space are important components of a medical center's ability to deliver employee education and training programs. This is particularly important in a hospital setting where staff such as ward clerks, food service workers, and nurses may not be able to use and synthesize some types of knowledge-based information at their duty stations.

Clinical and Allied Health Staff

The librarian's timely delivery of current, authoritative information from the health sciences literature is critical to the organization's ability to provide high-quality patient care. All VALNET libraries provide access to MedLINE, hold books and journals listed in basic selection guides, and participate in resource sharing networks; most offer expanded resources and services.

Clinicians and allied health professionals are the major users of VALNET library services and resources. The approximately 125,000 physicians, nurses, students, and allied health professionals, working in VA healthcare facilities, rely on VALNET libraries to identify, organize, retrieve and deliver information from the health sciences literature to support clinical decision making, research, and education (Department of Veterans Affairs, 1995a). The services and collections available through individual medical center libraries vary; however, all VALNET libraries offer access to database search service, provide reference and interlibrary loan services, hold core collections of clinical textbooks and journals (based on accepted selection guides and listings such as the Brandon Hill List, and Abridged Index Medicus), and provide current awareness services, enhancing users' ability to track and update key topics of interest (National Library of Medicine, 1970; Brandon & Hill, 1995).

Almost all VALNET libraries (91%) provide the healthcare professional with direct links to databases such as GraTEFUL Med, MedLINE, CINAHL, PsycLit, Micromedix, PDR, AHFS, and Health Reference Center. (See Table 21.1.) Most access is through stand-alone public

TABLE 21.1. Commonly accessed databases and decision support systems.

AHFS: American Hospital Formulary Service.

CINAHL: Cumulative Index to Nursing and Allied Health Literature.

DXPlain: Decision support system developed at the Laboratory of Computer Science, Massachusetts General Hospital, Harvard Medical School.

GraTEFUL Med: User-friendly interface to the National Library of Medicine's MEDLARS databases.

Health Reference Center: Consumer health information from selected journals, pamphlets, and textbooks.

ILIAD: Decision support system developed at University of Utah.

MedLINE: Medical Literature Online.

Micromedix: Databases synthesizing peer-reviewed pharmacy literature. Consumer and poison control information also available.

PDR: Physicians' Desk Reference.

PsycLit: CD-ROM database from the American Psychological Association.

QMR: Quick Medical Reference Decision Support System.

workstations and, increasingly, through the hospital's local area network. Some VALNET libraries have public workstations for access to diagnostic support systems such as DXPlain, ILIAD, and QMR, which are used primarily as teaching tools for VA students, trainees, and clinical staff.

Clearly, librarians share their users' interests in bringing external knowledge sources and databases to the point of care. Many are working with multidisciplinary information management planning committees to increase distributed access to these information resources, including helping users access information by navigating the Internet.

Targeted Outreach Activities

Clinical Librarianship

Librarians develop outreach services based on their users' requirements for timely, targeted information from the literature. Many VALNET libraries offer clinical librarian outreach programs where a librarian participates in ward rounds or teaching conferences (e.g., grand rounds, morning report, and tumor board), identifies case-related literature, and provides copies of the most relevant articles to staff, residents, and students. Clinical staff cite timely access to journal articles, the educational value of the service, and the discovery of new information as advantages of these programs.

Critical Assessment

Increasingly, librarians are combining their understanding of the literature, user requirements, and research methodology to quality filter bibliographic search retrievals, calling only the best articles to the user's attention. Much of the leadership toward this systematic, evidence-based approach to the practice of medicine and medical education is being provided by the Evidence-Based Medicine Working Group at McMaster University. This group publishes User's Guides for assessment of clinical and research oriented literature (Evidence-Based Medicine Working Group, 1992). The Medical Library Association and some academic institutions offer instruction for librarians who want to learn how to apply rules of evidence to assess the validity of articles. The ability to evaluate critically an article and apply standard rules to determine its validity is an important component of evidence-based medicine. VALNET librarians are encouraged to study the User's Guides, take continuing education courses, and use VA-provided audiovisual courseware pertaining to statistical analysis and research methodology, so that they can improve the precision and quality of literature reviews supporting clinical decision making.

User Mentoring and Instruction

Another way that librarians provide users with access to external, knowledge-based information is to teach them how to search bibliographic databases, access decision support tools, and find information by navigating the Internet. Most VALNET librarians help healthcare professionals learn to search the literature directly, either through NLM's GraTEFUL Med or other user-friendly versions of health sciences databases.

The mix of knowledge sources and services available to medical staff and patients should be dynamic. VALNET librarians are encouraged to pilot test new knowledge sources and introduce them to healthcare providers and other users. For example, VA librarians, information managers, and clinical staff recently participated in beta testing the Windows and Internet versions of NLM's GraTEFUL Med, and the web-based version of the DXPlain decision support system developed at the Laboratory of Computer Science, Massachusetts General Hospital, Harvard Medical School. Many VALNET librarians also provide user training for office automation software, computer-based training programs, and e-mail. Of the 73 VALNET libraries with Internet access, more than half provide Internet training. This is generally one-on-one rather than classroom instruction. A handful of librarians provide instruction supporting Decentralized Hospital Computer Program (DHCP) clinical packages. VALNET librarians view user instruction, for VA and external databases, Internet navigation, computer-based text and educational programs, as an area where they can make significant contributions. As VA medical centers improve and increase their access to Internet, networked databases, texts, and multimedia resources, the librarian's teaching and navigational roles will become even more important.

The types of activities described above should be part of a planned approach to management of information. More successful outcomes for integrating external knowledge sources will be achieved by including the customer (to identify needs), the librarian (to identify all appropriate data sources, evaluate search and retrieval capabilities, and train end users), and information systems staff (to determine how best to deliver the information resource based on the local environment).

Shared Information Services

One of the advantages of VALNET membership is the collaboration and sharing of professional expertise. VA librarians look for opportunities to leverage their investments of staff time and resources used to support their local needs. Often, the types of information identified by one group of users as a need will reflect needs at other medical centers. The following are examples of VALNET's responses to broad user needs.

Network Bibliographies

VA managers and healthcare professionals look to the external literature to learn about the rapidly changing healthcare environment. While individual managers may have specific questions and information needs relating to their local situation, they also share an interest in many broad areas relating to trends in health care. Monthly electronic bibliographies are produced by VALNET librarians in a number of sites: Washington, DC; Northport, NY; Portland, OR; Philadelphia, PA; Muskogee, OK; and Lexington, KY. Subjects include customer-focused healthcare delivery and satisfaction, marketing health care, organizing patient-care delivery in a managed-care environment, quality improvement, and women's health and women veterans. Citations are identified from a variety of sources, primarily databases in the Medical Literature Analysis and Retrieval System (MEDLARS). The citations are then compiled into subject-specific bibliographies and transmitted to interested VA staff via a self-enrolling FORUM mail group called Library Bibliographies. Some librarians modify these listings based on local needs and redistribute them to groups within their medical centers. Sharing electronic bibliographies reduces duplication of effort and improves access to information on high-interest subjects. In the future, these bibliographies will be available on VA electronic bulletin boards and internal and external World Wide Web pages.

AIDS Information Center

The VA's AIDS Information Center, located at the San Francisco VA Medical Center, was established in 1989 to meet the rapidly expanding information needs of those involved in VA AIDS/HIV patient care, education, and research. The center's activities complement local efforts to provide AIDS-related information. The center's librarian develops and transmits electronically a biweekly newsletter that provides bulletins, resource reviews of AIDS-related materials, updates from the Centers for Disease Control, perspectives of clinicians, and other information. Another electronic publication, the AIDS news service, includes news reports and journal article abstracts written by the center's librarian. The AIDS information center also supports educational activities of VA's national AIDS education task force and working groups, and the AIDS service, part of VA's Office of Public Health and Environmental Hazards.

Network Systems and Services

The Library Program Office, in VA Headquarters, manages and coordinates a variety of centrally provided services, such as shared database access, cataloging, and distribution of audiovisuals, to make it easier for VALNET librarians to identify and deliver information and educational materials to healthcare providers and patients. All of these activities involve resource identification and sharing among VALNET libraries. While many of the systems and processes described in this chapter facilitate the librarian's management of collections and knowledge sources, they are also the basis for making this same information accessible to users. As we move toward an electronic information environment, these activities must continue to be performed.

Cataloging of Books and Audiovisuals

Unless a book or learning resource is adequately described, in terms understood by librarians and users, it is difficult to locate and use. Cataloging and classification are the description of authorship, physical characteristics, and subject contents of bibliographic material. Cataloging is a specialized subfield of librarianship and conforms to national library information standards. Rather than duplicate this specialized expertise in individual VALNET libraries, VA has chosen to contract for centralized cataloging services. To ensure internal consistency and adherence to national library information standards, VA requires contractors to have experience and expertise in using Library of Congress (LC), NLM, and Dewey Decimal classification schedules, Anglo-American Cataloging Rules, Machine-Readable Cataloging (MARC) format, Online Computer Library Center (OCLC) databases, and the LC Rule Interpretations (LCRI).

Currently, VALNET library staff generate requests for cataloging locally and mail the requests to a centralized processing site for cataloging. When possible, the cataloging contractor locates a matching record on the OCLC database, orders catalog cards that are then mailed to the requesting library, and updates OCLC's master database. The contractor often modifies an existing record, based on the requesting library's preferred classification system or variables such as format of materials. If there is no existing record, the contractor performs original cataloging of the material and enriches the OCLC database. Centralized cataloging allows VALNET libraries to save time and money through volume discounts per unit cataloged, reduce administrative overhead (since much of the management is provided centrally), and have access to experienced catalogers. Another important benefit of centralized cataloging is the initial capture of library holdings (ownership) information that is the basis for resource sharing and improved public access.

Several frustrating aspects of VA's centralized cataloging program could be improved by using existing information technologies. For example, the next generation of contract specifications will include the requirement for the cataloging contractor to accept an electronic request form. This will reduce turnaround time for service by one workweek. Additionally, as more VALNET libraries move from manual card catalogs to automated

318   Clinical and Support Applications

public access catalogs, the centralized contract office will provide for electronic delivery of cataloging records.

Union Listings for Location of Library Resources

The ability to aggregate holdings records for all of the books and audiovisuals owned by VALNET libraries is an important byproduct of the cataloging process for resource sharing. Monthly batch tapes of OCLC MARC records for all items cataloged for VALNET libraries are provided to a commercial library database service company, Library Systems and Services (LSSI), for inclusion in a CD-ROM combined or "union" listing of VALNET holdings. The CD-ROM product, VA Library Public Access Catalog (VALPAC), is self-supporting through purchases by individual VALNET libraries as a subscription. Currently VALPAC is updated semi-annually; however, discussion is underway to develop a World Wide Web-based version that would be updated monthly.

VALPAC is a locator tool for the VA's 200,000 unique books and audiovisuals, providing easy and timely access to network library collections for VA staff and patients nationwide. In 1995, VA staff shared almost 20,000 books and audiovisuals held by other VA medical centers. This type of resource sharing, made possible by VALPAC, helps VA avoid unnecessary duplication of purchases and payment of interlibrary loan charges.

VALPAC's user-friendly search interface permits access by author, title, subject, unique identifiers (e.g., ISBN, ISSN, LCCN), and format (e.g., book, media). (See Figures 21.2 and 21.3.) It is also possible to limit a search to a particular library's holdings by selecting the library's unique five-letter

```
                       VALPAC LOCATOR
                          by LSSI                     Wed Mar 6, 96
                       BROWSE SEARCHING


         OCLC Number:

               Title:

              Author:

             Subject:

    ISBN/ISSN Number:

         Instructions:  To initiate a search
            . Locate the cursor on proper search criteria
            . Type the term to be searched
            . Press Enter key to start the search

Enter: Search
F3:  Clear Screen
F9:  Previous Menu        F10: Help
```

FIGURE 21.2. VALPAC entry screen.

```
     LC Call #:  ¤ RA399.A3 C66 1994
    NLM Call #:  ¤ W 84 AA1 M3685c 1994
   DEWEY Call #: ¤ 362.1/068/5
         Title:  ¤ Continuous quality improvement in health care : theory,
                 ¤ implementation and applications/  Curtis P. McLaughlin,
                 ¤ Arnold D. Kaluzny, [editors]
        Imprint: ¤ Gaithersburg, Md. : Aspen Publishers, Inc., 1994.
   Physical Des.:¤ xii, 467 p. : ill.; 24 cm.
          Notes: ¤ Includes bibliographical references(p. 435-447) and index.
        Subject: ¤ Total Quality Management.
        Subject: ¤ Delivery of Health Care--organization & administration--
                 ¤ United States.
    Added Entry: ¤ McLaughlin, Curtis P.
    Added Entry: ¤ Kaluzny, Arnold D.
                 ¤
    Location(s): ¤ CALBC
                 ¤ DEWIL
                 ¤ FLTAM
                 ¤ ILDAN
                 ¤ ILHIN

        F1: MARC Record            : Next Dn      ESC: Search Menu
        F2: Print Record   PageUp: Prev Page       F9: Return
                 : Line Up PageDn: Next Page      F10: Help
```

FIGURE 21.3. VALPAC record showing 5 holders for this title.

code as a search parameter. For example, a user could search for a book on quality improvement held by the library in Long Beach, California, by adding CALBC as a search parameter.

VALPAC can be mounted as a stand-alone or network database. Some libraries configure the database so that only local holdings are searchable by users. This capability adds a degree of automation, enhancing card catalog access; however, because VALPAC was designed as a locator tool rather then as an integrated library system, certain functions such as copy-specific status information are not present.

Some of the features that make VALPAC less useful as a local catalog result from the same policies and procedures that make centralized cataloging efficient and economical. For example, multiple libraries request cataloging for the same title at different times. The VA's contractors produce catalog cards and update holdings, using the most authoritative record in the OCLC database at the time of request. This practice can result in minor variations in the cataloging record and create duplicate records. To make it easier to identify holders, VALPAC developers merged different records for the same title, consolidating all the holdings on one record for convenient access. This serves the primary purpose of VALPAC as a locator tool, and represents the descriptive bibliographic information accurately, but may not reflect the exact OCLC cataloging record that was used for that library.

Online Public Access Catalogs

Whereas VALPAC is a composite of the various cataloging records that have been used on OCLC to describe VALNET holdings, the VALNET archival tape, also maintained by LSSI, is a repository of the exact cataloging records used for each individual library. VALNET libraries can purchase tape extracts of their individual holdings to begin the process of automating their card catalog and integrating library circulation functions. The fact that VALPAC developers have compiled and maintained the VA tapes from the larger OCLC database permits VALNET libraries to acquire holdings tapes at reduced cost. Additional tape processing, involving procedures such as removing duplicate records, deleting records for items no longer owned, and moving data from one field to another to meet local needs, is part of the retrospective conversion process necessary to make library holdings more accessible to users.

Currently, only 25% of VALNET libraries have automated public access cataloging and circulation systems. The majority of these libraries have purchased commercial library packages and added their holdings through tape processing from the VALNET tapes. A few libraries are using locally developed packages, created by using VA FileMan, a database management system that supports DHCP. The Library Expert Panel, the VALNET automation users group, has reviewed all of the locally developed library packages, and is discussing the development of class-three software (locally developed) that would combine the most useful features, for example, circulation, overdues, ability to print accession lists, bar codes, etc. Currently, libraries using locally developed software need to input a cataloging record for each title in their collections; however, the ability to convert OCLC MARC records to VA FileMan is also being discussed. Often the library staff abbreviate and modify standard information from OCLC MARC records for ease of entry and to suit local requirements. Local modifications to the software, such as changes to the way the bibliographic information is captured and displayed, would reduce the ability to merge or consolidate databases from several medical centers.

Management of Journal Collections

The VA staff and patients use information from health sciences journals more heavily than any other library resource. In 1995, VALNET provided almost 2,600,000 journal articles in response to VA staff and patient requests. If VA were unable to locate these items in VALNET, and had to acquire them (at $8 per article) from libraries outside of the system, costs for access to information would increase significantly. Understanding that management of journal/serial collections was VALNET's highest priority for automation, the Library Expert Panel began working with DHCP software developers at the Washington Information System Center, now called the Washington Information Resources Management Field Office (IRMFO), to develop a serials management package (McVoy, Leredu, Hunter, & Clark, 1987).

```
Displaying TITLE AUTHORITY file
TITLE: NEW ENGLAND JOURNAL OF MEDICINE    LOCAL SERIALS: ACTIVE LOCAL SERIALS
   PREDICTION PATTERN: S07.  THU/W/2V/Y,ST:JAN-JUL;ISRESET
   FREQUENCY: WEEKLY                    PORTION: NATIONAL
   SYNONYM: NEJM
INDEXED: INDEX MEDICUS
INDEXED: BIOLOGICAL ABSTRACTS
INDEXED: PSYCHOLOGICAL ABSTRACTS
INDEXED: EXCERPTA MEDICA
INDEXED: POPLINE
INDEXED: ABRIDGED INDEX MEDICUS
INDEXED: MAP NOTES
INDEXED: CANCERLIT
INDEXED: CANCER CORE JOURNALS
INDEXED: CHEMICAL ABSTRACTS
INDEXED: TOXLINE
   GENERAL NOTES:
CONTINUES BOSTON MEDICAL AND SURGICAL JOURNAL.
MONOGRAPHIC SUPPLEMENTS ACCOMPANY SOME ISSUES.
   SERLINE UNIQUE IDENTIFIER: N14600000  PUBLISHER: MASSACHUSETTS MEDICAL SOCIETY
   FIRST ISSUE: 198,1928--              ISSN/ISBN: 0028-4793

Enter RETURN to continue or '^' to exit:
   ALT-F10  HELP  ` VT-100    ` FDX `   9600 E71 ` LOG CLOSED ` PRT OFF ` CR    ` CR
```

FIGURE 21.4. Serials module: title authority file (TAF).

The serials module, released to VALNET in 1987, automated a number of tedious, recordkeeping tasks of a manual card system, and provided many varied listings and reports, previously unavailable, without hours of manual compiling and word processing. The module includes complete bibliographic information for each journal from a centrally maintained Title Authority File (TAF), as shown in Figure 21.4. Individual libraries use this file to build their record of local holdings. New or unique titles, added by one library, enrich the national TAF database. Other features of the serials module include journal check-in, routing of new issues, claiming issues not received, prediction of next issue, and reports such as holdings lists (see Figures 21.5–21.8). Recent enhancements to this module afforded

```
TITLE: NEW ENGLAND JOURNAL OF MEDICINE
   DISPLAY C.1 - RCI C.2-5

      JOURNAL                          COPIES    COPIES    COPIES
      DATE            V(I)             ORDERED   RECEIVED  COMPLETED
      ----------------------------------------------------------------
      FEB 8,1996     334(6)               5

         C                                                  LIBRARY
         A                                DATE              SITE
         T  STATUS           LOCATION     RECEIVED          LOCATION
         ----------------------------------------------------------------
   c1    M  NOT RECEIVED     DISPLAY                        VACO
   c2    M  NOT RECEIVED     STACKS                         VACO
   c3    M  NOT RECEIVED     STACKS                         VACO
   c4    M  NOT RECEIVED     STACKS                         VACO
   c5    M  NOT RECEIVED     STACKS                         VACO
   ToC      TOC NOT RECEIVED

DATE RECEIVED: TODAY//    (FEB 13, 1996)

Enter copy number/s to distribute separated by commas or a hyphen.
Copy number/s: EXIT// 1-5
   ALT-F10  HELP  ` VT-100    ` FDX `   9600 E71 ` LOG CLOSED ` PRT OFF ` CR    ` CR
```

FIGURE 21.5. Check in record showing ability to check in multiple copies.

322 Clinical and Support Applications

```
ALT-F10   HELP  * VT-100    * FDX  *  9600 N81  * LOG CLOSED  * PRT OFF  * CR     * CR
VA Library Network Serials Check-In                    FEB 13,1996
TITLE: NEW ENGLAND JOURNAL OF MEDICINE
   DISPLAY C.1 - RCI C.2-5

ID     JOURNAL                            DATE          CPY'S   CPY'S    DISPOSITION
NUM    DATE           V(I)                RECEIVED      ORD'D   RCV'D    COMPLETED
----------------------------------------------------------------------------------
56     JAN 25,1996    334(4)              JAN 31,1996   5       5        ALL
57     FEB  1,1996    334(5)              FEB  5,1996   5       5        ALL
58     FEB  8,1996    334(6)                            5
59     FEB 15,1996    334(7)                            5
60     FEB 22,1996    334(8)                            5

Do you want to check-in 5 copies of FEB 15,1996 issue today? Yes//




ALT-F10   HELP  * VT-100    * FDX  *  9600 E71  * LOG CLOSED  * PRT OFF  * CR     * CR
```

FIGURE 21.6. Serials module: check in record showing prediction pop-up.

each local site the ability to maintain consistent, accurate, and current information. This has been accomplished by linking a similar, constantly edited national file to the local site via network e-mail. A patch is now being developed to facilitate use of this package in integrated and multidivision facilities with separate physical locations.

```
COPY NUMBER OR ToC: 3//
GIFT:
VENDOR: READMORE, INC.//
PLACEMENT OF ITEM: STACKS//
CATEGORY OF COPY: MEDICAL//
COPY DISPOSITION: ROUTED AND RETURNED//
START DATE:
STOP DATE:
Select TO ROUTE TO: PEISER//
   TO ROUTE TO: PEISER//
   ROUTING ORDER: 4//
Select TO ROUTE TO: ?
   Answer with RECIPIENT(S) TO ROUTE TO
Choose from:
   PEISER                            4          011B
   PETERSON                          1          132
   HUMPHREYS                         3          011A1
   HEADQUARTERS LIBRARY - RCI        5          193A
   BOWEN                             2          08

      You may enter a new RECIPIENT(S), if you wish
      Choose who will be on the list to see this ISSUE/ToC.
   Answer with NAME
Do you want the entire 210-Entry NAME List?
ALT-F10  HELP  * VT-100    * FDX  *  9600 E71  * LOG CLOSED  * PRT OFF  * CR     * CR
```

FIGURE 21.7. Serials module: showing title set up needed to route journal issues to users.

```
Veterans Administration ROUTING SLIP
    ROUTE TO:
 132         PETERSON/AIDS PROGRAM
 08          BOWEN/OFF OF CONSTR MAN
 011A1       HUMPHREYS/BOARD OF VET APP, MED ADVISORS
 011B        PEISER/BOARD OF VET APPEALS SECT 02
 193A        HEADQUARTERS LIBRARY - RCI/HEADQUARTERS LIBRARY
 *****************************************
 NEW ENGLAND JOURNAL OF MEDICINE
 FEB 8,1996                334(6)
 COPY 3 ROUTING SLIP


 FROM LIBRARY, 142D          FEB 13,1996
```

FIGURE 21.8. Serials module: routing slip.

Sharing of Journal Collections

Department of Veterans Affairs libraries participate in borrowing and loaning arrangements within VALNET and with health sciences libraries nationwide. These arrangements are essential to an inidividual library's ability to provide VA staff and patients with access to the world's biomedical literature. The VA's Union List of Periodicals (ULP), including volume and issue-specific holdings for the almost 50,000 holdings statements for journal titles owned by VALNET libraries, is central to journal resource sharing. The ULP is resident as part of VA's FORUM e-mail system that supports a national communications network. Individual libraries send lists of their journal holdings to a librarian at the Cleveland, Ohio, VA Medical Center who updates and compiles the holdings information, and sends information concerning new titles, not previously part of the network collection to the TAF, currently maintained by library staff in Togus, Maine.

VALNET libraries need to share resources with their counterparts in the local community and with other health sciences libraries nationwide. In 1995, the Library Expert Panel, working with a commercial software developer, and the IRMFO responsible for FORUM, modified the ULP to ensure that data fields in the unit record for each journal title would include the information contained in NLM's SERHOLD database. SERHOLD is a national database of machine-readable holding statements of the serial titles held by biomedical libraries in the United States and Canada. A program, making it possible to reformat VA information from the ULP file so that it can be transferred to the NLM and uploaded directly into SERHOLD, eliminates the need for VALNET libraries to update their holdings in two separate systems. While the characteristics and functionality of VALNET's ULP and SERHOLD differ, the purpose for both of these databases is to facilitate resource sharing.

There is great interest within the health sciences library community, including VA librarians, in being able to provide users with electronic access to full-text articles, preferably with document images. Several commercial vendors are beginning to offer full-text documents to their subscribers. VALNET libraries subscribing to these services can direct the needed information from the literature to the user rapidly and conveniently. Issues relating to copyright protection, intellectual property, and equitable pricing strategies will need to be clarified and resolved before electronically published journals become more widely available.

Interlibrary Loan Module

The Interlibrary Loan Module uses VA's e-mail system, FORUM, to transmit standardized requests for loans between VA libraries. Users query the online ULP to identify holders for a particular title and issue and then complete a forms-based transaction to direct these requests to a location that can supply the material. (See Figure 21.9.)

The VA librarians like the module because all interlibrary loans within VA are filled without charge. Moreover, it links to VA e-mail, includes features that facilitate reserve bookings for audiovisuals, and provides ample capability for borrower and lender comments, linked in a message chain that improves tracking. Because VA staff and patients frequently need access to more than VALNET's 8000 journal titles, and because VA libraries also participate in other resource-sharing networks, most also use NLM's DocLINE, Document Delivery Online, an interlibrary loan and

```
SERVICE:
LOCAL ID:
DATE NEEDED BY:

Is this request ready for transmission? No//    (No)

Do you want to re-edit this request? Yes//    (Yes)
LENDER: MDPER,PERRY POINT//
SUBSEQUENT LENDERS: TXDAL//
FORMAT: PHOTOCOPY//
AUTHOR: GALLERANI//
TITLE: SUDDEN DEATH FROM...   Replace
JOURNAL TITLE: EUR HEART J//
VOLUME (ISSUE):PAGES: 13 (5):661-5 MAY//
YEAR: 1992//
VERIFICATION: MEDLINE//
CCG/CCL: CCG//
AUTHORIZATION: CSR//
PATRON: ROCK//
SERVICE:
LOCAL ID:
DATE NEEDED BY:

Is this request ready for transmission? No//
   ALT-F10   HELP  * VT-100      * FDX  *  9600 N81 * LOG CLOSED * PRT OFF * CR  *  CR
```

FIGURE 21.9. Interlibrary loan module: request form.

```
     1    DEPRESSION AND THE ELDERLY           WM 171 D424 1990
     2    DEPRESSION IN THE LONG TERM CARE SETTING      WM 171 D424 1993
   DEPRESSION IN THE LONG TERM CARE SETTING
CHOOSE 1-2: 1
TITLE: DEPRESSION AND THE ELDERLY        CALL NUMBER: WM 171 D424 1990
   DELIVERY LEVEL: REGIONAL              FORMAT: 1/2 INCH VHS VIDEO
   RUNNING TIME: 25                      COPYRIGHT DATE: 1990
   PRODUCER: FAIRVIEW AUDIO-VISUALS
   SUMMARY:    The prevalence of depression in the elderly is discussed and
   primary symptoms are described.  Common causes and examples of patient focused
   objectives are cited.
   AUDIENCE: PHYSICIANS
   AUDIENCE: NURSES
   AUDIENCE: GERIATRIC NURSES
   AUDIENCE: ALLIED HEALTH STUDENTS
   AUDIENCE: HEALTH CARE PERSONNEL
   AUDIENCE: GERIATRIC HEALTH CARE PERSONNEL
   GENERAL INFORMATION:   Narrator: Diane Knuth.
   SUBJECT HEADINGS: DEPRESSION
   SUBJECT HEADINGS: AGING
   SUBJECT HEADINGS: MENTAL HEALTH


Press return to continue, '^' to exit:
ALT-F10   HELP ˙ VT-100      ˙ FDX ˙   9600 N81 ˙ LOG CLOSED ˙ PRT OFF ˙ CR     ˙ CR
```

FIGURE 21.10. National audio visual (NAV) file listing.

referral system used by about 3000 health sciences libraries in the United States and Canada. DocLINE identifies holders of particular journal titles based on the SERHOLD database.

Dissemination of Audiovisual Information

The audiovisual software delivery program makes a core collection of commercially and VA produced information and training materials available to VA personnel. Materials are bulk purchased, centrally cataloged, and delivered to geographically dispersed sites in the VA Library Network. Based on recommendations by subject experts at Headquarters and in the field, materials are selected to support systemwide programs, not to replace heavily used titles purchased locally. The 550 titles now in the network collection cover subjects such as acquisition procedures, consumer and patient health information, customer relations, quality assurance, and clinical care.

Catalog cards for audiovisual acquisitions indicating the delivery site are sent to all medical center libraries, whether or not they actually hold a title. An online file, available through FORUM, and a print catalog, also facilitate identification of networked audiovisuals. (See Figure 21.10.) Library users can identify and borrow titles of interest from local Veterans Integrated Service Network (VISN) or Regional Audiovisual Delivery Sites.

In the past, 35-mm slide sets, audiocassettes, 16-mm motion pictures, and videocassette training materials were distributed to VALNET libraries, all of which had the equipment needed to offer these programs to users. Currently, almost all of the programs being distributed are videocassettes. CD-ROM and multimedia materials are just beginning to become part of the audiovisual software delivery program. Rapidly changing requirements for playback of this software make it a challenge to distribute.

Satellite TV Network

The VA satellite TV network is an analog, KU band, one-way video/two-way audio network. The network was planned to allow each of the 200 receive sites to receive programming from any source broadcasting over KU band. Half of the programs scheduled are produced by non-VA sources who often agree to offer the programming free or at a reduced cost because VA offers a single point of contact. Use of a single scheduling and technical information source ensures that the flexible nature of the network is preserved, while giving structure to VA-sponsored programming from many sources.

A VA satellite TV network management and scheduling program eliminates network satellite TV network scheduling conflicts and affords the VA, as a national network, the opportunity to participate in quality programming from non-VA sources. The monthly network schedule, along with program and technical information needed to receive the broadcasts, are disseminated over FORUM in a consistent, uniform package to the satellite mail group. (See Figures 21.11 and 21.12.) The network staff ensures print material, supporting VA and non-VA programming, meets network standards for format and accuracy before distribution (Tatman, 1995).

In fiscal year 1995, VA sponsored 53 broadcasts nationwide. Programs on emerging health issues like Persian Gulf veterans concerns and on new policies and legislation affecting VA health care have proven to be of great interest to VA staff.

```
PROGRAM: ETHICS, IN A CHANGING ENVIRONMENT
================================================================================
ID.........: 922.96.J0001              DATE START.: OCT 05, 1995 0100ET
STATUS.....: FINISHED                  DATE END...: OCT 05, 1995 0200ET
ENTERED BY.: CROWELL,JANET B.          MODE.......: SATELLITE TV
FACILITY...: OFFICE OF ACADEMIC AFFAIRS SCOPE......: SYSTEM-WIDE
NTP........: NO                        CONTENT....: LEGAL AND ETHICAL ISSU
SPONSOR FAC:                           CE CREDITS.: NO
SPONSOR OTH: OFC GOVERNMENT ETHICS     COURSE MTRL: NO Mail
LOCATION...: WASHINGTON, DC            PHONE TRBL.: 202-463-5912
CONTACT....: CROWELL,JANET B.          PHONE Q&A..: 800-368-5781
PHONE......: 202-565-7113              TEST SIGNAL: 1245ET

          C SATELLITE...: GALAXY 7     KU SATELLITE..: SBS 6
          C TRANSPONDER.: 11           KU TRANSPONDER: 8
          C DOWN LINK...: 3920         KU DOWN LINK..: 11896
          C POLARITY....: HORIZONTAL   KU POLARITY...: VERTICAL
          C AUDIO.......: 6.2 & 6.8    KU AUDIO......: 6.2 & 6.8
================================================================================
   AUDIENCE: DISTRICT COUNCIL
   <RETURN> to continue, '^' to quit:
   ALT-F10   HELP ° VT-100    ° FDX °   1200 E71 ° LOG CLOSED ° PRT OFF ° CR    ° CR
   REMARKS: PRESENTED BY THE GOVERNMENTWIDE SATELLITE BROADCAST NETWORK -
   ETHICS TRAINER'S PARTNERSHIP TASK FORCE
   REVIEWER REMARKS:
   STATUS: FINISHED//
      ALT-F10   HELP ° VT-100    ° FDX °   1200 E71 ° LOG CLOSED ° PRT OFF ° CR    ° CR
```

FIGURE 21.11. Satellite scheduling package.

```
Subj: ***February 1996 Satellite Calendar***   [#19157918]  07 Feb 96 15:19
  106 Lines
From: CROWELL,JANET - PROGRAM SPECIALIST (CHIEF INFORMATION OFFICE)
  in 'CALENDAR' basket.  Page 1
-------------------------------------------------------------------------------
SATELLITE CALENDAR                              FEBRUARY 1996
                                                (February 7, 1996)

           Chief Information Office Sponsored Broadcasts

Social Workers and   02/09   1:00-2:30    Social workers
  the Challenge of           (12:30 test)
  Violence Worldwide
(Credit Offered)

Acute Stroke         02/21   1:00-2:30    Health care professionals
  Diagnosis and              (12:30 test)
  Prognosis
(Credit Offered)
Enter RETURN to continue or '^' to exit:


ALT-F10   HELP °  VT-100    ° FDX °   1200 N81 ° LOG CLOSED ° PRT OFF ° CR    ° CR

Subj: ***February 1996 Satellite Calendar***   [#19157918]    Page 2
-------------------------------------------------------------------------------
Co-Construct-        02/28   1:00-2:00    Audiologists and speech
  ing Meaning in             (12:30 test) pathologists
  Conversations with a
  Man with Severe Aphasia
(Credit Offered)

Management of        03/13   1:00-2:30    Health care professionals
  Prostate: for              (12:30 test)
  Primary Care
(Credit Offered)

HRM: Information     03/20   1:00-2:30
  Highway FORUM 2            (12:30 test)

Family Responses     03/27   1:00-2:00    Audiologists and speech
  to Acquired Neuro-         (12:30 test) pathologists
  logical Injury
Enter RETURN to continue or '^' to exit:


ALT-F10   HELP °  VT-100    ° FDX °   1200 N81 ° LOG CLOSED ° PRT OFF ° CR    ° CR
```

FIGURE 21.12. Monthly satellite calendars.

Satellite broadcasts bring expert and timely information to large numbers of VA staff without having them travel away from their workplace. Almost all VA-produced broadcasts include a degree of interactivity, generally question-and-answer sessions, within the program. In 1996, the satellite TV network will explore the impact of a touch-pad viewer response system on interactivity. Incrementally, network sites are being retrofitted with the capacity to receive both digital and analog signals.

These plans are being developed in cooperation with the Chief Information Office, whose telecommunications support service is working to provide desktop video teleconferencing, and the Veterans Benefits Administration, which is implementing digital satellite communication within their network.

Summary

The library systems and network structure described in this chapter represent a continuum of service. In the past, librarians were heavily dependent upon large collections, located in close proximity to users. Locator tools and modes of information delivery required much time and patience. The value added by the librarian was primarily to organize physical collections of resources likely to contain needed information, and to give information to the user. Currently, physical collections and locations for library services and operations are becoming less important than access to knowledge-based information.

To meet the needs of local users, VALNET librarians are using new information technologies to expand access to their collections and services and to identify external knowledge bases maintained by others. The librarian's expertise is now needed to locate and assess the information that best matches the user's criteria, and to help the user access that information in direct and convenient ways. Librarians have long observed standard protocols for the description and exchange of resources. Today VA librarians need to change VALNET, to modify its systems, databases, and services to take advantage of emerging information technologies and make VALNET more accessible to its users.

Increasingly, VALNET librarians need to develop and use their professional skills to organize and manage information that they do not own and to help users to explore new technologies and navigate through familiar and new information pathways. They need to develop the skills to critically appraise research results from the literature and to build pathways that connect external knowledge-based information to patient-specific, aggregate, and comparative information. Using these skills, librarians will play a key role in the VA as it strives to improve the quality and delivery of health care.

References

Brandon AN, & Hill DR. 1995. Selected list of books and journals for the small medical library. *Bulletin of the Medical Library Association, 83(2)*, 151–175.

Department of Veterans Affairs. 1995a. *Computer output identification number (COIN), personnel accounting integrated data (PAID) report.* Washington, DC: Author.

Department of Veterans Affairs. 1995b. *Patient treatment file (PTF), outpatient clinic system, and extended care files.* Washington, DC: Author.

Evidence-Based Medicine Working Group. 1992. Evidence-based medicine. *Journal of the American Medical Association, 268(17)*, 2420–2425.

Joint Commission on the Accreditation of Healthcare Organizations. 1995. Section 2, pp. 53–63. Organizational functions/management of information. In *Accreditation Manual for Hospitals, Vol. I, Standards.* Oakbrook Terrace, IL: Author.

Joint Commission on the Accreditation of Healthcare Organizations. 1995. *An introduction to the management of information standards for healthcare organizations.* Oakbrook Terrace, IL: Author.

Kizer KW. 1995. *Vision for change: A plan to restructure the Veterans Health Administration.* Washington, DC: Department of Veterans Affairs.

McVoy JM, Leredu MB, Hunter JW, & Clark NA. 1987. Automating Veterans Administration libraries: I. National planning and development. *Bulletin of the Medical Library Association, 75(2),* 122–124.

National Library of Medicine. 1970. *Abridged index medicus.* Bethesda, MD: Author.

Tatman MA. 1995 (November). *VHA satellite TV network and scheduling program.* Paper presented at Chief Information Office Briefing, Washington, DC.

Wiesenthal D. 1987. VA library service: The VALNET system and resource sharing. *VA Practitioner, 4(5),* 49–51, 57.

22
Using Data for Quality Assessment and Improvement

GALEN L. BARBOUR

Physicians and Quality Measurement

For more than a quarter century, clinical data collection and analysis have been carried out in the name of quality; more often than not the analysis involved the retrospective evaluation of circumstances with outcomes that were less than anticipated. In most cases, the data were the basis of a "search for the guilty" and provided the basis for punitive actions taken to remove "bad apples" (Berwick, 1989). Physician activities were generally a focal point for these searches, and individual practitioners naturally developed a defensive attitude toward the entire data collection and analysis process. These pessimistic attitudes about the value of the quality process generally include a rather negative disposition toward the entire issue of measurement. There is a concept, instilled in us from early childhood, that measurement is always followed by some type of judgement like grades and report cards. Most of us feel we fall short, in one way or another, in some kind of measurement. Because we are reluctant to be judged inadequate, a reasonable defense is to avoid being measured. The human desire to avoid measurement has been amplified, for many physicians, by the punitive use of measurement and data obtained through the "quality process."

Nonetheless, even the most reluctant practitioners are aware of the many and varied pressures on health care to measure its processes and outcomes. As these pressures grow and the amount of data accumulated expands, there is an increased need for all practitioners and measurers alike to recognize an important additional function for measurement: to provide the basis and stimulus for improvement (DesHarnais, 1994).

In the early 1990s, the Office of Quality Management in the headquarters of the Veterans Health Administration (VHA) formulated a creed for its own actions regarding data collection, reporting, and analysis for quality-assurance activities in health care. That creed states

1. Quality is assured at the point of patient contact;
2. Quality is improved at the point of patient contact;

3. Quality improvement is data driven, so
4. Data belongs at the point of patient contact.

The concept that quality can be assured only by those involved in the process is nothing new. Neither is the idea that those involved in the activities are the only ones who can actually improve their processes. The theory that improvement is data driven requires that we understand that change, which may occur for a variety of reasons and without measurement of any kind, is not necessarily improvement. The idea of improvement implies that we know where we are and where we are going (i.e., have a goal) and only by repeated measurement can we determine whether we are moving toward that goal (i.e., improving). Therefore, the results of the measurements and the data analysis need to be in the hands of people involved in the process, not in some office of quality management or headquarters.

Consistent with this creed, diverse reports involving aspects of clinical care were sent from the national databases to field hospitals in the early 1990s. They were not greeted enthusiastically by practitioners, who were suspicious about the data and the motives behind sending out the reports. The greatest concern centered around the data elements themselves. Clinicians reported doubts about the validity of some of the data. Duplicate entries of some information created questions about the reproducibility of any report. Such concerns are not unique to the VHA information systems (Bean, 1994).

The discussions that ensued after the reports were circulated helped us to establish basic criteria for the best data and collection methods. VA practitioners believed that *encounter data* (the measurements needed to make a diagnosis or to follow treatment) and *outcome data* (to properly judge the effect of natural history or interventions) provided the most useful basis for improvements. They wanted these data to be *collected once* and stored for easy access and reuse. Further, they wanted the system for collecting, storing, and retrieving the data elements to be almost invisible and certainly user friendly.

An important issue about the value of outcome determination relates to the common measures used in medicine. In the main, quality-assurance activities focus their attention and screening measurements on the presence of untoward outcomes—mortality, morbidity, and the like. Obviously, focusing measurement in this area biases the collection and the results as far as representing the quality of the entire system. Such highly focused attention also fails to reflect the quality of care when dramatic outcomes do not occur or when our systems do not easily allow tracking and identifying intermediate outcomes.

Consider, if you will, a patient admitted to the hospital for a seemingly routine cholecystectomy. All appears to be going well until the second postoperative day when he suddenly develops chest pain and severe shortness of breath. Over the next few hours, in spite of heroic support and treatment for his acute myocardial infarction, he develops cardiogenic shock, a left-sided stroke and renal failure. His clinical condition stabilized, he starts on dialysis and is ultimately discharged to a nursing home. The current measurement system might well regard this hospitalization as composed of minor problems (largely attendant to the intensive care costs and long length of stay) because the patient was discharged alive. The fact that he can no longer "walk, talk, see, or pee" is not routinely captured in our clinical databases and deprives us of judging the merit of our interventions on the *functional* status of our patients. Proper outcomes measures must begin to include routinely collected functional assessments on all patients; the source of this information may be the patient or the healthcare system.

So, in the early 1990s, we turned to the Decentralized Hospital Computer Program (DHCP) in VHA to provide the necessary elements for improving our practice and outcomes. We found a powerful tool, containing an enormous amount of data—including repeat measurements over time of many important clinical parameters—with essentially the same data dictionary and file structure in every hospital and clinic across the country (Dayhoff & Maloney, 1992).

Unfortunately, the system did not contain certain data elements considered by many practitioners to be key to the proper determination of outcome. The lack of routinely collected functional assessments from patients discharged from hospital care or routinely followed in the clinics is one such omission. The Functional and Independence Measurement System (FIMS) provides this critical information for patients treated in VA extended-care settings but there is no such collection for other patients. The DHCP also did not routinely contain the common clinical measurements such as blood pressure and pulse or, with only a few exceptions, physicians' progress notes.

Defining Quality

What seemed to be needed at that time, to adequately assess quality and to obtain valid measurements on which to base improvements, was a useful working definition of quality health care. After some deliberation and consideration of published definitions, the Office of Quality Management developed and published the following definition of quality for use in the VHA:

Quality health care is that care which is *needed*, and delivered in a manner that is *competent, caring, cost-effective, timely* and which *minimizes risk* and *achieves achievable benefits*.

These attributes of quality, seven in all, should be understood to have a broad applicability. *Needed* implies the obvious provision of the right care at the right time ("doing the right thing"), but there is also the clear implication that every patient has access to the care that is needed. *Competent* means simply that the care delivered is done so in compliance with existing standards of care such as clinical practice guidelines ("doing the thing right"). The recent proliferation of guidelines published by nearly every major clinical society or organization has provided ample measuring tools for comparison. *Caring* is measured from the perspective of the patient and is intended to provide information about the degree to which the patient feels cared about in addition to being cared for. *Cost-effective* really means the proper use of resources to provide the most care for the available funding. The focus here is usually on trying to find and eliminate waste and unnecessary care, but equal attention should be paid to finding those instances where needed care is not being given, with resultant high costs occurring later on. *Timely* reflects both the practitioner's perspective (as when certain medications are necessary at prescribed times) and the patient's viewpoint (such as the timeliness of pain medication administration). *Minimizing risk* to the patient by controlling environmental factors is the most common of these measures but should also be extended to include control of risks to the practitioners and the community at large. *Achieving achievable benefits* may be the most difficult of all these measures. Actually, this measure may represent the summation of all the other measures. To achieve the achievable, medical science will have to increase our knowledge to determine what is achievable and practitioners will have to enlist their patients in a full partnership to reach those goals.

Evaluating Measures

Once this definition of quality health care was developed, we applied it to the measures then in use to determine whether there were underevaluated areas. Veterans Health Administration medical centers and nursing homes provide a definitive scope of services, namely, medical, surgical, and psychiatric, but not obstetrical or pediatric. By using the seven attributes of quality to assess coverage in those areas where we deliver the bulk of our care, including the extended care areas, we determined two important truths.

First, we had far too many measures in some areas. Hospital medical care, for instance, was evaluated by more than 100 separate measures such as mortality rate for each diagnostic group. We set some early goals to appraise each measure for value and to eliminate those not adding worth to the system. Second, certain areas were significantly undermeasured. In particular, we noted that our 1990 system of measures routinely glossed over the assessment of outpatient care, extended care, and psychiatric care and did not effectively provide useful information about the *caring* dimension.

Throughout the history of measurement of healthcare quality, a major concern for many has been the relative shortsightedness—and ultimately the ineffectiveness—of measuring only one aspect of quality. Overzealous attention to measuring outcomes as indicated by mortality rates has created new quality issues for practitioners and system managers. Cardiac surgeons become wary about operating on truly sick patients (often the ones that have the most to gain from a revascularization procedure) because the attendant higher mortality rate may threaten the physicians' certification. Even the addition of complex risk-adjustment protocols has not alleviated this issue. Often the drive to improve mortality rates leads to enormous expenses in personnel and machinery. The attempt to attain evermore difficult increases in survival can render the entire system suspect because of burgeoning costs.

Several institutions and gurus of quality measurement have suggested the obvious remedy: use a series of interrelated measurements, each focused on a different aspect of care, to reflect the whole picture. This concept, called a family of measures or the "dashboard approach" to monitoring progress, has gained some momentum in recent years. The VHA has also adopted the use of a balanced set of measures to reflect quality.

Specifically, VHA uses a value triangle of measures to study care processes and cohorts of patients. The three points of the triangle emphasize different aspects of quality and incorporate the various dimensions or aspects of quality from our definition. First, is patient satisfaction, measured from the perspective of the patient being treated. Second, are measures of the profession's outcome goals, such as risk adjusted mortality rate, infection rate, or complication rate; the functional status of the patient is used here as well. Third, are the costs of the care process for the cohort of patients, carefully reflected. All three dimensions of measurement, simultaneously reported in each area of service, are essential to ensure that concern about one aspect of care does not deflect the system from a proper and practical approach to management.

As a result of the review of measurement capability and the recognition of need for a balanced measurement system, the Office of Quality Management spent notable time and resources on the development of new tools and the improvement of existing tools to remedy the shortfall in measurements of each area of care using every attribute of quality. The office encouraged field personnel to develop local tools and to disseminate information throughout the system about these local efforts. During the first half of the 1990s, new tools were developed, tested, and implemented across the system. The concepts set forth in the creed operated to push for the local use of the resulting measurements to improve the processes of care and the outcomes we were producing in VHA.

Six basic patterns of data capture–storage–retrieval matured over the past 4 years. National and local programs developed capabilities in three separate and distinct domains: DHCP-based, totally or predominantly DHCP-based; totally or predominantly non-DHCP-based, using personal computer or other stand-alone capability including hybrid technology or

22. Using Data for Quality Assessment and Improvement   335

TABLE 22.1. Categories of data systems used for assessing and improving quality in the VHA (with examples).

Data system	National (examples)	Local (examples)
DHCP-Based	Quality Improvement Checklist	FileMan Routines, Performance Improvement Indicator Database
Hybrid	External Peer-Review Program	Ischemic Heart Disease CARE-GUIDE Project
Stand-Alone	Patient Feedback Program	PC-Based Clinical Indicator and Improvement Monitoring

shared databases between DHCP and another data system. Examples of programs are shown in Table 22.1.

DHCP-Based Programs

National Programs That Are Predominantly or Totally DHCP-Based

Several different data reports are available to field hospitals from national level searches of aggregated information. These searches provide the local facilities with their own data in comparative format and trended over time. Topics include clinically relevant information such as the incidence of pressure sores in patients treated in extended care beds, or the average length of stay (ALOS) for patients hospitalized for routine elective surgery that could have been done on an outpatient basis. Compiled quarterly, these reports display data for the facility over the past four or five quarters and compare them to the national VA average for the same activities. The data can also be compared to data for similar activities in other VA facilities. Used by facility directors and chiefs of staff to identify areas for improvement, these reports exemplify the usefulness of DHCP databases to local facilities. Other national databases can be used as a screening tool to assist system and local managers to identify potential areas for improvement (Goldman & Thomas, 1994; Iezzoni, 1990).

The most noted of the national programs is the Quality Improvement Checklist (QUIC). This program consists of nationally written routines that are exported to the facility level where they are run to extract data from local DHCP data files and create a new database (Barbour, 1994). Drawn from the facility's most recent experience, the information in this database relates to performance in several key clinical areas, such as mortality rates for common conditions. The newly created database is reviewed by the facility and, after approval by the director, is electronically transmitted to one of the information service centers in the system where it is aggregated into a national report with comparative displays of every facility report for each area measured. That report, in hard copy, is shared with each facility within 1 month after the completion of the data collection.

Several major points should be emphasized about the QUIC and its implementation. The concept of using an automated data evaluation system to pinpoint concerns in important areas is the idea behind the Indicator Measurement System of the Joint Commission for Accreditation of Healthcare Organizations (JCAHO) (Nadzam et al., 1993). Clearly, the JCAHO would prefer to have comparative data on important clinical issues to make decisions about actual site visits to individual hospitals; that same level of information from the QUIC regularly allows VA managers and clinicians to focus their attention on areas needing improvement when compared to the average activity in the system. The success VHA has had in bringing the QUIC into play—and in seeing the data from its implementation used for clinical improvements—should encourage the JCAHO to continue their efforts to implement a similar system in non-VHA hospitals. Meanwhile, VHA intends to augment the QUIC by using some of the more useful clinical indicators developed by the National Committee for Quality Assurance (NCQA) that are part of the Health Employer Data Information System (HEDIS) measures. Use of these monitors on a regular basis within individual VHA hospitals will not only lighten the burden of preparing for JCAHO surveys but will provide a good measure for comparing the quality of care in VA hospitals with that in nonfederal hospitals.

The QUIC set of indicators was derived by a task force of clinicians and managers in early 1991. The members of the task force used their experience and a thorough review of the clinical literature to determine key characteristics of the monitors they were developing. The decision to use a clinical indicator format for the individual monitors in the QUIC enabled them to develop a standard format for use in every question, making possible the ultimate use of the QUIC instrument as an educational tool. The clinical indicator format included five components: (1) a clearly worded question, (2) a brief explanation of the importance of the topic, (3) a set of definitions of terms, (4) a listing of the data sources and an explanation of the inherent calculations followed by (5) pertinent references from the current literature. These questions were edited to a maximum of one page in length and published as a handbook to accompany the electronic extract package into the facility.

The QUIC is collected and reported every 6 months, usually in May and November. The instrument has changed somewhat since its introduction in November 1991. That first version contained 53 questions that required a total of 91 different answers. Fifteen questions were answerable with a "yes" or "no" response, but the bulk of collected information involved direct measurement taken from the DHCP files and used to calculate rates of mortality, morbidity, lengths of stay, and the like. There were also some administrative questions dealing with waiting times, chart availability, and certain utilization measures.

In the second year of its use, the QUIC instrument was enhanced by the addition of local programmable flexibility in the questions themselves. This new capability allowed local clinicians to change parameters in the search extracts to use the system to search the DHCP for additional diagnoses and use different timeframes than those required by the national report. A graphing package was also added to the QUIC, allowing the local managers to display the data from their own files in a trended fashion over the past several collection periods.

At the local facility level, the quality manager is responsible for sharing the questions, the handbook, and the results of the indicator measurement with appropriate individuals on the staff. In many VA facilities, teams of involved individuals follow up on the biannual reports with further measurements and take actions to correct deficiencies. Several facilities have made striking improvements in clinical care processes or outcomes and have encouraged others to take similar actions.

The Long Beach VA Medical Center (VAMC) used QUIC data to document a significant time delay in the treatment of patients with chest pain who were eligible for thrombolytic therapy. An interdisciplinary team studied a flowchart of the treatment process and found 35 separate steps. The team introduced process changes that produced steady reductions in the "door to drug" treatment time but left the facility with a time that was still 50% longer than the community standard. The team revisited the issue, made additional changes, and further reduced the treatment time to the community level. Continuous monitoring with the QUIC helped the team and involved members of the treatment process to evaluate the effect of their actions and the changes being made in the overall system of treatment. Patient care was clearly improved.

The Durham VAMC exhibited the highest utilization of laboratory tests per patient day of hospitalization in 1993. There was no difference in usage between the two medical teams that provided coverage to all hospitalized patients. The clinical and administrative leaders arranged for weekly collection of information about test usage using the flexibility of the QUIC and provided weekly feedback to one of the teams during a controlled study period between August 1994 and May 1995. During that period, approximately 2600 patients received care from these two teams, and primary measurements were obtained regarding the laboratory usage rate and length of stay for all patients. At the end of the study period, the team receiving weekly feedback used 12% fewer laboratory studies than the other team, and its patients had a significantly shorter average length of stay. The reduced laboratory costs accounted for a savings of about $19,000 and the reduced days of care equalled savings of nearly $650,000. Adjusting for fixed costs, conservative estimates are that the routine use of QUIC data and weekly feedback could save $350,000 annually in that hospital without adverse effects on patient care or outcomes.

At the Washington, DC VAMC, the first QUIC data set rated the hospital's mortality rate for patients admitted for treatment of gastrointestinal bleeding to be higher than that in other similar hospitals. A clinical team of staff physicians, residents, and fellows studied their treatment patterns and identified practices where improvements were needed. They were able to effect those changes right away and were rewarded with improved patient outcomes almost immediately. By the time of the following QUIC collection and reporting period, the mortality rate for these patients (never exceedingly high) was about average for the VA system. The attention of senior clinicians to significant clinical measures promptly resulted in important improvements in patient outcomes.

According to a survey conducted in mid-1995, more than 80% of VA hospital staff directly involved in data collection and usage for the QUIC rated the QUIC as "above average" for usefulness in quality improvement activities compared to all tools at their disposal; they gave a similar positive rating to the impact that the QUIC was having on patient-care outcomes in their facility. Hospitals reported an average of nearly 10 individuals involved in the data collection. Almost half indicated they were not using the flexibility of the instrument to collect data of local interest. Their narrative comments were mostly laudatory, but some shared their skepticism about the continued usefulness of a centrally directed data collection. Nonetheless, the QUIC provides a credible and relatively prompt and painless way to collect systemic information and to provide local managers with important comparisons. Major concerns not yet resolved include confidentiality of the data and identity of practitioners and the lack of risk adjustment to the reported data.

Local Programs That Are Predominantly or Totally DHCP-Based

Although the DHCP is a national system, much of its power rests in its applicability to and usefulness in addressing local needs at individual medical centers. The "D" in the acronym—denoting the decentralized nature of the system—allows for local adaptations while preserving a basic systemwide architecture. That characteristic of systemness means that the system of VHA hospitals can actually benefit from locally developed software because the basic structure of the platform and files at each facility is strikingly similar to that at every other facility. This feature may be a modest motivator for local development: perhaps the local program will be used throughout the nation if it is developed in a manner that provides added value.

A particular local need exists at every hospital to collate and track many of the highly variable quality monitors they use to assess the clinical activity in the facility. In 1994, Bonnie Austin, the quality manager at the VA medial center in Grand Junction, Colorado, spent 1 month in the Office of Quality Management in Washington, DC, as the rotating quality resident. During that month, she was able to take one of the local packages developed at Grand Junction to a higher level of functionality and, in conjunction with the national programmers at the Information Service Center in Birmingham, Alabama, set in motion the coding and testing that allowed the program to be disseminated throughout the system in that same year.

The program initially conceived at Grand Junction is now known across the VA system as the performance improvement indicator database. This class 3 software package functions as a multiuser database for storing, recalling, collating, and reporting information and analysis regarding a wide variety of performance indicators under active tracking within the medical center. The program functions from a master database that contains every monitor in the system. Each monitor can be added, in standard format, to the database by any staff member. Once the indicator monitor is added to the database and assigned to some functional group for oversight, data can be added into the system from a variety of sources and locations (usually new data are entered manually, but demographic data can be addressed from other files as well). The aggregate data are visible to a variety of potential users; specific groups, including the sponsoring oversight group, can access data of interest and schedule a meeting to discuss them, as well as print an agenda and trend data for discussion at the meeting. Afterwards, the database will accept minutes from the meeting with discussion, conclusions, recommendations, actions and follow-up. The shared nature of the database allows management to know the scope of activities in quality assessment and improvement throughout the hospital at any given time.

A partial list of the reports available from the indicator database includes:

- A master list of every process being measured (including the oversight group and the specific indicator(s) in place);
- A master list of every current indicator;
- A list of indicators separated into the JCAHO dimension of performance measured;
- reports of oversight group conclusions and recommendations (this report is available to any service or function that contributes to the interdisciplinary indicator collection; and
- practitioner profiles for credentialling and privileging or competency assessment).

For the practitioner profiling to be effective and valid, both numerator and denominator data must be entered. The availability of these data helps each institution meet the requirements of the JCAHO to have available individual data to assess and analyze the competency of each licensed independent practitioner. The files containing these data are considered sensitive and are protected by menu access; FileMan access to these data is limited to programmers.

As of early 1995, more than 150 VA facilities have requested access to the performance indicator improvement database and the quality manager at Grand Junction VAMC has presented 12 sessions of a teleconference training module composed of five sequential hours; group size has varied from 9 to 30 facilities on each call. In addition, an e-mail users group of over 450 people has been established for less formal interchange of ideas and problem solving. Many facilities have reported highly positive experiences with JCAHO surveys using the tool as they are able to show the breadth and depth of programmatic and process assessment in the hospital as well as to show clear evidence of improvements made.

> *At the Grand Junction VAMC, a consistently high demand for orthopedic consultation had produced a waiting time of more than 90 days for an appointment to this clinic. The ambulatory-care oversight committee (comprised of primary-care teams of nurses, clerks, social workers, and physicians) established an indicator based on the national QUIC data and reviewed possible causes for delays in scheduling. The most apparent reason was that the orthopedic clinic was overcrowded with inappropriate referrals, for example, patients with low back pain where the primary-care physician had already tried most medical therapies. The team instituted a modification of the referral process so that all patients referred to the orthopedic clinic were first seen and screened by the therapist staff in physical therapy and occupational therapy. The therapists examined the patients and considered alternative treatment prior to referral to orthopedics.*
>
> *In the first 3 months after this modification was instituted, 192 patients were screened by the therapists, and only 11% were referred on to the orthopedic clinic. Three others were directed more appropriately to neurosurgery or podiatry clinics or for electromyographic studies. The remaining patients were placed in physical therapy regimens particular to their complaints and within 30 days of their beginning that regimen, the symptoms of 69% of the remaining patients had resolved. The most important improvements included the more timely attention to patient complaints, the addition of patient education regarding their role in caring for their injuries, and an increased satisfaction level for clinicians and patients. Importantly, the average waiting time for the next available orthopedic clinic appointment fell by more than 30% in less than 4 months and is still improving.*

Hybrid Programs

National Tools That Marry DHCP to an External System

A major strength of VA's databases is their comprehensiveness. Still, not all data that are desired by clinicians or that should be available for quality assessment are in the DHCP. In certain circumstances, the specific information is not captured or is too difficult to enter into the system. The current architecture makes it possible to meet these needs for usable data, through the creation of hybrid systems, where data from DHCP are shared with other databases, adding capability to the overall system. One such national system that aggregates information for use in quality assessment is the external peer review program (EPRP) (Walder, Barbour, Weeks, Duncan, & Kaufman, 1995).

Peer review of clinical activity has long been the gold standard by which the activities of practitioners could be assessed against an unwritten community standard. Unfortunately, the so-called gold standard has been tarnished in recent years and has lost a good deal of credibility in the eyes of the public and the profession. An evaluation of the value of peer review as reported in the literature (Goldman, 1992) exposed the difficulties and uncertainties that exist in a review system based, as the one in American medicine has been for so long, in subjective judgement grounded in implicit criteria and firmly established only in the mind of a single reviewer.

The VHA, like the Department of Defense (DoD), began to search for ways to provide credible outside review of its care without falling heir to the weaknesses of classical single reviewer systems. Using the program developed by the DoD as a model, VHA contracted for an external review of a 5% sample of its care nationwide on an annual basis. The program, EPRP, handled data in the following ways:

First, random sampling of the discharged cases was done using the DHCP databases for the patient treatment file (PTF). Specific diagnoses—ones that are high volume, high risk, or problem prone—were chosen for sampling. Each hospital's discharge records were searched and cases selected randomly from the cohorts of the targeted diagnoses. The records for these patients then underwent further onsite review by trained medical record reviewers.

Each reviewer arrived at the hospital with a laptop computer into which had been loaded a specific set of questions (that follow a treatment algorithm) for each of the clinical conditions under review. The record reviewer transcribed into the computer program the answers to the queries in the algorithm by finding the information in the hard copy of the medical record. The algorithm, actually a simplified version of a clinical practice guideline or pathway, was developed by non-VA physicians who were themselves board certified and in active practice in Joint Commission accredited hospitals (and representing a putative community standard of care and practice that would be judged as outstanding). The VA medical record information, residing in the laptop as part of the algorithm, was electronically transferred to the contractor's main database for further analysis.

Cases where the documented care met the explicit criteria of the algorithm were considered to pass review; in most instances the VA cases fell into this category. For those cases where the criteria were not met, a redacted record was produced and later reviewed by a panel of physician experts; after such review the vast majority (more than 98%) of VA cases were declared to have passed review.

The contractor's database, beginning with the DHCP information from a random sample of records in each diagnostic category and augmented with data taken directly from the medical record, provides a rich source of information for VHA to use in determining far more than a simple pass rate for the examined cases. The use of a hybrid system to amplify the DHCP information stores with additional data of high integrity and validity provides a powerful tool for the VHA's quality-assessment and improvement activities.

Local Tools That Marry DHCP to an External System

As noted, the VHA has set a prominent goal of continuously enhancing the value of the health care it delivers through the use of measurement and data analysis. Focused measurements are intended to reflect the status of outcomes of care, the resource utilization, and the degree of patient satisfaction with processes and outcomes. Melding all three measurements into a system for simultaneous use and interpretation calls for data collection that focuses on patient groups sufficiently homogenous to limit the amount of risk adjustment necessary to interpret the data. Another advantage of using such cohorts for analysis is that their small size allows the rapid accumulation of sufficient information on which to make decisions without having to wait years for statistical significance to be reached.

One of the more innovative and visionary local projects designed to use the strengths of DHCP and non-DHCP systems is the Ischemic Heart Disease CARE-GUIDE Project at the Denver VAMC. Driven predominantly by the cardiac staff, including internists and surgeons, the CARE-GUIDE Project is composed of two connected and somewhat overlapping elements: clinical practice guidelines and an automated clinical record for tracking care.

Using a model for developing and testing clinical guidelines instituted and supported by the Office of Quality Management, the CARE-GUIDE Project emphasized:

1. Choosing a well-defined patient population with opportunities for improvement (In this instance the cohort involved patients with ischemic heart disease ranging from angina pectoris to acute myocardial infarction through revascularization to rehabilitation.);
2. Using VHA seed guidelines prepared by VHA clinicians (Published guidelines from national organizations were used as a starting place when available.);
3. Recruiting pilot sites for testing and sending multidisciplinary teams to a convocation for guideline revision and adoption;
4. Agreeing upon an evaluation plan to measure the effectiveness of the implementation and the outcomes of the guideline;
5. During implementation at the pilot sites, developing careful documentation of barriers and ways of overcoming them for later systemic reference and export.

Today there is general agreement that the use of clinical guidelines or pathways can offer major assistance in reducing variation in medical practice, and virtually every major medical group is producing and circulating guidelines. Still, guidelines have not appreciably affected global resource utilization or patient outcomes. Significant variations in practice patterns persist in spite of the many guidelines available for nearly every major illness or procedure. The issue is not with the accuracy or clinical integrity of the guidelines themselves, but rather with the presentation. A specialty society generally communicates its recommendations to its members or to the profession at large in the form of a multipage document. Indeed, a recent guideline on the approach to unstable angina contained over 75 recommendations buried in 120 pages of text. The practical execution of these recommendations by busy practitioners seems overly difficult. The CARE-GUIDE Project is aimed at using computer-based records, coupled with step-specific guidelines, to place the most recent and useful information directly at the fingertips of the treating practitioners at the point in time where decisions about treatment options are being made.

The automated clinical record portion of the project is designed to incorporate clinical pathways based upon accepted guidelines and to collect and retain patient information regarding treatment preferences along with findings from the history and physical examination. This capability exists in a laptop computer available in the treatment area and connected to the DHCP. Each patient is enrolled in the CARE-GUIDE project by name and social security number; thereafter, information is automatically added to the patient's file, including laboratory results, medications prescribed, etc. The program uses clinical information about a given patient at the point of entry to classify the individual into a defined subset of patients; it then determines whether there is a known practice guideline, predicted sets of outcomes, and estimated costs of care for that subset and, if so, displays them directly to the practitioner. Processes of care, clinical decisions, and changes in clinical status may be entered at any time by the provider; outcomes may be entered at the appropriate time by either the provider or patients themselves.

The interface between the DHCP and the personal computer environment manifested by the laptop computer allows practitioners to enter (rather seamlessly) a patient into the system, access their considerable backlog of clinical information from the DHCP system, apply standard clinical rules that categorize each patient into a clinical cohort, and obtain guideline-based pathways and recommendations for treatment literally "just-in-time." Ultimately having the outcomes of those decisions—especially when those outcomes are evaluated by the patient under treatment—available for analysis holds significant potential for better understanding of our application of evidence-based medicine to the actual care of patients, in VHA and elsewhere in the profession. The CARE-GUIDE concept is a VHA prototype for the development of similar efforts in other clinical conditions. The marriage of DHCP power and PC flexibility can create major improvements in care for our patients. Veterans Health Administration researchers have already documented the value of using such data collection to monitor national and local surgical programs (Grover,

Johnson, Shroyer, Marshall, & Hammermeister, 1994; Hammermeister, Johnson, Marshall, & Grover, 1994).

Stand-Alone Programs

National Tools That Use Predominantly Non-DHCP Collection and Display

In 1992 the Office of Quality Management undertook drastic changes to the instrument by which VHA measured the satisfaction of its primary customers, its patients. This undertaking was largely driven by the importance of patient feedback to the assessment of overall quality. Since 1972, the VHA had measured the satisfaction of its patient population using a survey instrument that was hand scored, entered into a DHCP database at the local facility, and later rolled up into a national database. Although the VHA's scores using this instrument were comparable to those seen in the private sector (Rollins, 1994), there were shortcomings to the survey tool. Deficiencies included the lack of connection between the questions and patient priorities for quality care. In addition, because data were collected locally, some patients were concerned they would jeopardize their care if they criticized the healthcare personnel.

The new instrument, a patient feedback instrument, was developed from information gained from patient focus groups. Patients were interviewed to determine their priorities for quality health care and were asked to describe the behaviors of healthcare workers. Information from the focus group interviews was used to construct a series of questions that allowed patients to indicate whether appropriate behavior—appropriate to attaining the high quality health care they had defined—was regularly exhibited in the care setting where they had been treated. The survey instrument with about 45 to 50 questions was mailed to a random 5% sample of VHA patients; their answers were mailed to a central collection site, the National Customer Feedback Center in Boston, Massachusetts, and scanned into a stand-alone database.

The patient feedback database produced reports for each individual facility depicting the responses of their patients expressed as a "problem score" in the major categories of patient priorities, for example, timeliness, coordination of care, heeding preferences, continuity of care, emotional support, physical discomfort relief, and education. These reports were provided as hard copy. Each score was also plotted against the VHA average and when possible against scores from the private sector for comparison purposes. To create the national report, the survey is conducted once annually, questioning recently hospitalized and ambulatory patients. The questions and the methodology can and should be used more often at the local level to track improvements.

Increasingly, VHA facilities are recognizing the need to shift their resources and primary focus away from the expensive inpatient setting to the more economical outpatient arena. The patient feedback report reinforces their motivation and provides staff with measures reflecting how well they are meeting the needs of the patients in their care. Both the instrument and the results it provides are valuable resources to VHA in identifying key areas of customer satisfaction and determining where improvements are needed (Wilson, Cleary, Daley, & Barbour, 1994a, 1994b; Wilson, Daley, Thibault, & Barbour, 1994).

Local Tools That Use Predominantly Non-DHCP Collection and Display

Despite its powerful capabilities, DHCP presents difficulties for many individuals at the local level. They find the system complicated when they want to store, retrieve, and manage data not routinely collected and stored in the national system. Compared to many personal computer databases, the database capability of DHCP is not as easily used for entry, sorting, or display. Consequently, several facilities have used commercial database programs to store and collate data they collect for specific quality management issues.

San Diego Transfusion Review

The San Diego VA medical center developed an inventory/catalog for their performance improvement activities by using Microsoft's Access. Chosen over DHCP for its customization features and flexibility, the Access database was designed to use a standard format for data entry allowing multiple users access to the information. Part of the entry process called for annotating the rationale for the project (medical center priority, mandated activity, identified performance gap, etc.) and the dimension of performance addressed by the project (access, competence, caring, and minimizing risk), as well as the organizational function involved (assessment, education, and leadership).

An early use of the database involved tracking results of peer review performed by the Transfusion Review Committee and providing graphic displays of those results. In 1992, the number of cases failing screening criteria per quarter was over 21; almost 25% fell into the category of Level 3 practitioner error (i.e., most practitioners would have handled the case differently). The committee instituted a series of changes. In the fourth quarter of 1992, they implemented screening criteria from the National Institutes of Health. In the third quarter of 1993, they switched from retrospective to concurrent review.

Results were dramatic. In the quarter immediately prior to changing the timing of the review, there were 48 cases identified as not meeting the screening criteria (four were Level 3). By the first quarter of 1996, only 13 total cases failed screen, and there were no instances of Level 3 determinations. As the facility noted, the decrement in all forms of failed cases began after they changed to concurrent review, a change they believe was driven by the graphic display of the data and review by the Transfusion Committee, the review nurse, and the chairman of the committee.

Salt Lake City Ventilator Review

At the Salt Lake City VA hospital, a survey of ventilator patients disclosed a rate of nosocomial pneumonia of more than 59 per 1000 ventilator days. This rate was substantially higher than the national average. Data collected over the next 6 months confirmed the elevated rate; a specific database was established in Quattro Pro for tracking results. In April 1993, a multidisciplinary team, including nurses and respiratory therapists, evaluated the data and presented the information and a literature review to their colleagues in the intensive care unit. These bedside caregivers identified several problem areas: suctioning occurring at scheduled intervals without clinical assessment, high-volume lavage during suctioning, lack of coordination between nursing and respiratory therapy, and routine breaking of the closed-ventilation system for suctioning. They instituted changes in their procedures to reduce openings of the closed system; they instituted use of an inline catheter and an interdisciplinary flowsheet (to coordinate treatment). Later they reduced the ventilator circuit change to every 72 hours and ultimately to only once weekly in March 1994.

Over 3 years of tracking, the nosocomial pneumonia rate steadily declined; throughout 1995, it was less than 10 per 1000 ventilator days. Patients have exhibited fewer days of stay, and job satisfaction for caregivers was also improved.

Summary

Real quality improvement requires valid data and the ability to handle those data well enough to allow analysis and comparison to best practices. The VHA's DHCP, actively collecting and storing clinical and administrative information from the system's hospitals for more than 15 years, provides practitioners in the system with a virtual gold mine of information. Past difficulties in mining this lode—largely because of the interface with the Massachusetts General Hospital Utility Multi-Programming System (MUMPS) language—are being eliminated as new interfaces give faster and friendlier access. The belief that the struggle is well worthwhile is steadily increasing.

In the DHCP, the VHA has built a better mousetrap and is using it to monitor its quality of care. The resulting data demonstrate that VHA care is the equal of that provided by other sectors. The data also serve as the base for making continuous improvements. New approaches to using aggregate national data, including cohort analyses and risk-adjustment techniques, hold new promise (Khuri et al., 1995) for the VHA's quality-improvement program (Barbour, 1995) and for the services the VHA provides.

References

Barbour GL. 1994. Development of a quality improvement checklist for the department of Veterans Affairs. *Joint Commission Journal of Quality Improvement, 20,* 127–139.

Barbour GL. 1995. Assuring quality in the department of Veterans Affairs: "What can the private sector learn?" *Journal of Clinical Outcomes Management, 2,* 67–76.

Bean KP. 1994. Data quality in hospital strategic information systems: A summary of survey findings. *Top Health Information Management, 15,* 13–25.

Berwick DM. 1989. Continuous improvements as an ideal in health care. *New England Journal of Medicine, 320,* 53–56.

Dayhoff RE, & Maloney DL. 1992. Exchange of Veterans Affairs medical data using national and local networks. *Annals of the New York Academy of Science, 670,* 50–66.

DesHarnais S. 1994. Information management in the age of managed competition. *Joint Commission Journal of Quality Improvement, 20,* 631–638.

Goldman RL. 1992. The reliability of peer assessments of quality of care. *Journal of the American Medical Association, 267,* 958–960.

Goldman RL, & Thomas TL. 1994. Using mortality rates as a screening tool: The experience of the department of Veterans Affairs. *Joint Commission Journal of Quality Improvement, 20,* 511–522.

Grover FL, Johnson RR, Shroyer ALW, Marshall G, & Hammermeister KE. 1994. The Veterans Affairs continuous improvement in cardiac surgery study. *Annals of Thoracic Surgery, 58,* 1845–1851.

Hammermeister KE, Johnson R, Marshall G, & Grover FL. 1994. Continuous assessment and improvement in quality of care: A model from the department of Veterans Affairs cardiac surgery. *Annals of Surgery, 219,* 281–290.

Iezzoni L. 1990. Using administrative diagnostic data to assess the quality of hospital care: Pitfalls and potential of ICD-9-CM. *International Journal for Techology Assessment of Health Care, 6,* 272–281.

Khuri SF, Daley J, Henderson W, Barbour G, Lowry P, et al. 1995. The national Veterans Administration surgical risk study: Risk adjustment for the comparative assessment of the quality of surgical care. *Journal of the American College of Surgery, 180,* 519–531.

Nadzam DM, et al. 1993. Data-driven performance improvement in health care: The joint commission's indicator measurement system (IM System). *Joint Commission Journal of Quality Improvement, 19,* 492–500.

Rollins RJ. 1994. Patient satisfaction in VA medical centers and private sector hospitals: A comparison. *Health Care Supervision, 12,* 44–50.

Walder D, Barbour GL, Weeks HS, Duncan WH, & Kaufman A. 1995. VA's external peer review program. *Federal Practice, 12,* 31–38.

Wilson NJ, Cleary PD, Daley J, & Barbour G. 1994a. *Correlates of patient reported quality of care in Veterans Affairs hospitals.* San Diego, CA: Association for Health Services Research Conference.

Wilson NJ, Cleary PD, Daley J, & Barbour G. 1994b. Using patient reports of their care experiences to measure system performance. *122nd American Public Health Association Meeting.* Washington, DC.

Wilson NJ, Daley J, Thibault G, & Barbour G. 1994. Patient reported quality care in Veterans Affairs hospitals. *Brigham and Women's Hospital Symposium on Preventive Medicine and Clinical Epidemiology.* Boston, MA.

23
Developing and Implementing the Problem List

MICHAEL J. LINCOLN

Introduction

The Problem List software described in this chapter was developed by the Department of Veterans Affairs (VA) at the Salt Lake City VA Information Resources Management Field Office (IRMFO). The team of developers and clinicians included members of the VA's Clinical Application Requirements Group (CARG) and the Problem List Expert Panel (EP), a multisite advisory panel of VA clinicians appointed by the CARG. The Problem List is currently in release 2.0 (September 1994) and is presently installed in 145 clinical VA sites (April 1996).

The software was designed to support the documentation of patient problems in the Decentralized Hospital Computer Program (DHCP) system. A single, unified Problem List was designed to serve a variety of clinicians in ambulatory and inpatient settings. The clinicians who use Problem List include physicians, advanced practice nurses, physicians assistants, nurses, social workers, clinical psychologists, and others. Problem List has two main functions. First, the software can capture and display the current problem status and the historical evolution of patient problems. For this role, Problem List was designed to support direct clinician input, input from scannable encounter forms, and clerk entry from written or dictated clinician notes. Second, the software also supports a variety of output formats. These formats include direct manipulation of the Problem List interface on a terminal screen, output of summary data via the VA's Health Summary software, and screen display via the VA DHCP's newly developed, PC-based client, the Computerized Patient Record System (CPRS). Table 23.1 contains a summary of the Problem List's key features.

The first part of this chapter briefly reviews some of the work in the field of medical informatics that has provided a context and basis for the VA's work on electronic medical records. The second part of the chapter discusses how the VA's DHCP Problem List operates to support clinical decision making and how it is managed by local VA medical center (VAMC) sites. The third part of this chapter will describe the VA's Lexicon

350   Clinical and Support Applications

TABLE 23.1. Key features of the VA Problem List software.

Rich set of options for user data entry	Supports direct entry, clerk entry, scanned input
Linked to VA Lexicon Utility	Provides coded terminology with local user customization possible
Variety of output options	Terminal interface, PC-client system, links to Health Summary
Links with VA Encounter Forms	Problem sets on encounter forms can be linked to problem sets in Problem List
User customization	Users may choose from complete Lexicon Utility or subsets; local synonyms, links to local Encounter Form problem sets

Utility (LU) (clinical data dictionary) and its relationship to VA Problem List. Finally, the fourth part of this chapter examines the need for more advanced lexical data models to support the Problem List and other VA CPRS applications.

The Role of Problem Lists in the Electronic Medical Record

Weed (1993), in his early work on problem-oriented medical records, said that the problem list is a key feature of paper-based medical records. Weed and other researchers (1989) have subsequently stated that problem lists are also essential features of computerized patient records systems. In 1991, the Committee to Improve the Patient Record, convened by the Institute of Medicine (IOM) published a comprehensive report on computerized medical records. Their book, *The Computer-Based Patient Record: An Essential Technology for Health Care* (Dick & Steen, 1991), describes the potential scope and importance of the computer-based patient record (CPR). Many of the functions of electronic medial records described by the IOM Committee in this book are directly facilitated by problem lists. In my view, the medical record functions they listed that are especially facilitated by a problem list (whether viewed as individual patient records or in aggregates) include fostering continuity of care, describing diseases and causes, documenting patient risk factors, and documenting case mix. Furthermore, given that patient problems are a key index to further diagnostic work-ups and treatment, the Problem List probably also facilitates the functions of assessing and managing patient risk, facilitating care in accordance with practice guidelines, generating care plans, determining preventive advice or health maintenance information, supporting nursing diagnosis and treatment, analyzing severity of illness, managing risk, and assessing physician workloads. Therefore, a problem list provides one key index to diagnosis, treatment, and practice management for individual patients.

Despite the importance of problem lists, they have only recently been computerized. They have also only recently become an object of study in the general medical literature, apart from the pioneering work of Weed and a few others. A Medline search using the Internet Grateful Med system on the keywords "problem list" and "medical records" revealed that most of the relevant articles were published after 1989. Since then, the annual Symposium of Computer Applications in Medical Care has been a major source of information in this area. The fact that the work has been largely published in research symposia reflects the relative youth of the field. Problem list software has also been a late development for informatics vendors, coming long after such computerized stalwarts as admission/discharge/transfer systems, laboratory systems, billing packages, and pharmacy systems were in place.

The late development of problem list software may be related to the late availability of several fundamental technologies needed to make problem lists practical. For example, a truly useful computerized problem list must categorize problems using a controlled vocabulary. When problems are coded, their meaning is unambiguous to other programs, such as billing, quality assurance, decision support, research, and workload reporting software. Until recently, adequate coding systems have not been available. Another factor that limited early development of computerized patient problem lists was the need for clinician input. Recent developments in graphical user interfaces (GUIs) and faster computer systems have facilitated clinician interactions with computer systems. Also, the widespread adoption of personal computers, the advent of the online information services, and the increased use of computers in medical education and practice has produced a new generation of physicians who are more comfortable in using a computer for patient care. I believe these factors have combined to make the development and adoption of problem list and other CPRS applications much more feasible.

As a result of the advances described above, the VA has been building computerized problem lists. At this point, it is useful to review the work of some other investigators in this area. Their findings provide a context for the decisions and development work that the VA has undertaken. One important question the VA faced was how to determine for what audience the problem list is intended. For example, should all types of clinicians use the same problem list, or must each have their own? Henry and Holzemer (1995) examined this question by comparing problem lists generated by physicians, nurses, and patients (self-perceptions). While substantial overlap occurred among these problem lists, there were many unique problems (e.g., from nurses) that added substantial value to the medical record. Henry and Holzemer concluded that a multidisciplinary, unified, nonredundant problem list was required for CPRS. The VA, through its multidisciplinary Problem List expert panel, independently reached this same conclusion.

Another important question the VA faced was how to select a coding procedure for problem data. This coding system was designed to support computation by other DHCP programs while still providing a satisfactory clinical vocabulary. While the International Classification of Diseases, 9th edition, Clinical Modification (ICD-9-CM) has traditionally been used to code morbidity and mortality, it does not extend well to the types of narratives physicians wish to express when describing problems (Scherpbier, Abrams, Roth, & Hail, 1994). Some authors have proposed that meta-dictionaries such as the Unified Medical Language System (UMLS) or multiaxial coding systems such as the Standard Nomenclature of Medicine (SNOMED) may capture the clinical meaning of problem lists better than the International Classification of Diseases (Campbell & Payne, 1994). In our case, the VA Problem List expert panel also concluded that a metadictionary (a cross-referenced collection of source dictionaries) was desirable. Therefore, VA adopted a clinical lexicon that is UMLS based and supplemented with VA-specific sources. An alternative approach considered was to utilize free text narratives in the VA problem list. However, subsequent translation into coded forms would then be needed. Zelingher and colleagues (1995) looked at how well this translation could be done. They attempted to match free text problems that had accumulated in Beth Israel Hospital's Online Medical Record with a dictionary of 846 problems. In their study, they found that 80% of problem names could be mapped to three controlled terminology systems, namely, the UMLS Metathesaurus version 1.4, SNOMED version 3.0, or Read version 3.0 (O'Neil, Payne, & Read, 1995).

There were two reasons for the VA's decision to go directly with a coded system, the VA LU. First, the VA Problem List expert panel wanted to ensure that the clinician had seen and reviewed any coded terms associated with the patient's care (rather than allowing an automated program to pick them). Second, VA required more than 80% coding to permit efficient billing (yes, the VA bills for care in noneligible veterans), for national database roll-ups, for quality assurance, and for decision support. Despite using a coded dictionary, the VA system allows free text problems to be entered when the clinician cannot find a satisfactory code. Interestingly, Zelingher also found that physicians have a great affinity for attaching free-text comments and explanations to problems. Over 68% of problems had an attached comment, with a mean of 98 characters per comment. Therefore, the VA software also supports the addition of free text comments. Physicians use these comments to further explain or clarify any coded problem, to connect related problems, and to "jot down" important data relating to a problem.

An important motivation for developing the VA Problem List was to facilitate interpractitioner communication, given the VA's longtime emphasis on team care by primary-care practitioners and consultants. In this setting, communication is key. In the old days of private practice, when one practitioner was simply scribbling his or her own notes, communication between practitioners was usually formalized as a letter of referral or a consultation report. Today, as more medical systems move toward team managed care, a computer (and particularly a computerized problem list) offers a unique means to facilitate increased communication. The single copy of the paper medical record is not always adequate because it cannot be made available to all team members when they need it. For example, the billing department or another clinician may still have the record when the health educator needs it. In these situations, a computerized problem list can be simultaneously available to multiple practitioners. A computerized record can also save the money required to pull a paper chart and move it around to all of the interested parties. For example, Linnarsson and Nordgren (1995) studied the implementation of Swedestar, a CPR developed to support team-based primary health care. In Swedestar, all patient problems are recorded by the various healthcare team members and appear in the problem list. According to the authors, one of Swedestar's main practice advantages is that the problem list provides an integrated view, across provider categories, to facilitate a shared basis for primary-care team work, enhance quality improvement, and improve follow-up of individual patients.

In summary, the problem list has been found by most authors to be a critical component for patient care. The computerization of problem lists has been facilitated by the development of adequate coding systems, adequate clinician interfaces, and by the increasing computer literacy of physicians. A computerized problem list will become increasingly important as patient care is distributed among members of a healthcare team. A computerized representation of the medical record can best serve the diverse needs of such teams.

The VA Problem List

The VA Problem List has enjoyed widespread and rapid adoption among the VAMCs that use DHCP. While Problem List is not a mandated DHCP component, 145 of 170 VAMCs had adopted it by April 1996. The paragraphs that follow provide a detailed user's and technician's view of the Problem List package. The individual features discussed here were specified by the VA Problem List expert panel after considerable internal debate, research, and user feedback. As such, they represent a valuable repository of design information that may be useful to the reader.

Clinician and Clerk Use of the VA Problem List

The most commonly used clinician interface to the Problem List is the so-called List Manager version. This version runs on the American National

Standard (ANS) terminals commonly used in VA DHCP sites. The VA CPRS, now under development, will feature a GUI to the Problem List package. This new system has been alpha tested at one VA site and is scheduled for beta testing in July 1996 at three additional sites. However, this chapter focuses on the List Manager version, because it is currently released and is available for demonstration to interested readers at almost all VAMCs.

The List Manager is a VA DHCP software utility that permits Problem List and other DHCP programs to provide a "window-like" interface on ANS terminals. This interface features a fixed header and footer with a central, scrollable display area. An example of this interface to Problem List is shown in Figure 23.1. The header contains information that identifies the patient, describes the number and type of problems contained in the scrollable portion, and shows the current date and time. The footer contains a list of actionable menu options (with two-letter shortcuts) and a command line. The center pane, which shows the actual patient problems, is scrollable. When the problem listing exceeds the size of the central display area, the area can be scrolled with the arrow keys on the ANS terminal or by typing a + or − sign at the command line.

When the Problem List is installed at a particular site, the DHCP site manager will add the Problem List option to the menus of appropriate users. The actual DHCP program, which allows a clinician to review problems, add problems, inactivate problems and so forth, is called *GMPL CLINICAL USER*. Another routine, *GMPL DATA ENTRY*, allows clerks to perform similar functions, but adds an additional code, which requires

```
Select PATIENT NAME:cumquat,e  03-04-14 123432432     NSC VETERAN

Searching for the patient's problem list...

PROBLEM LIST            Apr 30, 1994 15:40:06     Page:1 of 2
CUMQUAT,EARNEST Q.   (C4567)                12 active problems
                        ACTIVE PROBLEMS
   Problem                              Updated   Clinic
1  Generalized aches, pain or           4/29/94   GEN MED
   stiffness, Onset 4/25/94
     Associated with nausea.
2  Dream anxiety disorder (Nightmare    4/26/94   PSYC
   disorder), Onset 11/27/92
     Lives on Elm Street.
3  Arm Pain, Onset 3/30/94              4/7/94    AMBULATORY CARE
4  Nocturnal Leg Cramps, Onset 3/30/94  4/1/94    AMBULATORY CARE
     Can't sleep.
5  Diabetes Mellitus,                   4/27/94
       + Next Screen   - Prev Screen   ?? More actions
AD Add New Problems   IN Inactivate Problems   VW Select View of List
RM Remove Problems    CM Comment on a Problem  SP Select New Patient
ED Edit a Problem     DT Detailed Display      PP Print Problem List
                      $  Verify Problems       Q  Quit
Select Action: Next Screen// AD   Add New Problems
```

FIGURE 23.1. A List Manager view of the Problem List application, with the user ready to Add New Problems. In this and subsequent figures, user responses are shown in bold.

23. Developing and Implementing the Problem List   355

```
Select PATIENT NAME:CUMQUAT,E  03-04-14 123432432    NSC VETERAN

Searching for the patient's problem list...

PROBLEM LIST         Apr 30, 1994 15:40:06        Page:1 of 3
CUMQUAT,EARNEST Q.   (C2432)                  12 active problems
                     ACTIVE CLINIC PROBLEMS
     Problem                          Updated     Clinic
1    Generalized aches, pain or       4/29/94     ISC
     stiffness, Onset 4/25/94
     Associated with nausea.
2    Dream anxiety disorder (Nightmare 4/26/94    PSYC
     disorder), Onset 11/27/92
     Lives on Elm Street.
3    Arm Pain, Onset 3/30/94          4/7/94      AC
4    Nocturnal Leg Cramps, Onset 3/30/94 4/1/94   AC
     Can't sleep.
5    Constipation, Onset 3/10/93      4/1/94      AC
+    + Next Screen   - Prev Screen   ?? More actions
AD Add New Problems  IN Inactivate Problems   VW Select View of List
RM Remove Problems   CM Comment on a Problem  SP Select New Patient
ED Edit a Problem    DT Detailed Display      PP Print Problem List
                     $ Verify Problems         Q Quit

Select Action: Next Screen// AD   Add New Problem
Clinic: Podiatry
PROBLEM: Diabetes Mellitus
Searching....
The following 10 matches were found
  1. Diabetes Mellitus * (ICD 250.0) (SNM D-2381)
  2. Diabetes mellitus without mention of complication (ICD 250.0)
  3. Diabetes Mellitus, Experimental *
  4. Diabetes Mellitus, Insulin-Dependent * (ICD 256.01) (SNM D-2385/D-241X)
  5. Diabetes Mellitus, Lipoatrophic * (SNM D-2402)
Select 1-5, '^(number)' to jump, '^' or (enter): 1
```

FIGURE 23.2. Adding the problem "Diabetes Mellitus" using the List Manager interface.

the clerk to first identify the responsible clinical provider. These are two of the main routines in the Problem List package. To use the Problem List, the clinician first invokes one of these routines on his or her main DHCP menu by selecting the Problem List option. As shown in Figure 23.1, the user can select a patient (e.g., Ernest Cumquat) by typing in the patient's name or social security number (and selecting from among any multiple matches). The Problem List can determine whether the user is a clerk by reference to the New Person File. In these cases, the system asks the clerk to enter the responsible clinician's name. The clinician or clerk can then take any of the actions in the footer pane, such as *Add New Problems*.

Figure 23.2 shows how the *Add New Problem* option would proceed during a typical interactive session on an ANS terminal. Because the figure illustrates an outpatient setting, the Problem List first asks which clinic is involved. The user then enters a problem keyword, such as "Diabetes Mellitus." This keyword is used to search the VA LU for potential matches. Considerable flexibility is possible for keywords. Other forms, such as "diabetes," partial words, such as "diab mel," or local and national synonyms, such as "NIDDM," may be used instead. The system then displays the potential matches, up to five at a time. The user can choose the appropriate match by number. After selecting a match, the program prompts the user to enter some additional problem data, as shown in Figure 23.3. Some of this additional data is VA specific, such as the relationship of the problem to chemical agent exposure, ionizing radiation, or "service connection" (i.e., the problem's relationship, if any, to military service). Other data is more generic, such as a free text comment, date of onset, activity status, and chronic or acute nature. When a user is finished entering problems and quits the Problem List package, he or she is prompted to print a new chart copy.

An alternative method of adding new problems is to pick the problem names from a clinic- or user-specific Problem Selection List ("pick list"). To use this option, at least one Problem Selection List must have already been created and linked to the user. Figure 23.4 shows a typical session using a Problem Selection List. In this example, the user picks a problem (Iron deficiency anemia, linked to ICD-9-CM code 281.91) and then fills out the additional data required to completely define the problem. If the desired problem is not on the pick list, then the user may choose the option *A: Add a Problem not on a menu*. In this way, any problem represented in the LU may be added. Clerks may use their clinician's pick list. As in the case of problems entered by keywords, several additional responses are required to completely define a problem. However, all of the responses are optional and may be bypassed with a return key. If problem status is bypassed, it defaults to *Active*, while the other responses, such as service-connected status, default to a null or negative setting.

Problems may be changed after entry. For instance, a free text comment may be added to the problem using the command *CM: Comment on a Problem*. Individual free text comments are limited to 60 characters

```
COMMENT (<60 char): Check foot sore
DATE OF ONSET: T-3       (MAY 19, 1994)
STATUS: ACTIVE// <RET>
   (A)cute or (C)hronic:  Chronic

>>>  Currently known service-connection data for CUMQUAT,EARNEST.:
            SC %: 50%
      Disabilities: NONE STATED
Is this problem related to a service-connected condition? Y  YES
Is this problem related to AGENT ORANGE EXPOSURE? N NO
Is this problem related to RADIATION EXPOSURE? Y
```

```
   Problem: Diabetes Mellitus
            <1 comment appended>
      Onset: 5/17/94              SC Condition: YES
     Status: ACTIVE/CHRONIC           Exposure: RADIATION
   Provider: GRIN,JOH
   Clinic  : Podiatry
   Recorded : 5/22/94 by GRIN,JOH

   (S)ave this data, (E)dit it, or (Q)uit w/o saving? SAVE// <RET>
   Saving......done

   Please enter another problem, or press <return> to exit.  <RET>
```

FIGURE 23.3. Completing the problem entry by filling out optional Problem List fields.

23. Developing and Implementing the Problem List   357

```
PROBLEM LIST              FEB 30, 1994 15:40:06      Page:1 of 3
CUMQUAT,EARNEST Q.  (C4567)                         12 active problems
              AMBULATORY CARE

   1  Endocrine & Metabolic ...

   2  Iron deficiency anemia (280.8)
   3  Macrocytic Anemia (281.91)
   4  Other Anemia (285.9)
   5  Anticoagulation

   6  Lymphatic
   7  Musculoskeletal
   8  Diabetes Mellitus (250.00)

    Enter the number of the item(s) you wish to view
AD   Add a Problem not on a menu    Q   Quit to Problem List

Select Item: Next Screen// 2

>>> Adding Problem #2 'Iron deficiency anemia'...
COMMENT (<60 char): Seems pretty rundown
ANOTHER COMMENT: <RET>
DATE OF ONSET: T-30  (January 30,1994)
STATUS: ACTIVE// <RET>
    (A)cute or (C)hronic: ??
You may further refine the status of this problem by designating it
as ACUTE or CHRONIC; problems marked as ACUTE will be flagged on the list
display with a '*'.

    (A)cute or (C)hronic: A
>>>  Currently known service-connection data for CUMQUAT,EARNEST.:
              SC %: 50%
      Disabilities: NONE STATED
Is this problem related to a service-connected condition? Y  YES
Is this problem related to AGENT ORANGE EXPOSURE? N  NO
Is this problem related to RADIATION EXPOSURE? Y

   Problem: Iron deficiency anemia
        <1 Comment appended>
        Onset: January 1994      SC Condition: YES
        Status: ACTIVE/ACUTE         Exposure: RADIATION
     Provider: GRIN,JOH
       Clinic: GENERAL MEDICINE
     Recorded: 2/10 by GRIN,JON

(S)ave this data, (E)dit it, or (Q)uit w/o saving? SAVE// <RET>
```

FIGURE 23.4. Using a pre-defined Problem Selection List to enter a selected problem.

(longer, multiline comments can be created by adding additional comments in sequence). Comments can later be deleted or edited using the *ED Edit a Problem* option. This option also allows the user to edit any other field, such as the service-connected status of the problem or problem activity (active or inactive). The command line also provides an option to completely *Remove Problems* from a patient's list. However, a permanent audit trail is created; if needed, a manager can later restore any removed problems.

The VA considered and rejected the possibility of having different clinician groups (e.g., physicians and nurses) each create their own problem lists. Instead, the Problem List was designed for joint use by all clinicians. However, the VA Problem List expert panel judged that inefficiency would result unless users could customize their use of a unified Problem List. Therefore, to better accommodate a diverse range of users, the Problem

List software permits customized views of patient Problem Lists, as well as temporary customized views for any one session. Temporary customized views are created by the option called *Select View of List*. The option makes only temporary changes to the user's preferred default view (usually *ALL ACTIVE PROBLEMS*). A prototype user session is shown in Figure 23.5. As seen in the figure, the selected attributes that can be used to filter the Problem List view include problem activity, originating provider, and point of care (e.g., outpatient and hospital). A user's Problem List view may permanently customized by means of the *GMPL USER VIEW* program. For example, a urology chief resident could ask the DHCP clinical coordinator to run this program and limit urology residents' default displays to *ALL UROLOGY CLINIC ACTIVE PROBLEMS*.

The expert panel realized that adverse decisions might be made if a user was not aware of a patient's full problem list. Sites are encouraged to set their users' default views to all active problems and to modify this default only after careful consideration (e.g., through the local Medical Records Committee and in consultation with service chiefs). The Problem List soft-

```
Select Problem List Menu Option: pl   Patient Problem List
Select PATIENT NAME: CUMQUAT,EARNEST    10-15-35 234034567   SC VETERAN
Searching for the patient's problem list...
PROBLEM LIST           Apr 30, 1994 15:40:06        Page: 1 of   3
CUMQUAT,EARNEST Q.  (C4567)              5 of 12 active problems
                           ACTIVE PROBLEMS
     Problem                          Updated      Clinic
1    Generalized aches, pain or       4/29/94      ISC
     stiffness, Onset 4/25/94
       Associated with nausea.
2    Dream anxiety disorder (Nightmare  4/26/94    PSYC
     disorder), Onset 11/27/92
     * Lives on Elm Street.
3    Arm Pain, Onset 3/30/94          4/7/94       AC
4    Nocturnal Leg Cramps, Onset 3/30/94 4/1/94    AC
       Can't sleep.
5    Diabetes Mellitus,               4/27/94
+       + Next Screen   - Prev Screen   ?? More actions
AD Add New Problems     IN Inactivate Problems  VW Select View of List
RM Remove Problems      CM Comment on a Problem SP Select New Patient
ED Edit a Problem       DT Detailed Display     PP Print Problem List
                         $ Verify Problems      Q  Quit

Select Action: Next Screen// VW  Select view of list

CURRENT VIEW: Outpatient, active problems from preferred clinics.
You may change your view of this patient's problem list by selecting one or
more of the following attributes to alter:

AT Active only          SC Selected Clinic(s)  PR Selected Provider
IA Inactive only        CL All Clinics         AP All Providers
BO Active & Inactive    IP Inpatient View      PV Preferred View

Select Item(s):
```

FIGURE 23.5. A default view of a Problem List (typically ALL PROBLEMS or ACTIVE PROBLEMS) can be reconfigured for a session using the VS ("View") menu.

ware provides a prominent alert whenever a user has selected (permanently or temporarily) a partial view of a patient's Problem List. In these cases, the Problem List's header area will state, for example, *ACTIVE PROBLEMS* or *ACTIVE PROBLEMS BY MICHAEL J. LINCOLN* instead of *ALL PROBLEMS*. Department of Veterans Affairs usability testing has confirmed that clinical users benefit from this type of user feedback and can become confused if it is not provided. To encourage an easy switch to a complete problem view, the main Problem List menu in the footer area contains an option *VW: Show All Problems*. To date, the VA expert panel and the VA software feedback system (the E-3R system) have not received reports of any adverse patient consequences resulting from partial problem views.

Another important function allows the user to use a subset of LU terms for problem look-ups. This option is implemented through the *GMPL USER SCREEN* program, which selects a filter for searching LU. For example, a user might choose to use a subset containing only ICD-9-CM and CPT terms. Users can choose from among predefined filters, such as ICD-9-CM CPT only, or create their own filter by choosing from among the source vocabularies contained in the LU. Site managers may preconfigure user filters for certain users or user groups. For example, some nurses might need to use only North American Nursing Diagnosis Association (NANDA) terms for a particular type of work. A NANDA-only filter can be created for them. Another option, *GMPL USER SCREEN* routine, allows a user to turn or off the display of associated numerical codes (e.g., ICD-9-CM numerical codes) shown in Problem List. A third option allows the user to either utilize the main LU file containing 92,241 terms, or a subset of this file by UMLS semantic types. Again, these subsets can be created in a custom fashion by sites.

Management Options for the VA Problem List

The Problem List package provides a variety of management functions. Many of these functions are optional. They allow the Problem List to be customized to better meet the specific needs of diverse VA practice environments and sites. These functions are listed in Table 23.2.

The site manager's assignment of menu access governs which users have access to the various program functions of the Problem List package. For example, as we have mentioned above, the menu option allowing users to run the program *GMPL CLINICAL USER* is typically given to physicians, nurses, clinical coordinators (clinician trainers), and other clinical users. Similarly, the *GMPL USER PREFS MENU* is assigned to clinicians. Other program functions, such as *GMPL CODE LIST* (allowing one to assign ICD-9-CM diagnoses to the Problem List) are typically assigned to clerks.

Four important site-specific parameters may be set within Problem List. The first parameter allows the site to choose whether or not clerk-entered

TABLE 23.2. Problem List management functions.

Assign menus and user options	required
Assign security keys	optional
Edit site parameters	optional
Set user preferences	optional
Create problem selection "pick lists"	optional
Create encounter form links	optional
Find patients with a selected problem	optional
Restore removed problems for a patient	optional

problems are flagged as not yet reviewed by clinicians. To remove the flag, a clinician must subsequently verify the problems. A second parameter allows the site to choose whether problems will appear on problem lists in reverse chronological order (most recent first) or chronological order (most recent last). When a third flag is set, the user is always prompted to have a chart copy printed after making changes to a problem list. A final flag can be set (it is the default) to utilize the LU as the source for problem selections. The alternative is to use only provider free text narratives for problem descriptions.

Problem List utilizes the general security mechanisms provided to DHCP users, such as access and verify codes required to start the DHCP program and display the main menu. In addition, Problem List uses one security key. This key is used to verify whether the user has been properly trained to code provider free text narratives in ICD-9-CM.

One of the most useful, albeit optional, functions in Problem List allows creation of Problem Selection pick lists. Problem Selection Lists can serve three important functions. First, they provide a quick means to enter commonly encountered problems. For example, a Problem Selection List might be created for clinicians working in the Home Oxygen Clinic, where a relatively homogenous group of patients with certain common problems is found. Second, the use of Problem Selection Lists can serve to standardize the LU codes, which clinicians use to represent problems. Third, Problem Selection Lists can be integrated with the encounter forms, which are supported by the VA's Integrated Billing package. When physicians have filled out Integrated Billing encounter forms, the forms can be scanned in and both Integrated Billing and the Problem List will be automatically populated with the appropriate data.

Problem Selection Lists are created with the *GMPL BUILD LIST MENU* program. This program allows the user to: (1) build Problem Selection Lists, (2) copy Problem Selection Lists from an Integrated Billing Encounter Form into Problem List, (3) assign or remove Problem Selection Lists for specific users, and (4) delete (from the site) Problem Selection Lists. To improve program usability for clinicians, the *GMPL BUILD LIST MENU* program allows the creation of hierarchical Problem Selection

Lists. For example, the Home Oxygen Clinic might have several hierarchical categories; for example, Obstructive Lung Diseases, Restrictive Lung Diseases, and Cardiac Diseases. Individual coded diagnoses can be associated with each category; for example, Obstructive Lung Diseases: Emphysema, Chronic Bronchitis, Asthma, and Bronchiolitis Obliterans. In this way, 100 or more commonly selected problems (e.g., 10 problems under each of 10 categories) can be conveniently displayed using the Problem List interface. Figures 23.6 and 23.7 illustrate how Problem Selection Lists may be built. Figure 23.8 shows a completed Problem Selection List. Problem Selection Lists may also be created by importing them from the Integrated Billing Encounter Forms package via *GMPL BUILD ENC FORM LIST*. This program allows the site to synchronize the Integrated Billing and Problem List packages without manually reentering Problem Selection Lists in the Problem List package.

The VA's need to efficiently generate patient bills (for veterans not eligible for free care) has affected the development of the Problem List package. Integrated Billing is a VA package into which patient problems can be entered in parallel with Problem List. However, while Integrated Billing must be populated with ICD-9-CM codes, clinicians often prefer to use non-ICD-9-CM terms to describe clinical problems. The UMLS Metathesaurus, which serves as a foundation for Lexicon Utility, provides a partial solution to this problem. As a "dictionary of dictionaries," the Metathesaurus provides a mechanism to navigate across different source vocabularies. However, the Metathesaurus content alone is not always sufficient to allow DHCP to link a problem code with a billable ICD-9-CM code. For example, perhaps a clinician has entered a Medical Subject Head-

```
Select Create Problem Selection Lists Option: B  Build Problem Selection List(s)
Select LIST NAME: Mirph's List
  ARE YOU ADDING 'Mirph's List' AS
    A NEW PROBLEM SELECTION LIST (THE 9TH)? YES
  PROBLEM SELECTION CLINIC: ?
  Enter the clinic to be associated with this list.
  Only hospital locations that are clinics are allowed.
ANSWER WITH HOSPITAL LOCATION NAME, OR ABBREVIATION
DO YOU WANT THE ENTIRE HOSPITAL LOCATION LIST? NO
  PROBLEM SELECTION CLINIC: CARDIOLOGY

BUILD PROBLEM SELECTION LIST Oct 07, 1994 16:42:17 Page 1 of 1
Last Modified: <new list>                                  0 categories
                         Mirph's List

No items available

        + Next Screen   - Prev Screen   ?? More actions
AD Add Category to List   SQ Resequence Categories   VW View w/wo Seq Numbers
RM Remove Category        CD Edit Category Display   CL Change Selection Lists
EC Enter/Edit Category    SS Assign List             SV Save List and Quit
                                                     Q  Quit
select Action: Quit// EC
Select CATEGORY NAME: Cardiovascular
  ARE YOU ADDING 'Cardiovascular' AS A NEW PROBLEM SELECTION CATEGORY (THE 1ST)?
YES
```

FIGURE 23.6. Building a new Problem Selection List.

362   Clinical and Support Applications

```
Searching for the list
EDIT PROBLEM CATEGORY      Oct 07, 1994 16:42:17   Page 1 of 1
Last Modified: <new category>                            0 problems
                        Cardiovascular

No items available

+       + Next Screen   - Prev Screen   ?? More actions
AD  Add Problems        SQ  Resequence Problems   SV  Save Category and Quit
RM  Remove a Problem    DL  Delete Category       CC  Change Categories
ED  Edit Problems       VW  View w/wo Seq Numbers Q   Quit
Select Action: Quit// ad   Add Problems
Select Specialty Subset: GENERAL PROBLEM// <RET>
PROBLEM: Hypertension
Searching....
The following 70 matches were found:
   1: Gestosis, EPH *
   2: Hypertension, Goldblatt (ICD 440.1)
   3: Hypertension * (ICD 401.9)
   4: Hypertension due to oral contraceptive
   5: Hypertension, Malignant *
Select 1-5, '^<number>' to jump, '^' or <enter>: 3
 DISPLAY TEXT: Hypertension// <RET>
ICD CODE: 401.9// <RET>
HYPERTENSION NOS
       ...OK? YES// <RET>
SEQUENCE: 1// <RET>
PROBLEM: obesity
he following 10 matches were found:

   1: Froehlich's Syndrome
   2: Hyperostosis Frontalis Interna
   3: Obesity *
   4: Obesity in Diabetes
   5: Obesity, Morbid *
Select 1-5, '^<number>' to jump, '^' or <enter>: 3
DISPLAY TEXT: Obesity// <RET>
ICD CODE: 278.0// <RET>    278.0    OBESITY
       ..OK? YES// <RET>
SEQUENCE: 2// <RET>
PROBLEM:   <RET>
```

FIGURE 23.7. Adding problems to a category in a Problem Selection List.

```
Select Create Problem Selection Lists Option: B  Build Problem Selection List(s)
Select LIST NAME: LIST 2
Searching for the list ...
BUILD PROBLEM SELECTION LIST  May 07, 1994 16:42:17 Page 1 of 2
Last Modified: April 20, 1994                     3 categories
                        LIST 2

           1 Diabetes
              Diabetes Mellitus, Insulin-Dependent (250.01)
              Diabetes Mellitus, Non-Insulin-Dependent (250.00)
              Hypertension (401.90)
              Myocardial Infarction (410.9)

           2 Miscellaneous
              Malignant essential hypertension (401.0)
              Lung Diseases, Obstructive (496. )
              Myocardial Infarction (410.9)
              Hypertension (401.90)
+       + Next Screen   - Prev Screen   ?? More actions
AD Add Category to List  SQ Resequence Categories  VW View w/wo Seq Numbers
RM Remove Category       CD Edit Category Display  CL Change Selection Lists
EC Enter/Edit Category   SS Assign List            SV Save List and Quit
                                                   Q  Quit
Select Action: Quit// AD
```

FIGURE 23.8. An already completed Problem Selection List, showing the manager preparing to add a new problem category.

ings (MeSH) term to describe the Problem "mitral valve prolapse." In this case, the Metathesaurus cannot easily provide a link to an equivalent ICD-9-CM term needed for billing purposes. The closest ICD-9-CM representation for the common clinical concept, "mitral valve prolapse" is ICD-9-CM 424.0, "Mitral valve disorders." This is the code under which VA must bill. However, in the Metathesaurus, the ICD-9-CM term "Mitral valve disorders" does not share the same concept unique identifier (CUI) as "mitral valve prolapse." Therefore, Problem List cannot make the link to ICD-9-CM 424.0 and assigns instead a default of ICD-9-CM 799.9, "Other unknown and unspecified cause." The VA has been tracking these 799.9 codes through the Problem List expert panel and other groups, and has added links between unbillable CUI codes and billable ICD-9-CM codes for the 5000 most common problems added to VA problem lists. Often, these links connect a more granular Metathesaurus concept (e.g., "mitral valve prolapse") to a less granular ICD-9-CM code (e.g., ICD-9-CM 424.0). Thus, these links do not indicate synonymy. Other 799.9 codes are also being reviewed to increase the number of linkages between these problems and a valid ICD-9-CM code. Until national changes are distributed, medical records technicians or clerks at individual sites may use a Problem List option to manually identify and link 799.9 problems.

Several other management options are also important. The first of these, *GMPL PROBLEM LISTING*, allows the user to search for patients who have any selected problem. The program requires the user to select the desired problem by name or ICD-9-CM code. This program can then display and print the identifying information for all patients in the system who have the specified problem. This function is useful for researchers, quality-assurance personnel, and administrators who wish to identify sets of patients with specific problems. A second, important function allows a manager to replace previously removed problems on a patient's problem list. The *GMPL REPLACE PROBLEMS* option is usually given only to the clinical coordinator and other site managers. The program displays a list of all problems that have been removed from the Problem List of the selected patient. Optionally, some or all of the removed problems that are displayed may be restored. This function is useful to replace accidentally deleted problems, or for creating a complete problem record for medical-legal purposes.

In summary, the VA Problem List interface provides a rich suite of user and manager functions. The Problem List has been designed by an expert panel of VA practitioners to facilitate direct clinician entry and review. The system also provides for clerk input, from either written clinician narratives or scanned encounter forms. Problem List has been developed using a List Manager interface that simulates some of the desirable functions of a windowing operating system. A GUI interface for DHCP's CPRS is under development and has already undergone alpha testing. A key feature of the

Problem List software is its ability to share data with VA's Integrated Billing package, Health Summary, and CPRS.

Lexicon Utility

The VA LU package is used by Problem List as a term file to populate patient problem lists. Although the LU is designed to be used with other DHCP applications, its main user is currently the Problem List package. The LU package is designed to provide

- A common terminology for identifying problems and other clinical data;
- Interrelationships between terms through use of the National Library of Medicine (NLM) UMLS content;
- The ability for local modifications to the code text and code definitions, and to provide local synonyms, shortcuts, and keywords;
- Compatibility with lexicon subsets for certain users.

Contents of the LU

The currently released LU[1] contains 92,241 terms (version 1.0). The LU was populated in 1993 from the then-current UMLS Metathesaurus and from VA sources. Currently, the LU contains terms from the sources shown in Table 23.3.

The deactivated ICD-9-CM codes in Table 23.3 are notable, because they are not valid for billing purposes. Because Problem List interacts with Integrated Billing, it is not desirable to use these codes to describe problems. These inactivated codes are included in the ICD-9-CM system because they are parents for lower level terms used for billing. For example, ICD-9-CM 250.0, which means diabetes mellitus without complication is a valid ICD-9-CM code but must be combined with a fifth-digit Clinical Modification extension to be acceptable for billing. Therefore, the VA has deactivated the codes that are not valid (e.g., 250.0), while retaining the corresponding fifth-digit codes that should be used instead (e.g., 250.01, diabetes mellitus without complication and not stated as uncontrolled).

A slightly different problem led to the deletion of a certain semantic classes of terms from the LU. For example, LU entries of semantic type "chemical" were deleted from the LU (and are not even shown in Table 23.3). This filtering was necessary to strike a balance between completeness of the LU and low precision of keyword searches (numerous false-positive results during look-ups). When the VA tested early versions of the LU,

---

[1] Originally called "Clinical Lexicon," this program's new version is called "Lexicon Utility." This paper refers to both the released version 1.0 and the new version under development as Lexicon Utility to reflect the new usage.

23. Developing and Implementing the Problem List 365

TABLE 23.3. Sources and terms contained in development version of Lexicon Utility 2.0.

System	Type	Total	Unique	Deactivated	Active	Description
ICD-9-CM	Diagnosis	21,507	14,990	2690	112,300	International Classification of Diseases
ICD-9-CM	Procedures	995	634	0	634	International Classification of Diseases
CPT-4	Procedures	364	235	0	235	Current Procedural Terminology
DSM-III-R	Mental Health	265	203	0	203	Diagnostic and Statistical Manual of Mental Disorders
SNOMED 2	Thesaurus	11,113	6,818	0	6,818	Systematic Nomenclature of Medicine
NANDA	Diagnosis	102	95	0	95	North American Nursing Diagnosis Association
NIC	Interventions	341	336	0	336	Nursing Intervention Classification
OMAHA	Diagnosis	78	76	0	76	Omaha Nursing Diagnosis Classification
ACR	Radiology Dx	119	118	0	118	American College of Radiology
AI/RHEUM	Thesaurus	757	753	0	753	National Library of Medicine
COSTAR	Ambulatory	1,397	1,392	0	1,392	Computer Stored Ambulatory Records File
CRISP	Thesaurus	5,123	4,588	0	4,588	Computer Retrieval of Scientific Projects (NIH)
COSTART	Adverse reactions	1,675	1,124	0	1,124	Coding Symbol Thesaurus for Adverse Reaction Terms
DXPLAIN	Diagnosis	488	487	0	487	DXPlain medical expert system
MCMASTER	Epidemiology	18	18	0	18	McMaster University Thesaurus of Epidemiology Terms
UMD	Medical devices	78	78	0	78	Universal Medical Devices
UWA	Neuronames	586	586	0	586	University of Washington Primate Information Center

which included chemicals, certain keyword look-ups returned far too many false-positive results that were not relevant to coding clinical records. For example, a search on "sodium" meant to code a patient with an abnormal serum sodium brought back hundreds of chemical names not relevant to clinical medicine.

User Options in LU

Users normally interact with LU through the Problem List application. However, users can be assigned two functions that are useful for direct manipulation of LU package. The first of these, provided by the *GMPL CLINICAL LEXICON UTILITY* program, allows simple look-ups in the LU and displays the information associated with the term upon request. This look-up utilizes the VA Multi-Term Look-Up (MTLU) package. The second function allows the user to change the defaults for look-ups. Users may change the search filters preferred vocabulary subsets for look-up, and the format for displaying the data.

Managers have additional options available through the *GMPT CLINICAL LEXICON MGT MENU*. The first option allows the manager to edit term definitions contained in LU. This editing process is illustrated in Figure 23.8. While these changes are local, any changes are placed into an e-mail message and sent to the development staff at the Salt Lake City IRMFO for review. Based on this review, future releases of LU could be modified. A second function allows the manager to edit the default filters, display features, or vocabularies associated with either single users or groups of users. A third related function allows the manager to list or print these user defaults for individual users or groups. A series of figures (Figures 23.9–23.12) illustrate the flexibility provided by LU for editing user and group defaults. While defaults are sometimes changed in practice for individual users, these changes are often most appropriate for particular groups of users (e.g., nurses, psychiatric technicians, physicians, and nurse practitioners).

Unresolved Narratives

So-called "unresolved narratives" are generated by the Problem List package when a user fails to choose a coded entry from LU and enters a free text narrative instead. When this happens, the unresolved narrative text is saved to evaluate and potentially improve the LU. After 50 narratives accumulate in a local site's unresolved narratives file, they are wrapped into an e-mail message and sent to the Salt Lake City IRMFO. In a period from April 1994 to August 1995, the IRMFO received 16,873 narratives from 86 service units in 30 hospitals. Subsequent to that time, an additional 55,584 narratives have been received (April 1996). Table 23.4 shows 10 randomly selected, representative narratives from among the initial batch of 16,873. The

```
Select OPTION NAME: GMPT CLINICAL LEXICON MGT MENU    Clinical Lexicon
Management Menu Option
Select Clinical Lexicon Management Menu Option: ?
        EDT     Edit a term ...
        DEF     Defaults ...
        UTL     Clinical Lexicon Utility ...

Select Clinical Lexicon Management Menu Option: DEF Defaults

Select Defaults Option: ?
        ED      Edit User/User Group Defaults
        LT      List User/User Group Defaults

Select Defaults Option: Ed Edit User/User Group Defaults

Select an application:

   1: Problem List
   2: Clinical Lexicon
   3: All of the Above

Select:  (1-3): 2

Clinical Lexicon Defaults:

   1: Filter              Unselected
   2: Display Codes       Unselected
   3: Vocabulary          Unselected

Select:  (1-3): 1
There are 7 filter options available:
     1    ICD/CPT Only
     2    Nursing Problems/Diagnosis (Less Nursing Interventions)
     3    Problem/Diagnosis (including ICD-9 and CPT-4)
     4    Problem/Diagnosis (including ICD-9, CPT-4 and DSM-IIIR)
     5    Unfiltered (include all, exclude none)

Select (1-5) or press <Return> for more:
     6    No filter - Delete current filter
     7    Create your own

Select (1-7): 3
con't next figure
```

FIGURE 23.9. Setting up Lexicon Utility lookup defaults, part 1.

clinical service of the users generating these narratives is shown in Figure 23.13.

Several Salt Lake City VA IRMFO staff (the author, two VA Special Fellows in Medical Informatics, developers, and a medical student) have been studying a sample of the unresolved narratives to determine how the Problem List and LU can be improved. One of our first studies involved classifying the causes of the unresolved narratives. Because of the large volume of the narratives (a consequence of the large number of VA sites that adopted Problem List), our study was undertaken partly to determine whether multiple physician raters (narrative evaluators) could be trained to produce reliable ratings. Multiple raters were deemed necessary because the volume of narratives required a "parallel processing" approach, where narratives could be parceled out among different reviewers. However, with such a design, we had to ensure that individual raters could produce compa-

```
Look-up filter will:
    Include:                          Exclude:
      Activities                        Chemical and Drugs
      Anatomy                           Geographic Areas
      Behaviors                         Groups
      Concepts/Ideas                    Occupations/Organizations
      Diseases/Pathologic Processes     Gov't/Regulatory Activity
      Physiology                        Machine Activity
      Procedures                        Manufactured Object
      Unknown/Untyped Concepts          Medical Device
      Fungus                            Conceptual Entity
      Virus                             Spatial Concept
      Bacterium                         Functional Concept
      Molecular Function                Intellectual Concept
      Genetic Function                  Language
      Cell or Molecular Dysfunction
      Medical Device
      Substance

Press <Return> to continue  <RET>

    Look-up filter will also include terms linked to:

       ICD   Intl' Class of Diseases, 9th Rev, 1989
       ICD   Intl' Class of Diseases, 9th Rev, Clin Mod, 1991
       CPT   Current Procedural Terminology, 1989
       CPT   Current Procedural Terminology, 4th Ed, 1992

Accept this as the default filter? YES//    <RET>     (YES)

Clinical Lexicon Defaults/User Groups:

 1 Filter            Problem/Diagnosis (including ICD-9 and CPT-4)
 2 Display Codes     Unselected
 3 Vocabulary        Unselected

Select default to modify:   (1-3):  2

Select/Create a display format
  1 All Classification Codes
  2 Codes commonly used by the VA
  3 ICD/CPT only
  4 ICD/CPT/DSM only
  5 Nursing codes only
  6 Delete the current display format, use the application default
  7 Create your own display format

Select (1-7):  5
cont'd next figure
```

FIGURE 23.10. Setting up Lexicon Utility lookup defaults, part 2.

rable results. The study was also designed to examine the specific causes of a random sample of unresolved narratives. We wished to study the causes of the unresolved narratives to understand how the DHCP LU and Problem List might be improved.

To contain the unresolved narrative mail messages received at the IRMFO, a Microsoft Access database was created. The individual narratives received to date were then inserted into this database in a random order. The narratives included in the study came from the 30 VA clinical sites that had installed Problem List and had generated at least one batch of 50 unresolved narratives by August 1995. A form, shown in Figure 23.14,

23. Developing and Implementing the Problem List 369

was designed for rating the cause of each unresolved narrative. The form had been designed in a preliminary study in which several small sets of ratings were made by three raters, and then the ratings were discussed to consensus and changes in the form were designed. In the present study, two physician raters used this form to separately rate 400 randomly selected unresolved narratives.

The raters were instructed to first rerun each narrative using a test patient in the Salt Lake City VAMC's clinical Problem List/LU accounts. This step was required to ensure that the narrative was truly unresolved. This step was necessary because, in some cases, the original clinician-user had simply

```
You have chosen to display the following codes during look-up:

      NAN  N Amer Nursing Diag Assoc, 9th Ed, Mar 1990
      NAN  N Amer Nursing Diag Assoc, 9th Ed, 1992
      NIC  Nursing Intervention Classifications, 1992
      OMA  Omaha Nursing Diagnosis/Interventions, 1991

Accept this as your default display? YES//    <RET>     (YES)

Clinical Lexicon Defaults/User Groups:

 1 Filter         Problem/Diagnosis (including ICD-9 and CPT-4)
 2 Display Codes  Nursing Codes only
 3 Vocabulary     Unselected

Select default to modify:  (1-3): 3

Default vocabularies are:

   1 Dental
   2 Immunologic
   3 Nursing
   4 Social Work
   5 Clinical Lexicon - the union of all available vocabularies

Select Default Vocabulary or press <Return> for more: 3

Clinical Lexicon Defaults/User Groups:

 1 Filter         Problem/Diagnosis (including ICD-9 and CPT-4)
 2 Display Codes  Nursing Codes only
 3 Vocabulary     Nursing

Select default to modify:  (1-3): <RET>

Set current user defaults for:

      1      Single User
      2      Group based on Service
      3      Group based on Hospital Location
      4      Group based on both Service and Hospital Location
      5      All Users in the New Person File
Select: (1-5): 2
Select users by service:  INTERNAL MEDICINE// NURSING    118
Replace existing user defaults? NO// Y       (YES)
Defaults will be set for:

  1. Clinical Lexicon
cont'd next figure
```

FIGURE 23.11. Setting up Lexicon Utility lookup defaults, part 3.

```
Look-Up defaults include:

  1. Filter      Added  Problem/Diagnosis (including ICD-9 and CPT-4)
  2. Display     Added  Nursing Codes only
  3. Vocabulary  Added  Nursing

Users in Service/Section:  NURSING

Replace existing data:Yes  Previously defined defaults will be altered
                           Filter will be replaced
                           Display will be replaced
                           Vocabulary will be replaced

Is this correct? YES//   <RET>    (YES)

Requested Start Time: NOW//   <RET>   (APR 12, 1994@15:57:09)

Task has been created to update user defaults

Do you wish to set the currently selected user
defaults to another user or group of users? YES//  N     (NO)

Select Defaults Option:  ^

Select Clinical Lexicon Management Menu Option:   ^
```

FIGURE 23.12. Setting up Lexicon Utility lookup defaults, part 4. The bold, underlined "up arrow" character answering the last two prompts is a DHCP convention for "exit and go back one menu level."

overlooked or mistakenly failed to choose a valid match. For example, the clinician might have retrieved the exact string he or she wanted, but failed to choose it because it was overlooked in a long list of potential matches. If this was the case, the narrative was classified as belonging to the first of four categories, as shown in Figure 23.14. The first category, *Lex Returns exact surface form* was selected for those unresolved narratives where the user had overlooked and failed to choose an exact match. The second category, *Lex Returns same concept different string* was chosen when the lexicon returned a synonym for the users keyword string, and the clinician did not select the synonym. *Lex Returns different concept different string*, the third

TABLE 23.4. Ten randomly selected unresolved narratives.

Narrative	Service
Angina-stable	Health Services R&D
414.00	Medical Administration
Advice or health instruction	Medical Administration
Gout versus Pseudogout	Medical Service
Possible allergy to penicillin	Medical Service
S/P CABG	Extended Care
Chronically elevated PSA	Nursing Home Care Unit
Pneumococcal vaccination	Nursing
hx-prostatic malignancy	Medical Administration
Influenza vaccine	Medical Administration

FIGURE 23.13. Sources of unresolved narratives by service.

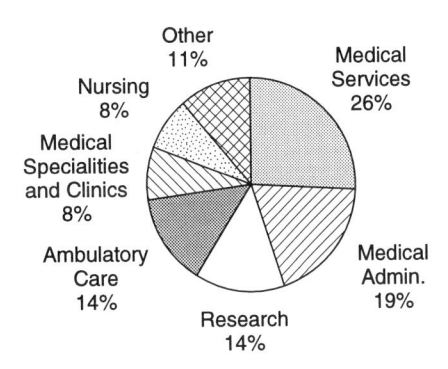

- Other 11%
- Medical Services 26%
- Nursing 8%
- Medical Specialities and Clinics 8%
- Ambulatory Care 14%
- Research 14%
- Medical Admin. 19%

category, was selected by the rater when the lexicon returned some matches to the clinician's keywords, but the matches were different concepts. The fourth category, *Lex Returns nothing* was selected when no keyword matches were returned by LU. Usually, these narratives were described by the clinician at a different level of granularity than the available lexicon matches. For example, the additional modifiers and qualifiers used by the clinician often prevented any match (e.g., keywords "large, right inguinal hernia" would not match any concept in LU, including "inguinal hernia").

Once the category for each unresolved narrative was assigned, each rater then selected the reason or reasons for failure to match (for categories 2 through 4). These reasons, as shown in rating form depicted by Figure 23.14, ranged from synonymous forms that did not match the user's free text input, to misspellings, numerical data (e.g., the user entered a number or code in a format or mode that the system could not recognize), modifiers, and granularity differences.

The results of the analysis by category showed that most unresolved narratives (about 81.5%) fell into category 4, *Lex returns nothing*. These results are shown in Figure 23.15. Inter-rater agreement measures were calculated on this data to determine whether the raters had been successfully trained to produce consistent categorical ratings. The coefficient of agreement, κ was equal to 0.80 ($p < 0.0001$), indicating a high level of inter-rater agreement, and *Finn's r* was 0.76, indicating that 76% of the inter-rater agreement was not due to chance. Bowker's extension of McNemar's test for symmetry of a 4×4 table showed a p-value > 0.10, indicating that there had been no systematic bias among the raters with regard to category selected. These results may have been dominated by the large size of the category 4 bin (*Lex returns nothing*). However, removing category 4 for both raters and looking only at inter-rater agreement for categories 1 through 3 gave similar results: $\kappa = 0.68$, $p < 0.0001$.

The reasons for failure to match in the LU were evaluated by each rater, and the data are shown in Table 23.5. The fourth column in the table shows

372 Clinical and Support Applications

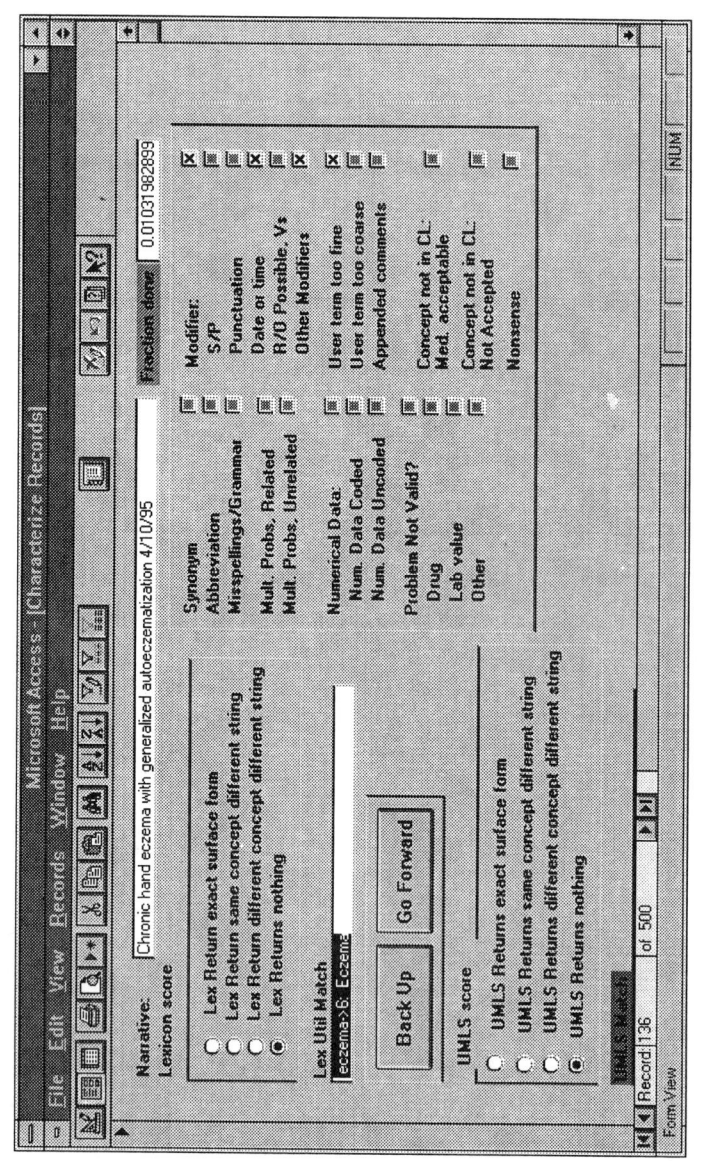

FIGURE 23.14. The form used to rate unresolved narratives.

FIGURE 23.15. Inter-rater narrative category agreement for a sample of 400 narratives. (1 = exact match, 2 = different string, same concept, 3 = different string, different concept, 4 = nothing.)

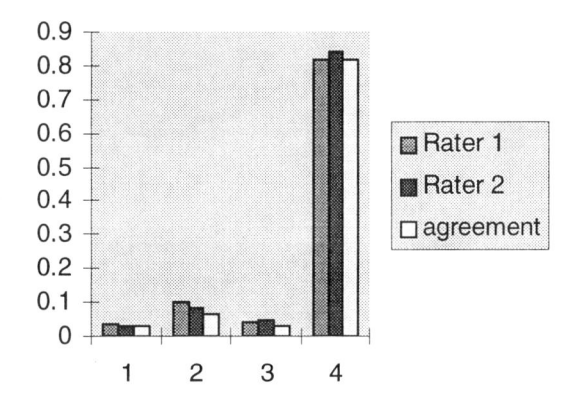

TABLE 23.5. Reasons for failure of unresolved narratives to match.

N = 400	Category selected by both raters n(%)	Category selected by at least one rater n(%)	Fraction of agreement (%)
Synonym	51(13)	145(36)	35
Abbreviation	3(1)	30(8)	10
Misspelling/Grammar	16(4)	37(9)	43
Multiple Problems, Related	24(60)	46(12)	52
Multiple Problems, Unrelated	0(0)	4(1)	0
Numerical Data	44(11)	50(13)	88
Coded	42(11)	47(12)	89
Uncoded	2(1)	3(1)	67
Problem not valid	20(5)	38(10)	53
Drug	0(0)	5(1)	0
Lab Value	1(0)	4(1)	25
Other	17(4)	31(8)	55
Modifier	160(40)	210(53)	76
S/P	38(10)	41(10)	93
Punctuation	0(0)	4(1)	0
Date/time	5(1)	13(3)	38
R/O, Possible, etc	5(1)	10(3)	50
Other	123(31)	186(47)	66
User term too fine	21(5)	75(19)	28
User term too coarse	0(0)	9(2)	0
Appended comments	2(1)	10(3)	20
Medically acceptable concept not in LU	15(4)	71(18)	21
Medically unacceptable concept not in LU	0(0)	4(1)	0
Nonsense	2(1)	9(2)	22
Totals	591	1082	
Average number per unresolved narrative	1.48	2.72	

the fraction of agreement between the two raters. Kappa (κ) statistics were not generated for this data because the reasons for disagreement were nondisjoint. The last row of the table shows the average number of reasons for failure to match per unresolved narrative for both raters in agreement (1.48) and for either rater (2.72).

Table 23.5 shows that the most frequent reason for failure to match was the use of modifiers (52.7% of cases). A diverse range of modifiers were used by clinicians. For example, 10.3% of narratives contained forms of the term "status post," often indicating a past procedural history (e.g., "s/p appendicitis"). An additional 3.3% of narratives used a date/time in combination with an otherwise valid lexicon entry (e.g., "appendicitis 1942"). In 2.5% of narratives, probabilistic adjectives were used as modifiers (e.g., "possible appendicitis"). Other modifiers were highly diverse and for this chapter, we have chosen to group them into an "other" cateogry. This category contained 46.7% of the total narratives, as judged by either rater. In many of these cases, multiple modifiers were used, for example, "appendicitis with sepsis and perforation."

The next most common reason for failure to match shown in Table 23.5 was the use of user-specific synonyms that were not contained in LU (36.4% of narratives). When this reason occurred for category 2 narratives (6.8%), the users probably wanted to find their exact preferred surface form, or they simply mistakenly failed to pick a good synonym. Most often, these were category 4 narratives, and an alternative term of equal granularity existed within the LU, but was not retrieved by the lexical routines utilized for lookup and thus was never presented to the user for consideration. This finding indicates that the look-up engine might be improved as one means of reducing unresolved narratives.

While modifiers and synonyms accounted for the bulk of the unresolved narratives encountered during our rater calibration study, other causes were also important. These causes are shown in Table 23.5, and include the use of abbreviations (7.5% of narratives), multiple, related problems (12.0%), misspellings (9.2%) and use of numerical data, such as ICD-9-CM or CPT codes (12.6%) as narrative "keywords." A few narratives were classified as "nonsense" (2.2%) because the researchers could not understand them. The cause "not valid" was assigned when the narrative seemed to indicate a nonproblem (e.g., a purpose of visit such as "health maintenance exam," which should have been recorded using DHCP's Integrated Billing 2.0 or Automated Information Capture System).

A less common, but still important, reason for unresolved narratives listed Table 23.5 is "medically acceptable, not in lexicon." Between 4% and 18% of narratives (depending upon the rater) fit into this category. This category means that the core narrative concept or synonyms, even when stripped of modifiers, misspellings, or numerical data, could not be found in LU. These terms represent an important source for updating and maintaining the VA LU and its Metathesaurus base.

In summary, while the LU was first designed to work with Problem List, the package is envisioned as serving a variety of more general coding needs in DHCP. The LU is based on the NLM's UMLS Metathesaurus. The VA has deactivated some UMLS codes and added additional VA content to better serve VA needs. Unresolved narratives were developed as a mechanism for reviewing the performance of the LU in Problem List. We have reported on our initial analysis of these narratives, which represent an important source to update the LU and Metathesaurus.

The Need for More Advanced Lexical Data Models

The Problem

The VA's initial work on LU has focused on how to encode patient problems in DHCP's Problem List. We are now also attempting to determine how the LU will evolve into a method for coding a variety of clinical information. The data we wish to encode are diverse and include items from histories, physical examinations, periodic progress notes, discharge summaries, health maintenance data, clinical practice guidelines, and data for monitoring clinical pathways. Unfortunately, as others have noted, there is no existing standard that is now adequate for this task (Campbell, Das, & Musen, 1994). Even in the case of Problem List, as shown by the unresolved narratives analysis, the LU cannot always include the full range of physician expressions. Some of the limitations we have encountered will be partly addressed by LU 2.0, which will add a complete set of Current Procedural Terminology (CPT) terms. The VA has a national site license for CPT, although for licensing reasons the complete CPT terminology cannot yet be included in the UMLS Metathesaurus. However, this approach will not address the other problems identified by the unresolved narratives analysis, including modifiers, appended dates, probabilistic assertions, and other factors.

Perhaps the largest problem is modifiers. In the study of unresolved narratives, 52.7% of users attempted to qualify the main problem term with additional information. For example, users would add one or more qualifiers describing the time of onset, laterality, or severity of the core problem condition. While many terms in the LU are "compound" in the sense that they include a core concept and some qualifiers, not all potentially valid, precoordinated combinations of core concepts and qualifiers are present. Campbell has pointed out that this problem is inherent in hierarchical vocabularies such as SNOMED, CPT, and ICD-9-CM (Campbell, Das, & Musen, 1994). Terms contained in hierarchies can have only one parent, but they may have many children. To provide an adequately diverse clinical dictionary for naming problems, one could simply add children at an appropriate location in the hierarchy. However, this approach can lead to combinatorial explosion. As a result, system developers must tightly control additions to the source vocabularies.

Developers of medical expert systems, such as Iliad, QMR, and DXplain, have also faced this problem. For example, Iliad version 4.5 (now undergoing beta testing) contains approximately 21,000 dictionary terms in its hierarchical data dictionary. During an Iliad knowledge engineering session, an expert may wish to add a patient finding to this dictionary to author a new frame or modify an existing diagnostic rule. Considerable discipline is required to search the existing dictionary, determine whether a suitable term is already present, and then perhaps add a new term as a child term at an appropriate level in the hierarchy. Subsequently, Iliad users must either remember or be prompted to map their diagnostic inquiries onto terms contained in the dictionary. Currently, in Iliad, there is no computable procedure for selecting the closest dictionary match to the user's term. A similar situation has also been faced by developers of QMR (Giuse et al., 1993). If Iliad or a similar system is to be interfaced with an electronic medical record system (EMRS), some computable method of mapping between the system's data dictionary and the data dictionary of the EMRS must be developed. If the mapping must be manual for each data item, then the usefulness of the system's interface will be limited. Even within a system as homogenous as the VA, differently configured local systems will need to communicate with other DHCP sites and with EMRS systems used by VA contractors. As a result of the VA's Veterans Integrated Service Network (VISN) initiative, sharing of patient care will occur between VA sites in a VISN region. As a consequence, medical records sharing between VAMCs will greatly increase, and there will be a need to accomplish this sharing in realtime. Therefore, the development of computable methods for modeling compound terms (core concepts plus modifiers) and for mapping compound terms to differing data dictionaries will be increasingly important for future VA EMRS efforts.

Approaches to the Problem

Other medical vocabulary systems, such as SNOMED version 3 and the Read Clinical Codes, have been proposed as the foundation for an EMRS system. These systems have been proposed because they contain advances, such as compositional extensibility and the ability to support compound terms. In this way, these systems are designed to overcome some of the limitations of strictly hierarchical systems. This chapter briefly reviews a sample of these systems, which have provided a context for our work.

The Systematized Nomenclature of Medicine was developed by the American College of Pathology to express the sorts of clinical concepts that their members encounter. Currently in version number 3, SNOMED is essentially a hierarchical system. However, it supports several different hierarchies, or "axes," such as "morphological," "disease," and "systemic regional and cellular anatomy." In SNOMED, compound terms can be constructed by creating linkages between individual terms in these hierarchies. However, the compositional rules governing the assembly of compound terms in SNOMED are not completely specified. The lack of compositional rules allows considerable flexibility, but this flexibility has several important consequences. Several different, equally valid compound terms can be constructed to describe the concept "acute appendicitis" (Evans, Cimino, Hersh, Huff, & Bell, 1994). If a system allows different surface forms to represent the same concept, computing on the concepts (as in an expert system) becomes more difficult. For example, the knowledge engineers and system designers may have to anticipate the different ways that users (or systems) may compose concepts and place these in their expert system rules and knowledge bases.

The Read Codes, versions 3 and 3.1, represent another approach to overcoming the limitations of strictly hierarchical coding systems. The Read Codes were designed to fit the needs of clinical practitioners in the United Kingdom (UK). Read arose out of three separate terming initiatives in the UK, the Clinical Terms Project, the Nursing Terms Project, and the Professions Allied to Medicine Terms Project (O'Neil, Payne, & Read, 1995). Version 3 of the Read codes allows a directed acyclic graph to replace hierarchical term structures. The system permits qualifiers (in version 3.1) and provides an informational model for computation. Under the Read system of qualifiers, a Read code can be linked to a "template file" that contains the relevant qualifiers. When a qualifying term has been entered, its attributes (e.g., its "anatomical site: bone") are inherited down to the level of the children of the parent term that is being qualified (e.g., the child "fracture fixation" of the parent "bone operation" inherits the "bone" site). This inheritance is not global as it is in the case of SNOMED axes. This approach allows the computation of more limited choices for users as they pick qualifiers to build a compound term. For example, a lookup on the keyword "fem" during coding of a "fracture fixation" procedure will retrieve "femur" but not "female" or "femoral hernia." As in SNOMED, Read allows multiple valid forms for a single term. However, because it has been created with extensive practitioner input, Read contains "precoordinated" compound terms for many commonly used clinical concepts.

Columbia Presbyterian Hospital's Medical Entities Dictionary (MED) system is another proposed solution for vocabulary representation (Cimino, Clayton, Hripcsak, & Johnson, 1994). The MED is constructed as a semantic network that uses directed acyclic graphs to represent multiple hierarchies. At Columbia Presbyterian, the MED developers have added terms from their laboratory and pharmacy systems, medical records diagnosis and procedural coding systems, and electrocardiographic terms. These terms have been added to other terms selected from the UMLS Metathesaurus system. In the MED system, a term represented in the acyclic graph can have more than one parent (e.g., "glucose" can be child of both "carbohydrate" and "biologically active substance," as Cimino et al. point out). Additionally, each concept at every node in the graph is actually a frame with slots that may or may not be instantiated (filled-out). These slots act as modifiers (for a complete discussion, see the Cimino reference above). Finally, concepts may have links other than parent–child. These "semantic links" to other MED terms represent membership information for the terms. For example, glucose is a "substance measured" by the plasma glucose test where "substance measured" is the semantic link between the two MED terms. The Columbia Presbyterian group describes some of their work on the MED project (now called InterMed) at *http://www.cpmc.columbia.edu/intermed_proj.html.*

Huff and colleagues (1995) have proposed an event model for representing medical information. Their model was developed from a manual and automated analysis of actual patient. The authors examined word patterns in paper medical records and dictated medical reports. They also examined the structure of the coded data in the HELP system (the LDS Hospital's clinical information system). Based on their analysis, they proposed that a representational model for medical data should include two components: a frame or template for each medical concept and a separate vocabulary containing elements that are used to fill slots in the frames. In this model, individual medical events are created by instantiating (filling out specific slots in) event templates. They have published samples of event templates for chest x-rays, laboratory tests, pharmacy, and physical examination events. They have used their event templates to map between the Iliad vocabulary system and the UMLS Metathesaurus (Rocha, Rocha, & Huff, 1993), to map between two hospital information systems (Fu, Bouhaddou, Huff, Sorenson, & Warner, 1990), and to construct coded records of echocardiographic finding from free text reports (Canfield, Bray, Huff, & Warner, 1990). Their model was developed to support expert system analysis, research and clinical queries on single patients or groups of patients, and natural language processing.

In summary, the VA has experienced some of the same problems experienced by other users of current vocabulary technologies. The problems of combinatorial explosion inherent in hierarchical systems may be soluble by utilizing concepts such as frames or acyclic graphs. These techniques allow core concepts to be combined with appropriate clinical modifiers without unacceptably increasing the number of terms in the data dictionary. The best of these techniques also provide a compositional grammar for compound terms. These methods may also allow a compound term to be computed and mapped to other coding systems. Some systems, such as the Read codes, provide precoordinated compound terms for common concepts while still allowing compositional flexibility. The VA is currently investigating how and where these approaches should be applied in the current DHCP framework.

Conclusion

The VA Problem List is an important application that is now installed at over 145 VAMCs (April 1996) and will probably be installed at most of the 170 VAMCs by the end of 1996. The Problem List was designed to allow problems to be entered and reviewed by a variety of methods: directly by clinicians, by clerks, and by reference from other DHCP packages (such as Integrated Billing and Health Summary). The Problem List was also designed to support a wide variety of clinical users, such as nurses, physicians, and therapists. Individual clinicians can customize their views of patient problem lists, their preferred methods for entering and reviewing problems, and their preferred vocabularies for naming problems. The Problem List is a key application in DHCP because it facilitates interprovider communication. This software also provides a database for research and identifies patients meeting who should be treated on certain clinical pathways. Finally, Problem List is also an important resource for billing.

The VA LU is the main vocabulary source used to populate patient problem lists. The Problem List package is closely linked in its functions to LU (for example, Problem List's user preferences can specify the subset of lexicons used for problem lookups). The LU is based on the UMLS Metathesaurus. As a "dictionary of dictionaries," the Metathesaurus content has allowed the VA to map automatically in most cases between terms clinicians pick to name problems and billable ICD-9-CM diagnoses. Additional VA content has been added to the LU to facilitate this task. The Problem List and LU packages provide for the return of so-called "unresolved narratives" (failed clinician searches) to the VA IRMFO in Salt Lake City. Analysis of these unresolved narratives will be used to understand and improve the functions of Problem List and LU.

The LU is well suited for naming problems in the Problem List application. However, analysis of the unresolved narratives indicates that the LU can continue to be improved. If the lexicon is to be used to code histories, physical exams, or other more complex clinical terms, some substantial changes will be needed. Including all commonly used clinician terms in a flat list or even in a simple hierarchy will incur the problems of combinatorial explosion and inadequate computability when translating to alternative vocabularies. Such translations are important to support quality improvement, billing, research, and decision support. Therefore, other researchers have been exploiting various approaches based on frames, acyclic graphs, semantic linkages, and other approaches. This research has resulted in some cases in concrete products that are ready for adoption, such as the Read version 3.1 codes. The VA will have to examine these approaches as it develops its vocabulary technologies to meet future needs.

Acknowledgments. The author is only one of many people who has contributed to the development of the VA Problem List and related technologies.

Therefore, he would like to acknowledge the important and innovative work of Melanie Buechler, developer of the Problem List, Kimball Rowe, developer of the Lexicon Utility, Debbie Price, and other VA developers. JoAnne Green, a VA clinical documentation specialist, originally prepared a number of the figures used in this chapter for the Problem List User Manual. The clinician members of the VA's Problem List expert panel, chaired by Michael J. Lincoln M.D. (past chairman) and now by Rolland Jenkins M.D., generated many of the functional specifications for the DHCP software. Three VA Informatics fellows, John Hurdle M.D., Ph.D., J. Christopher Eagon M.D., Robert Hausam M.D., and former University of Utah senior medical student John Lund M.D. have conducted evaluation studies of the Problem List and application. The author's work is supported by the Clinical Applications Requirements Group of the Department of Veterans Affairs, the Salt Lake City VA Medical Center, and the University of Utah Departments of Medicine and Medical Informatics.

References

Campbell KE, Das AK, & Musen MA. 1994. A logical foundation for representation of clinical data. *Journal of the American Medical Informatics Association, 1,* 218–232.

Campbell JR, & Payne TH. 1994. A comparison of four schemes for codification of problem lists. *Proceedings of the Eighteenth Annual Symposium on Computer Applications in Medical Care,* pp. 210–205.

Canfield K, Bray B, Huff SM, & Warner HR. 1990. Database capture of natural language echocardiology reports. *Proceedings of the Thirteenth Annual Symposium on Computer Applications in Medical Care,* pp. 185–189.

Cimino JJ, Clayton PD, Hripcsak G, & Johnson SB. 1994. Knowledge-based approaches to the maintenance of a large controlled medical terminology. *Journal of the American Medical Informatics Association, 1(1),* 35–50.

Dick RS, & Steen EB (eds). 1991. *The computer-based patient record: An essential technology for health care.* (Institute of Medicine.) Washington, DC: National Academy Press.

Evans DA, Cimino JJ, Hersh, WR, Huff SM, & Bell DS. 1994. Toward a medical-concept representation language. *Journal of the American Medical Informatics Association, 1(3),* 207–217.

Fu LS, Bouhaddou O, Huff SM, Sorenson DK, & Warner HR. 1990. Toward a public domain UMLS patient database. *Proceedings of the Fourteenth Annual Symposium of Computer Applications in Medical Care,* pp. 170–174.

Giuse NB, Giuse DA, Miller RA, Bankowitz RA, Janosky JE, Davidoff F, Hillner BE, Hripcsak G, Lincoln MJ, Middleton B, et al. 1993. Evaluating consensus among physicians in medical knowledge base construction. *Methods of Information in Medicine, 32(2),* 137–145

Henry SB, & Holzemer WL. 1995. A comparison of problem lists generated by physicians, nurses, and patients: Implications for PR systems. *Proceedings of the Nineteenth Annual Symposium on Computer Applications in Medical Care,* pp. 382–386.

Huff SM, Rocha RA, Bray BE, Warner HR, & Haug PJ. 1995. An event model of medical information representation. *Journal of the American Medical Informatics Association, 2(2)*, 116–128.

Linnarsson R, & Novdsven K. 1995. A shaves computer-besed probelm-oriented patient record for the primary care team. *MEDINFO 95 Proceedings*, 2, p. 1663.

O'Neil M, Payne C, & Read J. 1995. Read codes version 3: A user led terminology. *Methods of Information in Medicine, 34*, 187–192.

Rocha RA, Rocha BHSC, & Huff SM. 1993. Automated translation between medical vocabularies using a frame-based interlingua. *Proceedings of the Seventeenth Annual Symposium on Computer Applications in Medical Care*, pp. 690–694.

Scherpbier HF, Abrams RS, Roth DH, & Hail JJ. 1994. A simple approach to physician entry of patient problem list. *Proceedings of the Eighteenth Annual Symposium on Computer Applications in Medical Care*, pp. 206–210.

Weed LL. 1989. New premises and new tools for medical care and medical education. *Methods of Information in Medicine, 28(4)*, 207–214.

Weed LL. 1993. Medical records that guide and teach. *MD Computing, 10(2)*, 100–114.

Zelingher J, Rind DM, Caraballo E, Tuttle MS, Olsen NE, & Safran C. 1995. Categorization of free-text problem lists: An effective method of capturing clinical data. *Proceedings of the Nineteenth Annual Symposium on Computer Applications in Medical Care*, pp. 416–420.

Section 5
System Evolution

Chapter 24
Conceptual Integrity and Information Systems: VA and DHCP
Tom Munnecke   385

Chapter 25
Hybrid Open Systems Technology
Virginia S. Price   395

Chapter 26
Using DHCP Technology in Another Public Environment
A. Clayton Curtis   405

Chapter 27
International Installations
Marion J. Ball and Judith V. Douglas   426

24
Conceptual Integrity and Information Systems: VA and DHCP

Tom Munnecke

The hospital, said Peter Drucker (1973, p. 4), is "one of the most complex social institutions around." The intervening years have brought even greater complexity to the delivery of health care, the technologies involved, and the practice of medicine. The sheer size and complexity of the Department of Veterans Affairs (VA) healthcare system makes the Decentralized Hospital Computer Program (DHCP) one of the most challenging healthcare information system problems in the world.

Decentralized Hospital Computer Program

The DHCP has been coping with these organizational and managerial changes with the additional burden of having to keep up with a rapidly changing computer and communications environment. The computers on which DHCP was originally developed—PDP-11 minicomputers in specially constructed computer rooms—are less powerful than even the smallest laptop or handheld computers available today.

Given the enormous complexity of the problem, the continuous need for adaptation, and the constant technical change, installing any computer system is a tremendous challenge. Getting one to grow and prosper for over 15 years is a remarkable accomplishment. Other federal agencies, such as the Internal Revenue Service and the Federal Aviation Administration, have spent billions attempting to create large-scale systems, yet have seen little results from their efforts.

Despite the organizational complexity DHCP serves, it has managed to maintain a high level of integration. With a common database accessed by over 75 packages, it represents one of the most highly integrated hospital information systems in use today.

The underlying technical approach to DHCP has spread to nearly every healthcare provider in the federal government. Under contract to the Department of Defense (DoD), Science Applications International Corporation (SAIC) adapted the DHCP system for the Composite Health Care

System (CHCS). Competing with three other companies in 1986–1988, SAIC's approach cost about 55% of its competitors, while offering nearly double the functionality (General Accounting Office, 1988). After 8 years of operation in 1996, the General Accounting Office (GAO) has studied the cost-effectiveness of the CHCS system. It found benefits of about $4 billion, with costs of about $2 billion.

The technology has spread to the Indian Health Service (IHS), Finland and other European, Asian, and African nations. The low cost of DHCP has been proven repeatedly. A study by Booz Allen (1987) demonstrated that procuring a commercial version of the system would cost about twice as much as DHCP.

Conceptual Integrity in System Design

Frederick Brooks (1975) discussed conceptual integrity in architecture, using the example of cathedrals that were built over the generations by architects who each added their own style. The resulting cathedral was frequently a jumble of divergent styles and personal tastes, destroying the overall unity of the design. He singled out the Reims Cathedral as a design that shows remarkable conceptual integrity. The individual designers over eight generations sacrificed some ideas so that the whole might be of a consistent style. Brooks compared this to the software development process:

> I will contend that conceptual integrity is the most important consideration in system design. It is better to have a system omit certain anomalous features and improvements, but to reflect one set of design ideas, than to have one that contains many good but independent and uncoordinated ideas. (p. 4)

The DHCP was fortunate in that it had its origins with a small group of people who shared common goals and a common language within which to implement those goals. At an initial meeting in Oklahoma City in December 1978, the overall concepts of DHCP were mapped out. Over the years, these concepts proved to be remarkably valid. Some of the decisions at the meeting were:

- To use a single language, then called Massachusetts General Hospital Utility Multi-Programming System (MUMPS), now called M, for implementing the logic and the database of a decentralized healthcare system.
- To use general software tools as much as possible, to allow sharing and reuse between the developers.
- To use an active data dictionary to map out the data. This idea eventually became the FileMan system.
- To design the code to be portable across platforms. Although the microprocessor was just a curiosity at the time, the group discussed the eventual use of microprocessors. In 1990, VA converted its software to

24. Conceptual Integrity and Information Systems: VA and DHCP   387

Intel-based microprocessors, at considerable savings over the VAX minicomputers then in use. This effort required only minor changes to the software.
- To design the code to be robust and adaptive. For example, the group anticipated the need to use dates after the year 2000. This was built into the system from its inception. As a result, systems based on DHCP are able to move into the 21st century with little or no modification. This is in dramatic contrast to nearly every other system of its era, which will require expensive modification to cope with the change in millenium.
- To make the system portable across organizations. Decentralized Hospital Computer Program technology has spread to nearly every US federal healthcare provider. By 1996, it was in use by about 150,000 users in the DVA, DD, and the IHS.

Deployment and Evolution of DHCP

One of the most important achievements of DHCP is that it works. Although the number of packages activated at each site vary considerably, the software is installed and operational across the VA. Developing a hospital information system is a high-risk process. There are numerous examples in the industry of companies that spent millions of dollars and many years developing systems that failed.

Decentralized Hospital Computer Program is widely deployed within the VA. The equipment installed as of January 1995 includes:

CPUs	1924
PCs	40,000[1]
CRTs	85,116
Printers	41,596

Prior to DHCP, VA had a history of various prototypes that were not propagated throughout the system. It also had a variety of departmental systems and regional systems that were neither fully deployed throughout the system nor integrated with each other. The DHCP established a focal point for a hospital information system that made possible the deployment of an integrated patient database.

The DHCP has proven itself to be evolutionary. Starting with PDP-11 computers, it migrated to the VAX, then Intel-based platforms, then Alpha computers. Despite these changes in the hardware platform over the years, the software has continued to operate continuously. The VA has been able to move to lower cost, higher performing computer systems as the technology changed, without being locked in to a specific platform.

---
[1] This number is a rough guess and may underestimate the total number of PCs.

M technology has been a significant contributor to the success of DHCP. The reason that the software has been able to migrate across multiple generations of computer platforms is because the VA chose to rigidly adhere to the American National Standard (ANS). By choosing vendors who provided ANS M support, the software was ported with a minimum of effort. Basing a system as large and complex as DHCP on a single, standardized language was a bold step.

The FileMan data dictionary has served as a foundation for DHCP. Now containing over 20,000 data elements, it has grown from several hundred in 1982. This is an active data dictionary, in contrast to some systems that are after-the-fact paper trails of what is contained in the programs. Despite this growth, the FileMan data dictionary has not come up against any internal limitations.

Extensive use is made of M's indirection capabilities.[2] These capabilities, which blur the distinction between data and program, are shared with other languages such as LISP and APL and are used extensively in making DHCP adaptable to local needs.

This technology permits highly reentrant code to be written. For example, the original DHCP was written using only 19 commands and 22 functions of the ANS MUMPS language. This number may have increased by 50% over the years, but still represents a small number compared to other systems. This approach is somewhat subtle, and it is difficult to explain it satisfactorily to evaluators who are familiar only with compiler-based systems.

Indirection is missing in most commercially available compiled systems today. The closest concept that is popular on the market today is the family of scripting languages that can be found embedded into word processors, spreadsheets, and terminal emulators.

The software tools approach used by the VA also contributed to its success. The Kernel and FileMan software have given the VA a common infrastructure from which to work. This has been a crucial aid to the management of DHCP software. Because developers were given a strong set of tools to use, they focused their development efforts on the type of system that the tools support.

The DHCP system uses a layered model. The innermost layer consists of the native operating system (VMS, UNIX, DOS) coupled with an ANS M compiler/interpreter. The next layer is the Kernel toolset, consisting of such software as FileMan, MailMan, the Menu manager, Security system, and other utilities. The next layer is the database layer, containing the data dictionary and databases. Beyond that are the applications that customize the system to the needs of the users.

---

[2] Indirection permits late binding of variables. It also allows data to be turned into a program, and vice versa.

Public Domain

The VA software has remained in the public domain. Distributed under the Freedom of Information Act, it has been adopted by a variety of organizations around the world, and VA has benefitted from technology sharing with groups from Finland, Germany, and the IHS.

The DHCP model can be compared with the model used by the National Center for Super Computer Applications (NCSA) and its distribution of the MOSAIC World Wide Web (WWW) browser. This software, developed by NCSA staff, is freely available to the Internet community, and was a major trigger for the recent interest in WWW. The software has been licensed to commercial companies for further development and incorporation into their products. Limited by federal regulations, the VA has not entered into any such commercial arrangement.

Adaptability

The DHCP system has proven itself adaptable to the diversity of healthcare settings in the VA. The old saying "If you've seen one VA hospital, you've seen one VA hospital" is just as true with regard to implementations of DHCP. Although they use the same base software, the site parameters and usage patterns vary widely with the needs and skills of the local users. One of the reasons for the success of DHCP at any given site was the fact that local managers were able to customize the software to their needs.

In 1982, there was some fear that DHCP would lead to "helter skelter" development and unreliable data. This has not proved to be the case. For example, during the Gulf War, the VA was able to modify its systems to prepare for a possible influx of casualties in a very short time.

Many of the original goals of DHCP were validated over the years:

- Information systems have to be relevant to the end user.
- It is best to collect workload information by streamlining the work process and unobtrusively measuring it.
- It is best to provide flexible tools to the end users.
- Extremely complex systems can be managed by "creating a path of least resistance."
- Users have to feel that they have ownership in their system.
- Creating a simple initial condition and letting the system grow "organically" with the organization is a powerful approach to managing complexity.
- Complex, integrated systems can be created without the use of the linear "waterfall" development approach.

Examples

The DHCP has made some dramatic, systemic changes in a very short time. For example:

- *Operation Desert Storm.* A casualty tracking system was designed, built, distributed, and integrated into the nationwide system in less than 3 weeks.
- *Pharmacy Copay.* Retroactive congressional legislation prompted VA to produce a phamacy copayment system that impacted the Registration, Billing, Accounts Receivable, and Outpatient Pharmacy systems. The product was "in the channel" in 13 weeks. It was delivered to the field for operational use on the same day a contractor delivered a proposal to develop like functionality for the Philadelphia integrated hospital system.

Adoption by Other Organizations

The relationship between the IHS and VA seems to have been one of great benefit. The IHS has done a good job of deploying their system in the face of budgetary, staffing, and geographical challenges. They have contributed substantially to the current and future patient-oriented features of DHCP.

The DHCP system approach was adopted by the IHS in the early 1980s. They have continued to work closely with the VA via an interagency sharing agreement. This sharing is bidirectional: their IHS's Patient Care Component system (PCC) is being used as a model for the VA's Patient Care Encounter (PCE).

The DHCP software has been used extensively in other state and local government settings. Because the software is acquired through the Freedom of Information Act and distributed freely by the M Technology Association, there is no accurate accounting for all the places in which it is used.

Informaticians in Finland, from Kuopio University and Helsinki University, were among the early adopters of VA FileMan. They have installed a DHCP-based system in about 30% of the hospitals in Finland, and have exported the system to a hospital in Ile-Ife, Nigeria. The Finns have offered software back to the FileMan development team, to allow graphic representation of FileMan files.

The DHCP has been adapted by a number of other sites in Europe, and has been translated into German. A volunteer from Germany worked with the Kernel development team to give it tools to be localizable in different languages.

In 1986 SAIC was one of three competitors for the CHCS for the DoD Health Care system. It adapted the DHCP software for the CHCS and won the competition to install the software at DoD facilities world wide.

M Technology

The DHCP has been very closely associated with M technology and the M community. The VA is a frequent participant at M-Related meetings, a sponsor of the M Development Committee and the ANS standardization of M.

M is a highly portable language. It includes a database management system, programming environment, operating system, and other support to manage its entire runtime environment. The DHCP has been able to migrate through several generations of hardware without reprogramming.

This minimalist nature of M has served VA well, however. For example, M programmers are not able to deal directly with specific addresses in the computer's memory. They deal with names only, not addresses.

It is difficult to compare M with other computer programming technologies. It is a programming language, command line interpreter, database manager, linkage editor, debugger, communications manger, and (sometimes) an operating system all combined into a single language.

Tools Versus Applications

M, the Kernel, and FileMan were essential tools for the development of DHCP, and gave it much of its flexibility within the constraints in which it operates. The use of the proper tools can dramatically decrease the need for customized application software development and stovepipe applications.

For example, before the advent of electronic spreadsheets, budget analysts were dependent on systems analysts to retrieve their data. They would tell the analyst what they wanted, who would write requirements describing their needs. These requirements were turned into program specifications, which were turned into programs. The program would be run, and the analyst would present the results to the budget analyst. At that point, the budget analyst would invariably see something and ask for changes to the report. This would continue, and use up much of the analysts' and programmers' time, and eventually become a frustrating and expensive process for all concerned. This might then become subject to coordination by a committee to ensure that information systems (IS) resources were expended in the most efficient manner possible.

With the advent of the spreadsheet, however, this process has changed. The IS department can deliver the raw data directly to the budget analyst's desktop. Analysts can manipulate the data as they see fit, without incurring the costs of the systems analyst or programmer. If they want to see their information as a pie chart instead of a line chart, or have one column calculated as a percentage of something else, they can do this on their own. This type of tool encourages *What if* or browsing types of activities, which are expensive in the older paradigm of application development. (Browsing has been defined as the art of not knowing what you want until you see it.)

The addition of such tools into the organization changes the way it thinks about its systems. The introduction of MailMan, for example, changed the way that a hospital organization thinks about itself.

The Kernel, MailMan, and FileManager were designed to be the basic toolset of DHCP. Over time, however, the role between tools and applications has blurred somewhat. The Order Entry/Results Reporting (OE/RR) team, for example, is developing general sets of tools and architectural elements that are applicable to other applications. The exact role and focus of messaging in DHCP seems to be distributed between several sites and designers. This is a critical component for the future architecture, and needs special attention as a tool, not just a component of an application.

Co-Evolution of the Organization and Its Information Infrastructure

The use of a software tool can change the way an organization works. As the VA becomes more proficient with a given toolset, new employees readily find others to help them learn the tool. They may demand more sophisticated tools and add-in modules to do their work. They may develop templates and scripts and share these among themselves on a peer-to-peer basis. This type of "organic" growth cannot be predicted. It is the result of a unique interaction between work groups, their tools, and their computer and communications networks.

Rumbaugh (1991) discussed a high-level object model of a system that differentiates between aggregates and associations. Aggregates are collections of classes, in which the whole is the sum of its parts. A relational database is an example of this kind of thinking, in which all of the components are formally structured. Relationships between the rows and columns of a table, and between tables, is well defined. An association, however, can have a more specific meaning, that of specific link relationships between individual objects. This type of relationship attaches meaning to the specific link, independent of the class relationships expressed in their aggregations. The FileMan variable pointer is an example of this, as are the links found in the WWW.

One of the reasons DHCP has been successful is its use of links and associations. There is a tendency to think that a hospital IS can be designed from the perspective of aggregates; that all information can be expressed in the form of normalized tables. This view, however, ignores the associative links necessary to design a medical record system. The associations (Hypertext links) in the WWW could never be cast as a normalized relational database system; they map an entirely different type of database problem. Clearly VA needs to understand the balance between aggregations and associations with regard to the medical record toolset.

The traditional requirements-based design process breaks down when attempting to address this type of organic system. This happens for several reasons:

- Requirements are written from the perspective of users who are not using the tools. They cannot be expected to know what they want from the tool without using it.
- The requirements process is one sided. It presumes that the organization is static, and does not expect the organization to evolve as a result of installing the system.
- The requirements are usually one evolutionary step behind the reality of the organization. The organization is evolving to the next generation system, while the requirements deal with the last one.
- Different users may use the tool in different ways. Thus, a single point of view described by the requirements document cannot reflect all the possible uses of the system. The organic vitality of the tool is lost when it is restricted by rigid requirements.
- Requirements are used to create specifications, which by their very nature produce specific things. Tools are general things. The greater the generality of the tool, and the greater the diversity of its use by the organization, the more successful the tool will be. The more specific an application becomes (a typical evolution in stovepipe applications), the less general it becomes.

The Organicity of a System

Organic systems exhibit the property that the whole is greater than the sum of their parts. The toaster, a nonorganic system, can be disassembled and reassembled to become the same toaster. A cat, being an organic system, cannot be dissected and reassembled to become the same cat. The difference between the whole and the sum of the parts can be called the emergent properties of the system. (This term was introduced to DHCP at the first Oklahoma City conference in 1978 by Dr. Richard Davis.)

The DHCP must be considered as an integral part of the VA organization: "Orgware" is one term that has been used. Both the software and the organization have evolved together, providing a platform for the next generation of technology and organizational evolution.

References

Booz, Allen, & Hamilton. 1987 (February 23). *Decentralized hospital information system and integrated hospital system compatibility study*. Washington, DC: Author.

Brooks F. 1975. *The mythical man-month*. Reading, MA: Addison-Wesley.

Drucker P. 1973. *Management, tasks, responsibilities, practices*. New York: Harper and Row.

General Accounting Office. 1988. *Composite health care system procurement: Fair, reasonable, supported*. Washington, DC: Author.

Rumbaugh, et al. 1991. *Object-oriented modeling and design*. New York: Prentice Hall.

25
Hybrid Open Systems Technology

VIRGINIA S. PRICE

Getting Ready: A Brief Historical Perspective

In 1983, the Veterans Health Administration (VHA), part of the Department of Veterans Affairs (VA), began to develop and implement a modular system of M-based applications across its nationwide healthcare delivery system. These applications addressed areas where automation was vitally needed to help with backlogs and heavy workloads. Few automated systems were in use at VA medical centers at this time, and the VHA desired to install systems at all 171 hospitals throughout its healthcare delivery system. Based on the state of the technology available at the time, the VHA decided that the size of the implementation and the degree to which automated solutions would need to address VA-centric workflow requirements and information needs warranted establishing an inhouse development effort. The resultant product, the Decentralized Hospital Computer Program (DHCP), is still in operation at all VA medical centers, and forms the foundation of the present VHA healthcare information environment.

In 1984, the VHA instituted the Integrated Hospital System (IHS) program in response to a Congressional mandate to identify commercial healthcare information systems that could be used in VA medical centers. The VHA installed commercial hospital information systems in three VA hospitals. Consistent with the industry approach of the day, IHS attempted to use commercial systems to support all aspects of medical center operations. A monolithic, all-or-nothing approach did not prove to be the best avenue for infusing commercial technology into the VHA information environment. Experience showed that it was extremely difficult for vendors to keep pace with adaptations required by changing VA regulations and newly mandated requirements for gathering VA-specific data. In addition, the cost of maintaining these systems restricted the use of IHS as a means of exporting commercial technology to the larger VHA healthcare community.

Meanwhile, VA medical centers had started to augment DHCP capability with commercial products on their own. At first, the use of commercial technology focused on research databases and tools. As medical centers gained experience with automation, sophisticated users saw great potential for using commercial products to supply additional functionality. The installation of automated instruments in VA laboratories opened up new possibilities for the efficient and accurate transfer of medical information to DHCP systems.

The VHA began to see in a new way the wider importance of standards. Within DHCP, a standards-based approach made it extremely easy for modules to communicate internally. However, interfacing to external systems did not prove to be so easy. The many worthy niche-system candidates, such as intensive care, biophysical monitoring, oncology, anesthesiology, had to run in a stand-alone mode. After some time, unidirectional connectivity was established. Clinical data could be transmitted to DHCP, but DHCP data was not returned to the application. As DHCP clinical functionality grew richer and standards matured, bidirectional interfaces were achieved. Each interface still had to be written independently and adjusted for the environment.

Over the years, VHA's vision for providing automated support to enable delivery of the best health care for our nation's veterans has evolved in response to a changing healthcare industry and the maturation of information technology standards and products. Commercial technologies are now widely used throughout the VHA healthcare delivery system, and have been provided by VHA initiatives (decision support systems, or DSS), agency initiatives (e.g., PAY VA, financial management system or FMS), and independent procurements by healthcare facilities to satisfy local needs, as well as through the avenue of the official VA Hybrid Open Systems Technology (HOST) program.

The VHA has recently embraced a new healthcare delivery paradigm, based on a networked care model. The regional structure of VHA has been replaced by 22 Veterans Integrated Service Networks (VISNs). Each VISN will have significant latitude in organizing its treatment settings and delivery patterns to meet the healthcare needs of veterans in the most effective manner. The VHA is committed to achieving the best, most cost-effective mix of inhouse and commercial off-the-shelf (COTS) software to provide automated support for VISNs, and has enlarged the scope of HOST accordingly.

This chapter provides a brief description of the goals, processes, and successes of HOST. The chapter focuses on the steps HOST has taken to achieve an open architecture approach, maximize flexibility for the system as a whole, and support networked capabilities. The goal of the chapter is to describe how HOST is changing in concert with VHA and the healthcare industry to enable VHA's healthcare information environment to take the next step forward.

Taking Aim: Establishing HOST

Hybrid Open Systems Technology is the successor to the VHA IHS Program. Hybrid Open Systems Technology provides VHA with a strategy for technology management, and the mechanism to evaluate the cost-effectiveness, practicality, and benefits of using commercial information technologies to improve veteran patient care.

Based on lessons learned through the IHS experience, HOST was designed to infuse commercial technology to improve the capability of the information environment at a wide cross section of VA healthcare delivery settings. Initiatives were to be chosen to meet defined VHA business needs, and to support the patient-care mission of the VHA. To maximize the benefits and minimize the need for customization, HOST was to marry commercial products that offer potential value-added benefits with proven components of VHA's existing DHCP. The VHA could evaluate innovative commercial technologies and migrate them to meet the needs of the larger VA healthcare delivery system. A key feature of HOST was compliance with maturing open systems standards, which would enable VHA to achieve a "hybrid" approach where commercial off-the-shelf products and DHCP could effectively exchange information.

Implementing the HOST Concept

HOST Field Projects

In choosing which automation initiatives to pursue, VHA has always placed a high priority on assessing the needs of the grassroots user population. The VA medical centers (VAMCs) were invited to submit projects for consideration during an annual submission cycle. To select commercial technologies that would best satisfy business needs at the widest range of field facilities, the VHA Information Resources Advisory Council (IRAC) was requested to rate these proposals. During 1994 and 1995, 33 commercial technologies were chosen to be piloted at 23 VHA healthcare delivery sites, as shown in Table 25.1.

Other HOST Activities

Originally, the methodology for testing and assessing commercial technologies paralleled the classical research model. Recommendations to nationally migrate commercial technologies were to be predicated entirely on the data gathered and results achieved in prior phases of the HOST process. In real-world practice, however, the time required to select and implement HOST projects, mature the implementation, establish an assessment cycle, and develop a comprehensive migration strategy often meant that events

398   System Evolution

TABLE 25.1. Commercial technologies piloted at VHA healthcare delivery sites, 1994 and 1995.

Pilot Sites 1994
(Site: Project Title)

Asheville, NC: VA/DoD CHAMPUS Project
Baltimore, MD: Evaluation of Radiology Picture Archiving Communications Systems (PACS)
Big Spring, TX: Alternative Forms of Data Entry and CD-ROM Libraries
Bronx, NY: Evaluation/Integration of Anesthesia Systems with DHCP
Durham, NC: Ambulatory Care—A Managed Care Workstation
Houston, TX: Voice Operated Interactive Clinical Equipment
Minneapolis, MN: CD-ROM Libraries
Philadelphia, PA: Commercial Integrator (Integrating seven commercial technologies)
Saginaw, MI: Commercial DHCP—Dental Interface
San Diego, CA: Development of a DHCP-Electronic Anesthesia Record Keeper Interface
Tucson, AZ: Information Acquisition & Management Point-of-Care (POC) Test

Pilot Sites: 1995
(Site: Project Title)

Albany, NY: Correspondence Management System
Brockton, MA: Integration of Disparate Databases
Fayetteville, NC: Telepathology—Imaging
Iron Mountain, MI: Remote Telepathology Consultation & Imaging System
Lebanon, PA: Mobile Point-of-Care Device with Delphi and KERNEL Request Broker
Long Beach, CA: Mediphor Computerized Medical Record System
Muskogee, OK: Voice Recognition Interface with DHCP
Muskogee, OK: Computerized Telephone Triage
Oklahoma City, OK: Zip Plus Four Mail Integration
Northampton, MA: Teleradiology for Rural Medical Centers
San Francisco, CA: Automation of Patient Intervention Data
Togus, ME: Electronic Integrated Patient Assessment
West Haven, CT: Regional Digital Echocardiography Network

and technology were overtaking the assessment process. Other VAMCs were identifying and implementing promising technologies before the similar technologies could be stabilized and assessed at HOST pilot sites. The VHA was not achieving the full benefits of standardized implementation, common interfaces, and economies of scale that the HOST process promised, nor was it able to leverage corporately the lessons learned through the experience of individual healthcare delivery sites.

To better meet this challenge, HOST began to refocus resources, redesign processes, and open other avenues of identifying and assessing innovative technologies and making these assessments available to the VHA healthcare delivery network.

The HOST Laboratory

The HOST Laboratory was established to enable VHA to keep up with technology changes and enhance the capability to introduce COTS software into the VHA information environment. The laboratory monitors emerging technologies, determines which of these technologies appear to be advantageous to the VHA clinical and management environment, tests these technologies for compatibility with the clinical systems infrastructure, and supports the introduction of promising technologies into VHA healthcare facilities. The laboratory is helping to define the hybrid DHCP/commercial, open systems-based environment and to promulgate standards that commercial technology can be expected to meet for inclusion in the VHA hybrid information systems mix. The VA HOST Laboratory also provides expertise in integrating certified products at VHA healthcare facilities.

Currently, 13 studies and evaluations are in process at the HOST Laboratory, including network-based security, desktop video teleconferencing, VHA telemedicine assessment evaluation, telephone triage, WWW access to appropriate VHA medical data, and investigating security implications in accessing VA's DHCP over the Internet.

The HOST Clearinghouse

The central clearinghouse has been established to collect, coordinate, and share information on HOST, VHA, and VA field-based HOST-like projects. The clearinghouse will coordinate studies and market analyses on leading industry technologies and products as well as industry trends, and will provide VHA healthcare facilities and other interested parties with access to information on state-of-the-art and innovative products and technologies. Current information is available electronically through the WWW, VHA Bulletin Board Service, fax-on-demand, and telephone.

The clearinghouse maintains a full file on each certified technology including the accumulated test results, problem reports, resolution actions, installation and support techniques and advice, standardized implementation/project plans, and training materials. The VA healthcare delivery sites interested in COTS capabilities can match their needs with descriptions of the functionality delivered by certified HOST technologies.

Comprehensive COTS Baseline and Ongoing Assessment

The VHA medical facilities have been using local funds to implement commercial healthcare information technologies. Hybrid Open Systems

Technology has undertaken the identification, review and publish a detailed listing of COTS technologies in use at all VA healthcare delivery sites. This establishes a baseline of the current information system capabilities and aids in determining the interests in certain commercial technologies. The survey assists in identifying the potential for COTS integration, the likely level and extent of COTS utility within the target population, the likely impacts, benefits and costs, and the level of effort required to accomplish migration.

The COTS survey addresses three main areas: (1) administrative and clinical systems; (2) office automation; and (3) local area networks and hardware.

A supplemental survey that focuses on a comprehensive baseline of DHCP functionality at each facility will also be conducted. The surveys address the following key areas:

- User information needs and functional requirements;
- Targets of opportunity for COTS;
- Impact on existing procedures;
- User types;
- Personnel training (users, managers, and administrators);
- DHCP interfaces;
- Data stores;
- Operating systems;
- Networking capabilities;
- Data access mechanisms;
- Hardware requirements;
- Security issues;
- Physical facilities;
- Environmental conditions.

Information derived from these comprehensive inventories will serve as a valuable resource to VHA facilities when making automation decisions.

Successes

As shown in Table 25.1 above, 33 commercial technology pilot projects have been commissioned at VA healthcare delivery sites. These projects are in various stages of implementation. Although only four assessments have been completed, many of the specific technologies promise to be useful in a wider implementation.

Baltimore, MD: Evaluation of Radiology

The Baltimore VAMC has implemented a "filmless" radiology department through the use of the DHCP digital imaging system and Picture Archiving Communications Systems (PACS) commercial technology. To establish a standards-based link between the two systems, the Baltimore facility worked with Siemens-Loral. The result was the first link in the world that uses healthcare standards to exchange text and image information between a hospital information system and a PACS, and marked the first time these standards have been mapped to each other. This ground-breaking achievement overcame a major obstacle to telemedicine information exchange on the information superhighway.

Philadelphia, PA: Commercial Integrator

The Philadelphia HOST project is perhaps the most ambitious of the current VA HOST projects. The scope is equivalent to eight projects, including seven clinical application products and the commercial Message Routing and Translation System (MRTS). Four of the COTS products have been installed in the medical center (intensive care monitoring, nurse scheduling, anatomic pathology, and blood bank), and three COTS systems (anesthesiology, teleradiology, and voice recognition) are in various stages of implementation. The project is evaluating the feasibility of using an open-systems approach to integration, while simultaneously assessing the clinical benefits of these seven COTS technologies. In many ways, this project may serve as the model for future multi-COTS implementations.

Saginaw, MI: Commercial DHCP-Dental Interface

Saginaw has completed assessment of a windows-based dental information system that runs on a local area network, in conjunction with a commercial dental imaging system that captures x-rays and intra-oral color images. The DHCP information is displayed in a third window on the workstation. The implementation offers a significant improvement in operations over the standard film-based system due to the nearly immediate availability of digitized images at the patient chair. Both black and white x-rays and intra-oral color images are captured. Being able to view archived digitized images concurrently with reviewing current images offers significant clinical advantage. X-ray radiation is reduced by a factor of four for the patient. The provider's exposure to the patient's bodily fluids is reduced because film does not have to be handled. Dentrix includes a powerful feature for scheduling patient visits that offers advantages. The commercial products enables the medical center to capture billing information in greater detail (approximately 200 dental codes). Synchronization of the three systems required a great deal of effort, the project has received a positive reception at Saginaw and throughout the VISN.

Coming Attractions

The VHA healthcare delivery network is embracing the use of commercial technologies in the field of telemedicine. Enhancing telemedicine capabilities will enable VHA to reduce travel costs for clinicians and veterans alike without adversely affecting quality of care. Several HOST technology assessments are underway to investigate telemedical capabilities.

Muskogee, OK: Computerized Telephone Triage

Voice response applications that provide appointment, billing, drug, and prescription status information have been integrated with an advanced voice messaging and telephone transfer capability. The end result will be a user-friendly interface that allows veterans at the Tulsa out patient clinic (OPC) and the Oklahoma City VAMC to obtain information available on DHCP, leave a voice message for a specific department, or transfer to the appropriate department, via his/her touch-tone telephone keypad and minimum amount of user intervention. Although the formal project assessment is not complete, this technology has caught on and is in various stages of implementation at many VA healthcare delivery sites.

Iron Mountain, MI: Remote Telepathology Consultation and Imaging System

The Iron Mountain and Milwaukee VA MCs recently activated a Remote Control Dynamic Telepathology system. This is the first of its type in North America and only one of a few in the world. The system utilizes robotics technology to allow pathologists the ability to have full control of microscopes located hundreds of miles away.

The system also uses videoconferencing to allow the grossing process (preparation of the specimen) to be monitored from a remote location. Through robotics, the pathologist in VAMC Milwaukee controls the microscope located in Iron Mountain, and is able to view a live image via the VA wide-area network (WAN). The live image can be digitized to high resolution, should greater detail be needed for diagnosis or consultation. The system has many features that will expedite the remote diagnosis process. Consultation between pathologist and the patient's physician (through the videoconferencing component) is occurring in realtime. Clinicians are able to discuss diagnosis and provide more productive and expedient healthcare to veterans. The project protocol calls for the digitized images to be stored in the electronic patient record, utilizing components of the DHCP Imaging System.

Although the project is in its infancy, the ability to provide Iron Mountain VAMC access to the pool of clinical specialists at the Milwaukee VAMC appears to be progressing smoothly, and represents a leap forward in rural health care. The end result of this dynamic telepathology network will be to provide immediate clinical information to the primary-care physician, and to expedite treatment and quality patient services.

A Look to the Future: How HOST Can Move the VHA's Healthcare Information Environment Forward

The VHA has recently embraced a new healthcare delivery paradigm, based on a networked care model. The regional structure of VHA has been replaced by 22 VISNs. Each VISN will have significant latitude in organizing its treatment settings and delivery patterns to meet the healthcare needs of veterans in the most effective manner. Veterans, like other members of US society, will increasingly wish to gain access to VHA facilities and health care via electronic communications media (e.g., voice, data, and video) as much as possible rather than traveling to VHA facilities. Veterans Health Administration employees and contractors will increasingly use voice, data, and video connections to conduct business with the VHA from remote locations. Current overall trends in healthcare utilization among veterans will continue over the planning horizon (e.g., hospital-based tertiary care will continue to decline while outpatient and long-term care will steadily increase). Pooling and sharing of healthcare resources will be increasingly common as part of overall efforts to improve efficiency and control costs in the VHA system. The US healthcare system as a whole will continue to embrace the use of information technology at a steadily increasing rate, adding impetus to VHA's efforts to incorporate information technology (IT) more effectively.

The scope of the HOST concept will also need to be broadened because VHA has determined that interconnectivity is needed among all ITs currently in use, and being planned for use, within VHA. The VHA must upgrade its information infrastructure and adopt a technology architecture. This will enable VISN partners to coordinate clinical and administrative information needed to deliver high-quality, cost-effective health care, and to communicate electronically with each other no matter where in the VHA system they are located. This will also promote the identification and dissemination of "clinical, managerial, and technical best practices" (i.e., practices worthy of adoption throughout the VHA system) no matter where such practices are developed. Connections with other medical centers, educational networks, healthcare providers, and local governments will permit VHA medical centers to develop telemedicine projects that provide realtime medical care to patients at remote locations, to share scarce medical specialties with other institutions, and to serve veterans better by providing ready access to information on available programs.

When fully implemented, this information infrastructure will provide a suitable mechanism for implementing proven IT innovations on a VHA-wide basis. This will be a key factor in accelerating the rate with which commercial technologies can be introduced into the VHA information environment. The implementation of this expanded HOST concept will result in a large infusion of standards-based commercial solutions to assist VISNs in delivering cost-effective patient care and making information-driven management decisions.

26
Using DHCP Technology in Another Public Environment

A. Clayton Curtis

Introduction

Background

The Indian Health Service (IHS) is part of the US Department of Health and Human Services and, like the National Institutes of Health (NIH) and the Centers for Disease Control (CDC), is an agency of the US Public Health Service. It is responsible for providing comprehensive health care to approximately 1.2 million American Indians and Alaskan Natives, often from birth to death, in rural and generally remote regions of the country scattered across the continental United States and Alaska. This is accomplished through a system of IHS direct-care facilities (47 hospitals and 122 outpatient centers), IHS funded tribal programs (another 7 hospitals and 149 outpatient sites), and services contracted for from the private sector. The organization has approximately 15,000 employees and a budget of about $1.7 billion. Its programs range from installation of basic sanitation infrastructure, such as water and sewer systems, to dental and medical care delivered in facilities, communities, schools, and homes.

IHS Initiatives in Information Systems

Development of IHS information systems began shortly after the formation of the organization in 1955. The earliest applications were primarily administrative and centralized in a mainframe-based data processing center. The first decentralization of computing capacity occurred in 1974 with the inception of the Remote Access Data Entry Network (RADEN), a minicomputer-based distributed data entry system with some local processing capability. Remote Access Data Entry Network was initiated to address problems with centralization; these included general inflexibility, inability to respond adequately to local needs, untimely distribution of data, and low return on the efforts of individuals supplying source data.

Efforts to bring information systems support to primary care in the IHS began in 1968 as a project of the IHS Office of Research and Development, located in Tucson, Arizona. The Health Information System (HIS) used clinically oriented encounter forms to capture a wide range of patient data in ambulatory-care settings. It was mainframe based, written in COBOL, and did not use a database management system (DBMS). A later version, the Patient Care Information System (PCIS), expanded on the HIS while retaining its essential nature (Garratt, 1979; Brown, 1980). These systems eventually served the Sells service unit (southern Arizona) in an online mode, and Billings (Montana) and Alaska areas through microfiche. While advanced for their time and highly productive clinically, their reliance on mainframe and non-DBMS technology made the HIS and PCIS too expensive and inflexible for widespread implementation, as well as constant targets for cutbacks.

In 1983, the IHS developed its first 5-year strategic plan for information systems, which made radical changes in the organization's approach to computing, calling for

- Decentralization of computer capacity to points of patient care and
- Program management in the field with increased local control of and responsibility for data systems.

As a result of previous experience, namely the inability to move the RADEN system off of its aging platform without a massive and expensive conversion effort, the IHS made vendor independence and portability across hardware platforms major considerations. The plan also resulted in the Resource and Patient Management System (RPMS) project, which targeted the integration of management and clinical data systems and the use of information systems resources in multiple functional areas.

At the same time, there was considerable pressure on IHS by its parent agencies to purchase commercial systems for cost accounting and any clinical applications. The survey carried out in response to these directives failed to identify appropriate systems in the marketplace but served to make the IHS aware of the Department of Veterans Affairs (VA) Decentralized Hospital Computer Program (DHCP). In 1984, the IHS decided to adopt DHCP's underlying technology as the basis for development of the RPMS and to make selective use of DHCP applications while converting clinical applications such as the PCIS (as the Patient Care Component, or PCC, of RPMS) to operate in the DHCP setting to support ambulatory and longitudinal care.

Veterans Administration Initiatives

With 159 medical centers, approximately 200,000 employees, and a budget for medical programs of over $16 billion, the Veteran's Health Administration (VHA) is a large operation by any standard. For almost two decades, the VHA has been engaged in a massive implementation of information systems to provide automation support for clinical and administrative activities at VA medical centers. This project is the DHCP. Initial core applications included patient registration, admission/discharge/transfer, clinic scheduling, and outpatient pharmacy. Later phases have included inpatient pharmacy, clinical laboratory, nutrition and dietetics, radiology, medicine (diagnostic procedures), surgery (operative reports, etc.), and others. Applications have also been developed in areas such as fiscal and acquisitions management and engineering services. The DHCP is now comprised of over 70 integrated packages.

To support these applications, the VA implemented its own database management system, known as FileMan, and a comprehensive structure for security, menu management, e-mail, and so on, colloquially known as the "Kernel." The FileMan, which offers powerful data dictionary capability, was used to provide the standardized data definitions necessary for applications making multiple uses of the databases shared by multiple groups.

Massachusetts General Hospital Utility Multi-Programming System (MUMPS) was chosen as the implementation vehicle to maximize hardware and vendor independence. Efficient and highly portable, MUMPS is recognized by the American National Standards Institute (ANSI). The ability of MUMPS to support applications development on different types of equipment without idiosyncratic extension of the language, together with implementation of the FileMan in MUMPS alone, has conferred a degree of software portability seldom (if ever) seen in a project of the magnitude of the DHCP.

Historically, despite significant opposition from the commercial sector, emphasis has been placed on the following:

- Self-directed evolution, inhouse development, and dedication to user-driven specification and design;
- A high degree of vendor independence and platform portability;
- Standardization in the areas of programming language (MUMPS), data exchange protocols (Health Level 7, or HL7), and graphics standards (X-Windows).

These considerations have had significant impact on every phase of DHCP, from application design through hardware and software procurement.

Initial DHCP releases focused on rapid-payoff vertical departmental applications. Now DHCP is embarked on a major effort to provide direct support for front-line providers. Applications in this category include ones developed de novo by the VA as well as ones that borrow from their IHS counterparts. The infrastructure provided by existing, widely deployed departmental systems is an asset.

A strong technology focus differentiates the VA's approach to computational support for patient care from that of the IHS. Major initiatives currently underway across the VA system include imaging, intelligent workstations, and point-of-care data collection and verification. These have significant implications for resource requirements and design strategies for new applications. In addition, the VA explicitly espouses the goal of a complete electronic medical record, which IHS has not.

Rationale for Use of DHCP Software in RPMS

The RPMS project was mandated to develop support for four major areas: (1) cost accounting, (2) administration and management, (3) hospital operations and ancillary services, and (4) delivery of patient care. An unsuccessful attempt was made in 1983 to identify affordable and functionally acceptable systems approaches to this goal in the private sector. A key issue was the general absence of support for the evolutionary growth of information systems as contrasted to complete specification of all future requirements in advance.

Shortly thereafter, the IHS became aware of the VA's DHCP, which offered significant benefits in terms of availability at low cost, operation on more than one vendor's systems, and an open architecture. As a result, it was decided to give serious consideration to building on the VA DHCP for the hospital-centered components of RPMS. While it was clear that the merits of portability and economy were moot unless a considerable match of functional requirements existed between the VA and the IHS, this indeed proved to be the case.

It is worth noting that VHA charges only for the cost of reproduction and distribution; the software is otherwise free, having been developed by a federal agency and placed in the public domain. This alone resulted in considerable cost savings to IHS. In addition, the open architecture of DHCP and the infrastructure supplied by the Kernel provided IHS with a foundation for development of major clinical applications that complemented the functionality of DHCP.

Implementing the RPMS

Hardware and Software Strategies

Following a year of experimentation with MUMPS, the MUMPS language and the VA FileMan were adopted as standard development tools in April of 1984. Initial copies of the VA Kernel were obtained shortly thereafter. By October of 1984, somewhat more than 100 computer systems, varying in size from single-user microcomputers to multiuser minicomputers, had been procured. These systems—all capable of supporting MUMPS and the VA software environment—were purchased from three vendors representing distinctly different hardware and operating system software architectures.

Using the existing IHS data processing center and its staff as a hub for coordination and software distribution, a structure based on voluntary participation was implemented to support distributed development of applications as well as testing and adaptation (where necessary) of the VA software prior to installation in IHS sites. This effort was founded on standardized data dictionaries, standard software tools, and emerging standards for development approaches.

Paralleling the overall structure of RPMS, responsibility was assigned for development of cost-accounting applications, testing and adaptation of VA applications, and rehosting of the existing IHS PCIS from a COBOL and mainframe environment to MUMPS and the FileMan in a minicomputer setting.

Key components of the RPMS project were the desire to avoid reinventing systems and to avoid incurring future maintenance obligations wherever possible. As a result, the basic VA data dictionaries, which define essential data structures, were left intact. A few were given extensions to reflect differences in the way the IHS and VA healthcare systems function. This was critical to maintaining compatibility with current and future releases of VA applications software.

Current Status of RPMS

The IHS has kept up with changes in Kernel software, including the FileMan, and has been supplied by the VA with new releases of Kernel and applications software as they have been distributed within the VA. Department of Veterans Affairs software is currently being run at more than 200 sites, both hospitals and outpatient centers. Over the years, the DHCP-RPMS amalgam has been on radically different equipment from at least four vendors (Plessey, DEC, Altos, IBM, and multiple PC sources), each under a different operating system (native DSM, two flavors of UNIX, AIX, and DOS).

A number of the core VA packages are in widespread use in IHS, including outpatient pharmacy, clinic scheduling, and admission/discharge/transfer. More limited use, primarily in larger facilities, is being made of the laboratory, radiology, and nutrition and dietetics applications. The Anchorage Native Medical Center utilizes nearly all DHCP modules.

Indian Health Service development has resulted in a large set of non-DHCP components of RPMS. Significant administrative applications include fiscal management, acquisitions management, and time and attendance. A number of hybrid administrative/clinical packages exist, most importantly for patient registration. These hybrids provide a major core data set of data idiosyncratic to IHS, contract health services management, and referral management (incorporating significant elements of managed care). Major clinical applications include the PCC, clinically oriented dental care, activity reporting for community health representatives, immunization, obstetric clinic tracking, and women's health (tracking of preventive and therapeutic procedures, such as pap smears or biopsies). Rehosting of the Patient Care Information System (PCIS) as the Patient Care Component (PCC), a large and multifaceted task, required almost 5 years; major effort was put into conversion of the Sells (Tucson, AZ) clinical database, which at the time covered 1.7 million visits involving 31,000 patients over 15 years. Numerous applications, both clinical and administrative, have been implemented by end users using the VA FileMan database system. The availability of end-user tools has been a major advantage to IHS.

Cooperative VA-IHS Activities

Liaison has been maintained for almost a decade now between designated VA and IHS development centers, and informal meetings between VA and IHS developers have taken place at professional meetings. Input from IHS had major influence on new features in FileMan and other portions of the Kernel. A formal interagency sharing agreement was negotiated in 1986, which enhanced communications between corresponding groups of developers and members of professional specialties.

The Patient Care Component

Background

The PCC is a product of a mixed heritage, melding functional attributes of its predecessor, PCIS, and technology drawn from the DHCP. Its objectives are surprisingly similar to those set down 25 years ago, although the circumstances of implementation have changed dramatically.

Common barriers to the delivery of comprehensive health care faced by the IHS include:

- Patient mobility, partly as a result of a hierarchy of healthcare facilities;
- Distance to health care, and the related opportunity cost of not taking full advantage of the patient's visit;
- Limited time with the patient, perhaps as little as 6 to 10 minutes;
- Provider turnover, typically on the average of every 2 to 5 years;
- Difficulty of keeping track of periodically performed tasks.

In the attempt to overcome these barriers, the PCC was structured to:

- Support providers of front-line primary care with generic capabilities as well as specialty-oriented applications;
- Meet the comprehensive needs of longitudinal care and ambulatory settings;
- Integrate patient data from multiple disciplines and sources, even across sites;

- Record core aspects of every encounter of a patient with the healthcare system, whether services are provided directly or through external contracts;
- Provide managerial and administrative data as a byproduct of the patient-care process.

Clinical data systems have developed in the IHS primarily as a result of grass roots interest rather than through top-down management direction or as a follow-on to the automation of ancillary systems. Thus, the PCC evolved in a climate that emphasized ways to improve the quality and effectiveness of direct patient care in a setting characterized by small rural sites, a mobile patient population, and community-based public health.

From an organizational perspective, this has meant establishing design guidelines and operational policies that emphasize:

- Independence of clinical data from data management policies of ancillary departments;
- Accommodation of data originating outside the site;
- Patient-centered data organization;
- Access to data along multiple axes;
- Minimizing ownership and control issues characteristic of vertical applications.

To date, the IHS has not aspired to a complete electronic record, viewing the PCC as a summary and index of the traditional record rather than an eventual replacement; this has had a strong influence on application design. However, major emphasis is placed on PCC's role as a focal point for communication and data sharing among members of the healthcare team, even in the face of security and confidentiality constraints.

The IHS has had to approach the development of information systems in general, and medical systems specifically, from a much lower level of technology than the VA. Beyond geographical remoteness of facilities, as well as a generally lower level of resources for development of computing infrastructure, this is a reflection of the disparity in size of facilities, a consequently lower level of clinical services offered, and extraordinary difficulty in recruiting and retaining competent Information Resources Management (IRM) staff. While working toward increased sophistication in the long run, IHS has traditionally had to develop less technology-intensive approaches to problems such as data capture in ambulatory-care settings, for example, using clinically oriented encounter forms rather than direct provider-system interaction.

Patient-Centered Data Integration

Technical Integration

Patient Care Component was designed specifically to integrate tightly with DHCP. This relationship is represented schematically in Figure 26.1, which

412 System Evolution

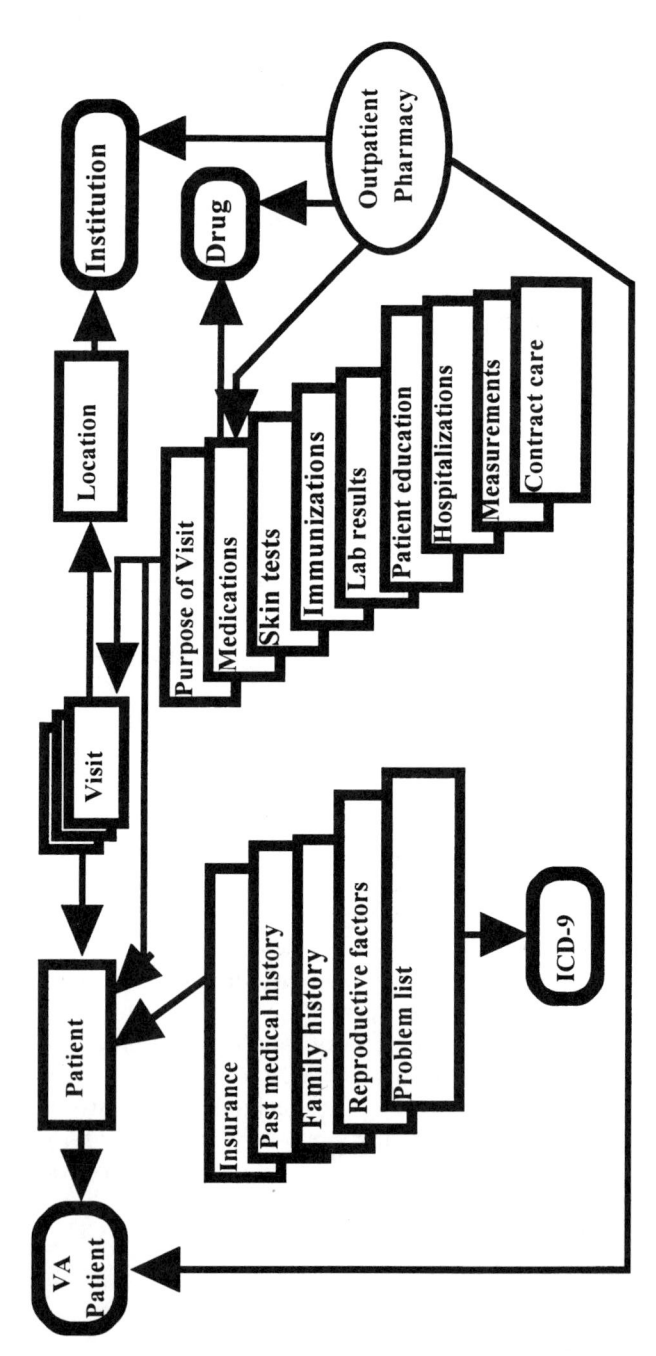

FIGURE 26.1. Relationship of patient-care component and Decentralized Hospital Computer Program.

illustrates major PCC and DHCP files (squared and rounded rectangles, respectively) and the sharing of core DHCP files (patient file, drug file, etc.) by PCC through pointer linkages. As can be seen, the PCC file structure is patient and visit centered, and designed for ease of access along axes of patient, visit, time, or class of data. For example, the patient-specific Health Summary follows links from patient file to data file. Population-based epidemiological reports follow links from visits to data; and program-specific reports (such as immunization statistics) focus on individual data files. The flat, normalized structure combines aspects of relational and network database models, and simplifies addition of new data classes.

Like all RPMS applications, PCC relies on the VA Kernel for services such as database management, menuing, messaging, etc. Unlike another notable comprehensive system, the Department of Defense Composite Health Care System (CHCS), RPMS and PCC are designed specifically to coexist with DHCP. This has made it possible to attempt bidirectional transfer of technology between IHS and VA (Curtis, 1986). The PCC is currently in test at the VA medical center in Tucson and has required comparatively minimal adaptation to cope with VA-specific healthcare practices.

Operational Integration

In operation, the PCC supports three major types of integration: cross-application, cross-discipline, and cross-facility, as discussed below.

Cross-application

As shown in Figure 26.2, the PCC receives data from a number of sources, including active links from VA and IHS applications (e.g., lab and pharmacy), entry of encounter form data supplied by providers, and automated external data sources. This repository structure isolates clinical data from dependencies on data management policies of ancillary systems and provides the ability to store data arising from other sites without contaminating data used for workload measurement, etc., in ancillary packages. In addition, it allows generic tools, such as the PCC query manager (QMAN) and the clinical reminder system to access data without issues of ownership or complex data structures (Curtis, 1989).

Cross-discipline

Health care in the IHS is truly a multidisciplinary activity, involving a spectrum of staff including physicians, nurses, community health workers, etc. The PCC is the team's system, rather than a "physician's system" or a "nurse's system."

Cross-facility

As noted above, the PCC database houses patient-care data regardless of the site at which they were delivered. This has been exploited in the IHS

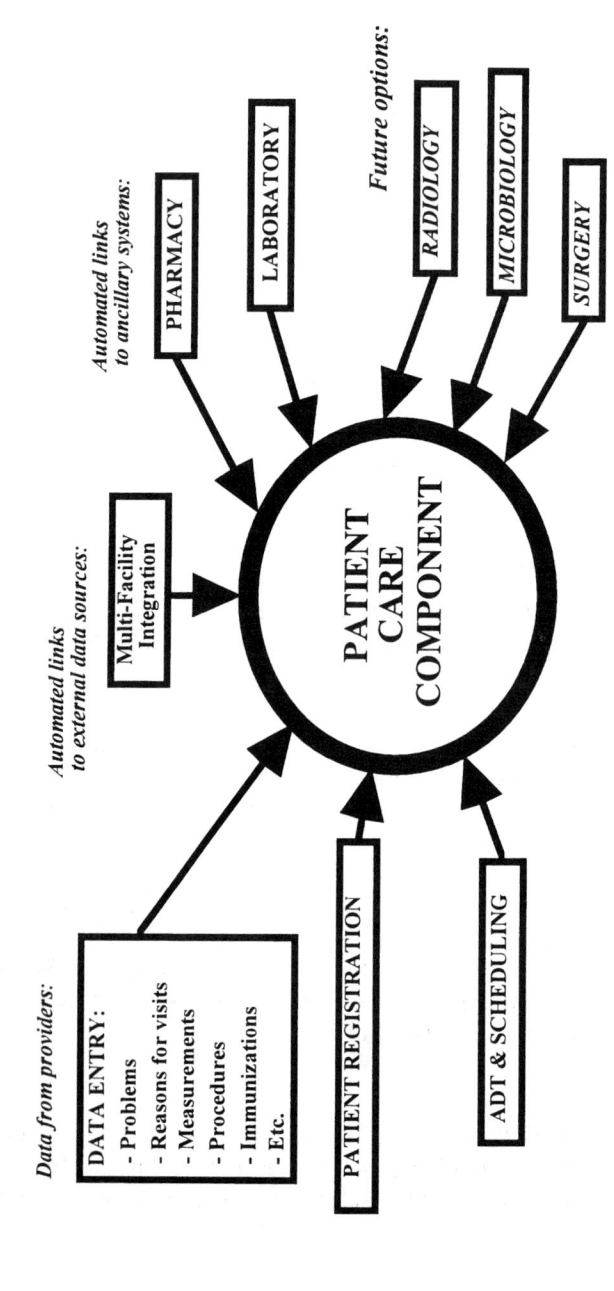

FIGURE 26.2. Patient-care component as an integrating focus.

Multi-Facility Integration (MFI) project (Curtis, 1992), which uses e-mail to route transactions reflecting clinical activity and alteration of demographic data to all sites at which the patient has records. The design and implementation of MFI are discussed in more detail later in this chapter.

Achievements

Resources for formal evaluation of PCC have been nonexistent. However, outcomes observed as a result of site reports and surveys of provider perceptions include improvements in:

- Integration of patient information across sources;
- Communication among healthcare team members;
- Follow-up of high-risk patients;
- Provision of preventive services;
- Performance of tasks related to chronic conditions.

The PCC has proven its value in production use as a system that is patient centered from the perspective of front-line providers, but that is simultaneously capable of supporting views of the database focused on cohorts or selected data classes for purposes of quality assurance and program management. Beyond its clinical roles, PCC has replaced earlier statistical data collection systems with clinically relevant data capture, and now provides the basis for third-party billing, an increasingly important aspect of patient care in the current fiscal climate.

Future Challenges

To remain a long-term success, the PCC must successfully address a number of critical issues:

- Requirements for storage of additional types of data as interests and priorities change, for example, health status and risk factors, radiology, microbiology;
- Requirements for new modes of integration with specialty systems, for example, maternal and child health, which imposes a strong programmatic framework on a constellation of independent classes of data (such as procedures, anatomic path, and pharmacy);
- Incorporation of decision support capabilities, for example, enhanced clinical reminders and surveillance, which entail development of more powerful inferencing facilities with tighter integration into mechanisms through which data enter the system;
- Complexities of intersite data exchange, for example, interfacility transfer, and installation in the database, of patient data; for IHS this currently means adaptation of the MFI project to HL7 as a part of working with the VA;

- Position and funding limitations impacting the use of clerical-based data entry, which may dictate shifting data entry workload to clinicians.

The common factor underlying all of these matters is the requirement to adapt to changing circumstances. This need for flexibility was anticipated early on, and PCC was designed with an open architecture. The design has enabled PCC to accommodate changes such as the recent addition of microbiology results, tests in a live DHCP environment at a VA medical center, and (perhaps the ultimate role shift) use as a "patient" information system in a veterinary school.

Current Status

The PCC has been in operation in the IHS for 6 years, expanding from the initial sites of development and testing to support daily patient care at some 220 hospitals and clinics. In fiscal year 1993 (the most recent period for which statistics are available), 2,912,500 ambulatory visits were recorded in PCC at 85 IHS direct-care sites. For contrast, the Tucson VA medical center pilot project captured data at that single facility from some 186,200 visits in PCC in the 12 months beginning February 1993. The system has also had an impact outside IHS; it is the basis of the VA's Patient Care Encounter (PCE) application, now in the early stages of release within the VA DHCP system. Patient Care Component forms the outpatient component of the hospital system of Saipan, and has informally been reported to be in operation as far away as Siberia.

Despite its "school of hard knocks" flavor, the PCC experience provides evidence that technology integration can be successful given the right circumstances, and that an integrated, patient-centered clinical database can play an important role in an environment of sophisticated vertical applications.

Multi-Facility Integration

Background

As discussed previously, the major objectives of PCC included:

- Maintenance of an electronic summary of the critical elements of patient care (versus implementation of a total electronic medical record);
- Integration of data from multiple disciplines and sites of care; and
- Support for convenient and flexible access to data by noncomputer professionals.

The primary motivation for an integrated multisite database is availability of at least a summary record wherever an individual patient is seen. The need for individual patient records at multiple facilities stems from the mobile nature of the IHS patient population and from typical referral patterns, which may result in a patient receiving care from an ambulatory-care site, a small hospital, and a tertiary-care referral center.

In addition, the geographically dispersed nature of IHS creates a significant requirement for a consolidated database to support wide-area applications such as immunization programs and managed-care programs. Other important uses of the database include program management, quality assurance, and applied research.

The Multi-Facility Integration Project

A side effect of RPMS decentralization was the loss of the integrated clinical database, which occurred when all data no longer resided in a single set of files on the IHS mainframe. While unfortunate, this was anticipated, and reintegration of the database through exchange of data between facilities was a goal from the start. Even so, pressure to regain the previous level of integration became considerable over the course of implementing the Patient Care Component and, later on, Multi-Facility Integration (MFI).

Alternative Approaches to Data Integration

In general, there are three ways to approach the integration of medical databases across multiple sites: maintenance of records on a centralized data system, transmission of data between sites, and patient-carried records.

Constraints resulting from the remoteness of IHS facilities and lack of a sufficiently powerful and reliable communications network precluded use of a centralized data system for RPMS itself, and thus for MFI. It was recognized that optical card technology holds great promise for implementing patient-carried records, and that such records could be a useful adjunct to other data systems; indeed, VHA has actively investigated such applications. However, it was felt that it would be unrealistic in the IHS to rely exclusively on availability of such records as an essential component of the integrated data system.

The two major alternatives involving transmission of data are distinguished by whether or not data generated outside a site and data generated at the same site are integrated and treated equivalently. An example of a nonintegrated design is VHA's PDX project, in which database records are converted to text and transmitted; once received, data are readable, but not capable of being installed in the database. Multi-Facility Integration typifies the integrated approach: other than visit-related data being tagged with a location, remotely generated data are indistinguishable from those which are generated locally once installation is complete.

Implementation Framework

An absolute requirement for MFI was that it be implemented using core RPMS/DHCP MUMPS-based technology. This included FileMan (basic database functionality), Network MailMan (communications and messaging), Menu Manager (message servers), and Task Manager (server and task control)—virtually every element of the VA Kernel.

The need to transport DBMS data in a system-independent manner led to development of "filegrams": a method of representing individual or linked collections of database records as structured, human-readable, ASCII messages. While filegrams are an intrinsic component of FileMan, they can be generated and interpreted by arbitrary programs, regardless of the processing environment.

Because of the diversity of hardware and software environments in which MFI might be required to operate, independence from special characteristics of those environments (e.g., operating system file or communications facilities) was essential. It was recognized that inefficiencies that might result from this decision were a necessary trade-off for portability.

System Architecture

Overall flow of data

In MFI, data flow between three classes of sites: parent sites, satellite sites, and integration sites (see Figure 26.3). The integration site (IS) is the hub of activity: all transactions pass through it for routing as well as addition to the database located at the integration site, which contains a full demographic and medical database. All subsidiary (i.e., nonhub) sites have a complete demographic database; a satellite site has medical data only for patients with records established at that site, while parent (or regional) sites have a full medical database for patients registered there as well as at associated satellite sites.

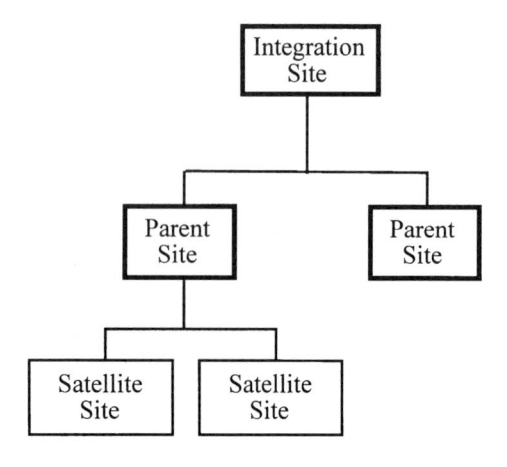

FIGURE 26.3. Multi-facility integration schema.

In the IHS, satellite sites are individual facilities (hospitals or ambulatory-care centers). Parent sites are generally service unit offices (the next level of aggregation for administration and patient care, typically consisting of a hospital and one or more associated ambulatory-care centers) and may be located at the largest facility regardless of type. An IS is located at an area (regional) office.

Generation of messages

Messages may contain database updates (sent as filegrams) or commands (e.g., "Merge patients A and B"). They are generated as a result of activity in patient registration or the PCC. The message generation process is run periodically, at an interval determined by characteristics of the site, such as size and level of activity. Patient registration activity sets a flag indicating that registration data have changed and should be transmitted. Only visit-related data are exchanged through MFI, and thus all visits not previously transmitted are candidates for transmission.

Transmission of messages

All MFI transactions are transmitted as network mail messages. Commands are sent as formatted text. Database updates (registration data and medical data) are created by a filegram generator and encapsulated in a message with a header that permits routing to be determined without examining the entire message.

Messages are addressed to processes, referred to as servers, rather than human recipients. The network mail system allows system-to-system contact to be initiated as soon as a message is generated for a site or for contact at scheduled times. Messages are not currently encrypted or secured; although this has been discussed, general inaccessibility of messages by users has been considered sufficient thus far. Encryption of telecommunications may be implemented in the future, although the relative cost and risks are still being evaluated.

Installation of messages

When a message is received at a site, the mail system stores the message and activates the appropriate server (identified by the recipient name). At the IS, the initial recipient is the message router, which forwards the message to the message processor task at all sites where the patient is known, including the IS itself. At satellite and parent sites, the initial recipient is the message processor itself. The message processor dispatches tasks (command processor or filegram installer) appropriate to the type of message.

Problems and Challenges

Stimulus for transaction generation

The original system design called for use of the database audit facility to detect changes to the database and facilitate transmission only of the changed portion of a database record. However, it was discovered that the audit facility of the host database management system was flawed and could not be used as intended; this remains true to date. As a result, messages must contain entire database records, and this substantially increases the size of the average message. While the audit facility has since been fixed, shifting to selective data transmission would now require significant changes to filegram generation, and the impetus is not sufficient to warrant embarking on a significant revision.

Entity uniqueness

Two attributes of the operating environment—lack of unique identifiers for patients and visits, and inability to guarantee complete standardization of reference tables (such as medications, lab tests, etc.) across all sites—have had a major impact on the PCC and MFI. The primary effect is an increase (although not to unacceptable levels) in the estimated rate at which errors will occur at typical sites.

This has led to implementation of complex mechanisms for carrying auxiliary identifying information ("specifiers") along with primary lookup values to disambiguate such values on the receiving system; the VA was responsive to IHS needs in incorporating this capability into VA FileMan. It has also complicated the linkage of visit nodes and dependent items (e.g., prescriptions and measurements).

Asymmetric demands for data

Multi-facility integration is based on an assumption that the relationship among all sites where a patient is seen justifies maintenance of consistent data at all of those sites. However, this is not always the case. For example, a referral center may not want all medical data for every patient ever seen there (which may have been a one-time occurrence), but the facility referring the patient may want to receive data on events while the patient was at the referral center.

Because this is an actual situation, the system provides a means for specifying whether a site will receive demographic data only or demographic and medical data, and from what facilities. This provides flexible control over patterns of data distribution.

Consistency of reference data across sites

The host database management uses a field type called a "pointer" to implement linking references from one file to another. Pointers are internal entry numbers, and thus are not dependable across systems. As a result, de-referencing pointers (i.e., converting them to symbolic values) is an integral part of filegram generation, and conversion back to pointers must occur on the receiving system. Thus, successful installation of a filegram rests on matching elements of incoming data with entries in files to which they refer (e.g., provider of service, International Classification of Diseases, 9th ed., Clinical Modification [ICD-9-CM] code). Failure of such lookups can result from true errors (e.g., a required entry is missing from a "standard" table at a site) or from a desire to keep data tables small (e.g., substitute a generic provider name for all references to outside providers).

To cope with such situations, a capability known as "conditionals" has been provided that enables a site to specify an algorithm for resolving lookup failures. A conditional consists of a set of sequentially processed tests that can examine the input data and optionally return a substitute value (pointer reference). If no value has been determined when the conditional terminates, an error is registered and the filegram fails.

Current Status

Multi-facility integration is operational in the healthcare delivery area surrounding the Tucson center where it was developed. Multi-facility integration links a computer serving as the integration site with a small hospital, three ambulatory-care centers, and a research system that requires an integrated database. (See Figure 26.4.) During the first phase of the project, routine data transmission was limited to patient demographic data; testing of medical data exchange was being initiated at the time this paper was written. Several tens of thousands of messages have been sent to each of the five systems involved; traffic rates vary by site from approximately ten to approximately 300 messages generated per day.

Historically, the principal problem affecting routine operations has been the communications environment, which has evolved significantly over the life of the MFI project. Telecommunications facilities were initially limited to 2400 baud dial-up lines and a 7200 baud multiplexor channel to the hospital. Reliability was enough of a problem that large collections of messages were frequently filed off onto tape, transported to the receiving

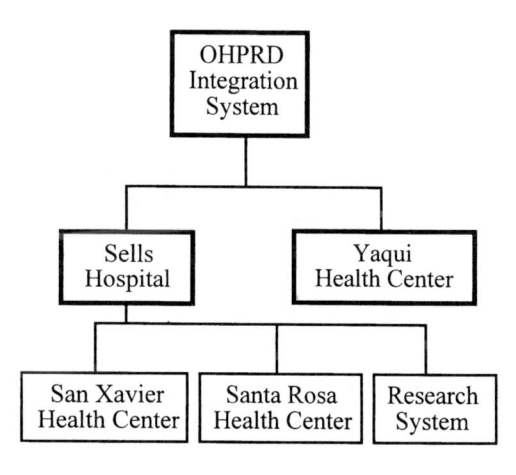

FIGURE 26.4. The multi-facility integration pilot test.

site, and reinstalled, whereupon the mail system picked them up as though they had arrived via telecommunications. (That this worked at all is a tribute to the generality of the underlying VA MailMan software.) In addition, the two smallest sites conducted backups during the day and shut off their computers overnight. The resulting limited system availability sometimes conflicted with the amount of time required to transmit large message batches. Over the lifetime of the project, IHS installed an X.25 network, similar to VA's data network, but implemented via FTS2000, the Federal telecommunications system. This made all the difference in the world in reducing the system's failure rate.

Adopting a message-per-transaction model and using network mail software that already existed in the Kernel exacted a price. Inefficiencies of transmission in the current setting are considerable. The mail system, not originally intended to support applications like MFI, does not incorporate capabilities sufficient to support management of a complex messaging application. This can become critical. In the event of a system crash, which results in loss of data at a receiving site, it may prove impossible to determine which messages must be regenerated and reinstalled. Such problems will be a focus of future development.

Future Effects of External Factors

The MFI project and its fundamental design were motivated primarily by internal requirements and influenced relatively little by external factors. However, events in the wider community may have a significant effect on the project over its lifetime. The Institute of Medicine's report on electronic medical records and an emerging Health Care Financing Agency legislative initiative (which proposes that electronic medical records be required at sources of health care receiving reimbursement from the US government and that development of appropriate standards be undertaken by the Department of Health and Human Services) demonstrate national trends that may provide strong incentives for adherence to standards for data interchange. The relative immaturity of official and de facto standards (e.g., IEEE, ASTM, HL7, HICCS) at the outset of the project resulted in their having little impact on MFI (Rishel, 1989); it is clearly desirable, however, that they be incorporated into the transmission protocols if at all possible. However, the structuring and linkage capabilities of current proposals (e.g., compared to the visit orientation of much MFI data) are still fairly rudimentary, and need refinement before they serve as the formal basis of an MFI design.

Conclusion

Experience in the development and early operation of the MFI system has resulted in a number of insights that are worth sharing.

Multisite Effects

Extension of data systems beyond a single site introduces new twists on old themes. Some attendant problems can be addressed through standardization, yet variation among sites in a large organization remains a challenge, as does the need to cope with the introduction of potentially inconsistent or unwanted data from external sources (Curtis, 1989). Organizations considering interfacility data exchange should consider carefully what can be done to the type and structure of data involved to simplify installation of data and reduce potential conflicts.

Technology Effects

Multi-facility integration is a "meta-application" that combines multiple tools, including a database manager, task manager, and e-mail, in unusual ways. This combination, while synergistic, tests the limits of the component technologies and tends to expose boundary condition problems that remain undetected in less complex settings. As a result, close cooperation with the team supporting those tools is essential. An organization considering this type of application must evaluate available fundamental technologies and the degree to which they can be enhanced if necessary.

Implementation Effects

Conflict between theory of design and reality of operations occurs constantly, often confounding apparently appropriate technical approaches. Organizations thinking of implementing an MFI-like application should be able to respond with a strong affirmative to the following questions:

- Is the implementation environment (database management system, communications facilities) really robust enough to support the application?
- Do site and user attributes (availability of the site's computer, degree to which the staff can be taught enough to use the system) match the requirements of the application?
- Can crucial events or entities (visit date and time, patients) be identified uniquely and consistently?

Validity of Basic Goals

In spite of the difficulties encountered and described in this chapter, it is evident that the project's basic goal—an integrated multisite clinical database supporting decentralized health care in the IHS—remains valid. This assertion is supported by continued demand from clinical users as well as interest expressed by other multi-facility organizations, including the VHA, as their clinical information systems mature. Nevertheless, large organizations considering such a project should think twice before taking it on, and be sure of management commitment and adequate resources.

Summary of Experience

Costs

The initial phase of the RPMS project, representing about a third of total systems requirements, costs approximately $4 million, primarily in hardware and software purchases. Subsequent phases brought the total to nearly $20 million over the next 10 years. These costs do not include personnel. Most of the work so far has been done by 20 or so developers at five sites (provisions are being made for contracting out some development), but by some estimates approximately 400 positions (not counting end users who are not working with computers full time) are devoted to RPMS, primarily in site operations capacity.

Benefits

The author of this chapter and others associated with the early planning of the RPMS believe that the approach taken is leading to lower overall costs than would have been incurred through alternative approaches. The VA's core applications have been demonstrated to meet IHS requirements, and the technology has provided an evolutionary growth path as well as demonstrated hardware and vendor independence. In addition, end users with access to reasonably powerful tools and appropriate databases have proven to be capable of implementing significant applications, and comprise an invaluable resource for the organization. There is no doubt that IHS experienced a decade of explosive growth of clinical information systems that would not have been possible without leveraging the VA's investment in DHCP.

Drawbacks

The approach taken in the RPMS project is not without its drawbacks. Key among these is the inability to fully meet user expectations and difficulties in the management of a large systems project, especially one requiring cooperation between organizations of disparate size.

For the sake of economy and simplicity, the VA's applications are based on relatively primitive technology: serial terminals, scrolling displays, etc. While this is beginning to change (the first PC-dependent client–server applications were to be fielded in 1996), monochrome ANS terminals are still the norm. It is clear that this results in less elegance than can be achieved with color, graphics, windowing, non-keyboard input, and so on. In addition, FileMan lacks a truly powerful query and data manipulation language, although this may be remedied in the future. As a result, it has been difficult for IHS to meet the expectations of users who have been exposed to the current crop of microcomputer-based commercial products.

The absence of "sizzle" often acts to obscure the fact that at least equal functionality is being provided in a setting that integrates numerous basic "bread and butter" applications.

Management of any really large information systems project is difficult at best. For the IHS, management structure, budgeting, and regulatory mechanisms affecting data processing procurements made managing the development and operation of the widely distributed IHS information systems program a heroic task. This was complicated by incomplete understanding of the VA's operations, major differences in size and activities of the two organizations, and consequent uncertainties in estimating the resources required for implementing similar systems in the IHS. An additional factor has been the (thus far) rather dependent nature of the IHS-VA relationship and the somewhat superficial understanding of the RPMS project by VA staff who perceive only its DHCP-related aspects. We always believed that improved communication would alleviate this.

Conclusion

The IHS experience has supported the VA position that hardware and vendor independence can be achieved. It has not proven that the tools used in the RPMS and the DHCP are optimal for this task; few, however, are as inexpensive initially. Experience has also demonstrated that portability and interoperability of applications—even those developed by separate organizations—are feasible given sufficient dedication to standardization of an underlying software interface and consensus uniformity of development approaches.

References

Brown G. 1980. The patient care information system: A description of its utilization in Alaska. *Proceedings of the 4th Annual Symposium on Computer Applications in Medical Care*, pp. 873–881.

Curtis AC. 1986. Portability of large scale medical information systems: The IHS-VA experience. *MedInfo '86: Proceedings of the 5th International Congress on Medical Informatics*, pp. 371–373.

Curtis AC. 1989. Knowledge-based systems in an imperfect world: data-based decision support systems in ambulatory care. *MedInfo '89: Proceedings of the 6th International Congress on Medical Informatics*, pp. 248–252.

Curtis AC. 1992. Multi-facility integration: An approach to synchronized electronic medical records in a decentralized health care system. *MedInfo '92: Proceedings of the 7th International Congress on Medical Informatics*, pp. 138–143.

Garratt A. 1979. An information system for ambulatory care. *Proceedings of the 3rd Annual Symposium on Computer Applications in Medical Care*, pp. 856–858.

Rishel R. 1989. Pragmatic considerations in the design of the HL7 protocol. *Proceedings of the 13th Annual Symposium on Computer Applications in Medical Care*, pp. 687–691.

27
International Installations

MARION J. BALL AND JUDITH V. DOUGLAS

Principles

A rapidly evolving field, enlisting some of our best academicians and scientists, health informatics holds great promise. The challenge now is to deliver on that promise, to move informatics developments from theory into practice.

The International Medical Informatics Association (IMIA) has worked to encourage that move, through its strategic planning, its 1995 triennial congress, and its many working groups. As far back as the early 1980s, IMIA informaticians agreed that existing information technologies were sufficient to support healthcare applications.

The reality is that we already have the tools and technologies to change health care for the better. Within the private and the public sectors, we can find excellent examples of systems that have been in use for a period of time and that could serve far greater needs if they were only to be adopted more widely. The case in point is the Decentralized Hospital Computer Program (DHCP), which offers proven capabilities in multiple facilities and diverse applications. The Veterans Health Administration (VHA) began development of DHCP in 1982 using Massachusetts General Hospital Utility Multi-Programming System (MUMPS), now known as M language and acknowledged by the American National Standards Institute (ANSI). The system has evolved over time, opening up its architecture, adding applications, and enhancing its functionality. Winner of a 1995 award from the Computer-based Patient Record Institute (CPRI), DHCP is a robust system, capable of supporting the VHA and its national service network.

Testimony to the degree of robustness that DHCP possesses comes not only from the VHA, but from other sites that have chosen to acquire and implement DHCP software. This chapter documents installations of VHA-developed software in healthcare settings outside the United States and shares informal survey results, gathered in the attempt to measure user satisfaction.

Utilization

Database of Purchasers of VA Software

The M Technology Association (MTA) prepares printouts listing persons and/or organizations that have paid for Department of Veterans Affairs (VA) software through the MTA. For the year identified as 1995–1996, this prototype application library listed 35 organizations. The item most frequently purchased was FileMan; other purchases ranged from utilities like Kernel to VA CD-ROM and a range of clinical applications. Of these 35 organizations, 20% (7) were based outside the United States. A review of the library lists for years 1992, 1993, 1994, and 1995–1996 cites organizations in 17 different countries. Of these, some are cited several times and for multiple packages.

This count is a clear underestimate, for countries where DHCP components are known to be used are not cited. However, the listing does not indicate whether these are healthcare installations; some are clearly so, others are ambigious. Table 27.1 shows the 17 countries listed. Egypt, Finland, and Nigeria are all known to be using VHA software. Both Finland and Nigeria obtained the software in 1980s, which may account for their not being listed. Still other countries are thought to be using or intending to use VA software, including Colombia and Rwanda.

E-mail Survey of Satisfaction with DHCP

An informal e-mail survey conducted in December 1995 to obtain some measure of the satisfaction of non-VA users of DHCP software. Those responding included non-VA users in the United States, those aware of international users, and several international users themselves. The comments of the third group are summarized below.

TABLE 27.1. Countries acquiring VA software.

Australia (6)	Austria (1)
Belgium (3)	Brazil (1)
Canada (3)	China (3)
England (2)	Germany (2)
Italy (1)	Japan (2)
Korea (1)	Malta (1)
Netherlands (2)	Pakistan (1)
Romania (1)	Singapore (2)
Spain (1)	

Source: The countries listed above appeared on the MUMPS Users Group Prototype Application Library List, 1992–1996. Numbers in parentheses are the number of times a country appeared on the Library List during the years given. It should be noted that the listing is not limited to healthcare use.

Finland

Mikko Korpela, Senior Researcher at the University of Kuopio in Finland, wrote Catherine Pfeil, the VA's Assiociate Director for Software Production, stating that "the VA software is one of the very best American things Finnish health care has ever got." Korpela continued to state that "VA FileMan and Kernel are the cornerstone of the hospital information system that has more than 50% of the installed market base in Finland. Because of the language and other factors, the DHCP Core Applications are not in use in Finland."

Others responded regarding Finnish use of VA software. In a separate communication, in February 1996, Hellevi Ruonamaa, also from the University of Kuopio, reported that a total of 34 hospitals, laboratories, and universities use VA software.

Germany

Marcus Werners, at the German Heart Institute in Berlin, where DHCP technology is used for its hospital information system, responded to the same query. Werners reported that "The decision to use DHCP was not a strategic decision, but a gradually developing process, not without its share of Pro- and Contra-MUMPS debate. The decision to go along with a DHCP solution was based on 'desired functionality,' 'easy to integrate' and 'costs.'" He continued to state that "By using DHCP we found that we spend less money on software (and hardware) purchases but invested more in training of people and exchange of information with the VA developers and users." According to Werners, "in our experience with vendors of Hospital Information Systems in this country no system came even close to the level of integrated functionality that DHCP offers." Werners further reported that healthcare legislation in Germany was requiring many changes, and stated that the VA technology allowed the hospital to make necessary adaptations "much faster" than commercial systems.

Quantitative Survey

Survey Form

As mentioned above, the informal e-mail survey also served to identify international users, who were subsequently contacted and asked to respond to a brief survey describing their experience with DHCP. To ensure rapid turnaround, the survey was conducted by fax and e-mail, as well as by hand at a 1996 international conference held in Boston, Massachusetts, for M users. Attempts to reach possible users in Colombia, China, and Rwanda were unsuccessful. The form was limited to two pages.

Part 1 of the survey gathered descriptive information on the respondent, who was then asked in Part 2 to specify the year in which VA software was first acquired and which components (e.g., FileMan, Kernel, or Nursing System) were in use at the time of the survey. Part 3 asked for a ranking of seven key attributes on a scale of 1 to 10 (10 being highest). Part 4 asked for yes or no responses to three questions and brief descriptions if the answer was yes. Part 5 asked the respondent to state what he liked most and least about the VA software being used, and to rank the impact of software use on five areas.

Respondents

Those responding to the form included:

Abdebowale Adewusi, on behalf of Mrs. H.A. Soriyan, Ile-Ife Project, Nigeria;
Wolfgang Giere, JW Goethe, University Klinikum, Germany;
Omar El Hattab, National Cancer Institute, Cairo, Egypt;
Mikko Korpela, on behalf of the Ile-Ife Project, Nigeria;
Hellevi Ruonomaa, University of Kuopio, Finland; and
Marcus M. Werners, German Heart Institute, Berlin, Germany.

Results

Use of VA software dated to 1983/4 in Finland, 1988/9 in Nigeria, and 1990 in Egypt. Wolfgang Giere translated Version 14 into German in the early 1980s; the Heart Institute first acquired VA software in 1992. Thus, the respondents represented established installations, although at least two (Finland and Heart Institute) were installing, testing, or awaiting delivery of new components and/or releases at the time of the survey.

All respondents reported using FileMan and Kernel. They did not, for the most part, make extensive use of applications. The National Cancer Institute in Egypt reported using five applications (Admission/Discharge/Transfer, Laboratory, Nursing, Surgery, and Radiology). The German Heart Institute reported using Radiology and having Laboratory in production and Record Tracking in test. Finland and Nigeria reported use of MailMan. Giere of Germany stated his institution used no applications but rather developed their own. In Nigeria, an early version of the Admission/Discharge/Transfer package (3.3) was modified for their Medical Records purposes.

In rating the software, respondents gave highest marks to its transportability, with an a average score of 9 out of 10. Next highest were Applicability (7.8), Documentation (7.5), and Ease of Use (7). These rankings fall well above the midpoint of the scale and thus speak well for the software in what are critical areas to implementation and ongoing operations. The next group of rankings includes Ease of Modification (5.8), End User Support (5), and Technical Support (4.2). In this last group, with the lowest rankings, responses varied widely, in two cases from 1 to 9, and in one case from 2 to

10. In looking at this last group, it is important to note that VA's mission does not include providing support services to users such as those responding here.

Three respondents reported retrofitting the software, including translation of portions into German and Arabic. Four reported making improvements, changing formats, developing a dictionary, adding an instrument to the Laboratory package, adding text windowing capabilities, and otherwise modifying the package to local use. Two reported that the English-language base was a problem; one respondent stressed the necessity of translation for end users while stating English posed no problem for information management staff.

Responses to the more open-ended portion of the form were consistent with the rankings. Replies to the query "What do you like most about the VA software?" cited the following attributes and features:

- Integration;
- Robustness;
- Portability;
- Comprehensive;
- Continuous Improvements;
- Open System;
- MailMan;
- FileMan Cornerstone;
- Flexibility;
- Framework; and
- Price.

Responses to "What do you like least?" targeted the following:

- Effort required for modifications;
- Old version of ADT and outdated user interfaces;
- Some code in terms of routines (e.g., upper/lower case, etc.).

Finally, those surveyed were asked how the software had changed the way in which they do their work and to rank its impact from positive (5) to negative (1). Aspects identified on the form and the rankings given are as follow: Efficiency (3.66), Cost-Effectiveness (4.83), Productivity (3.83), Quality of Outcomes (3.16), and Organizational Structures (2.5). It should be noted that responses to the fifth item, Organizational Structures, had the widest range, of from 1 through 4; other items had ranges of only 3 (e.g., from 3 to 5) or less.

Conclusion

The number of respondents appears to be small in absolute numbers, yet in relative numbers it represents a reasonable sample of international health institutions known to be using VA software. The responses to the individual items tend to confirm what is known anecdotally (i.e., from e-mail).

Relevance

As international health informaticians, we applaud the work done in Finland, Germany, and elsewhere. Users such as Wolfgang Giere have made substantive contributions to biomedical computing. Giere's comments (1996) at the Tenth Annual Meeting of the MUMPS Users Group-Europe in 1985 helped to define the challenge for the M development over the next 10 years when he stressed operating systems, standards, interfaces, and multilayer development—in essence, open hybrid systems.

We take special note of the work done in Nigeria, in the Ile-Ife Project, with the help of Mikko Korpela of Finland. Well documented in the literature (Makanjuola, Daini, Soriyan, & Korpela, 1991; Daini, Korpela, Ojo, & Soriyan, 1992; Makanjuola, Korpela, Soriyan, & Adekunle, 1995; Soriyan, Akinde, Farewo, Adekunle, Orisatoberu, & Korpela, 1996), this project can serve as a model for future collaboration in developing nations, with populations yet to be served by health informatics.

Key participants in this project attended the 1996 Helina conference in South Africa, which drew attendees from across Africa, including English- and French-speaking countries. Among them was Dr. Amal Ibrahim, who spoke on the hospital information system he has installed in Cairo, Egypt, discussed the VA aspects in detail. (Salamon, in press.) The Ile-Ife Project exemplifies many of the challenges involved in such settings and speaks to the benefits that international collaboration offers. Participants clearly identify the low cost of the M-based VA software as a main factor in its use.

Special Considerations

As the survey results document, transportability and portability are prime considerations. So too are appropriateness and ethics. Mikko Korpela (1992), who has been and continues to be a key player in the collaboration with Ile-Ife, addresses the latter issues. For Korpela, "the real issue is that of practical fit to concrete requirements" (p. 285). He comes to "the conclusion that if computers can be used to benefit African health care, then that is precisely *the most ethical form* [Korpela's italics] of informatics" (p. 286). Korpela stresses (p. 293) that if systems can be maintained, benefits follow, "in day-to-day applications like text processing, 'card boxes' and statistics, which are the most fundamental uses of computers in any country." He continues to state that "more indigenous systems development capabilities" make possible, he calls, "more ambitious uses ... where African people can benefit the most from computers" (p. 293).

The Helina 96 Conference in South Africa, sponsored by the International Medical Informatics Association (IMIA), confronted many of the issues Korpela and his Nigerian colleagues address. It is the hope of IMIA that Helina will foster the creation of a regional group for health informatics in Africa, which can then serve as a catalyst for further collaborations of the type described briefly in this chapter.

Whatever the future holds, it is clear from the experiences reported by international users that the VA software, most notably FileMan and Kernel, is capable of supporting healthcare computing in a multiplicity of environments, ranging from Finland to Nigeria and beyond. Perhaps at a future date the difficulties of gathering information on efforts reported to be ongoing in China, Colombia, and elsewhere can be overcome. In any event, the limited evidence we have assembled here speaks to the VA software—exceptionally robust and distinctly serviceable.

References

Daini OA, Korpela M, Ojo JO, & Soriyan HA. 1992. The computer in a Nigerian teaching hospital: First-year experience. *Medinfo 92 Proceedings*, pp. 230–235.

Giere W. 1996. The tenth anniversary of a challenge: Software development in MUMPS. Reprinted from proceedings of the 10th annual meeting of the MUMPS Users Group-Europe Travemunde 1985. In W Kirsten and R Klar (eds): *Dokumentation und infomationsauf bereitung fur den artz: beitrage zur medizinischen informatik von Wolfgang Giere*. Hocheim, Germany: Epsilon, pp. 403–406.

Korpela M. 1992. Health informatics in Africa: Ethics and appropriateness. In SC Bhatnagar and M Odedra (eds): *Social Implications of Computers in Developing Countries*. New Dehli: Tata McGraw-Hill: pp. 285–294.

Makanjuola ROA, Daini OA, Soriyan HA, & Korpela. 1991. Low-cost hospital informatics in Africa: the Ile-Ife experience. In *Medical Informatics Europe 1991*. Berlin: Springer-Verlag, pp. 111–115.

Makanjuola ROA, Korpela M, Soriyan HA, & Adekunle MA. 1995. Making the most of a hospital information system: A Nigerian case. In MC Sosa-Iudicissa, J Levett, SH Mandil and PF Beales (eds): *Health, Information Society and Developing Countries*. Amsterdam: IOS Press, pp. 163–168.

Salamon R, et al. (eds). In preparation. Special issue: proceedings of Helina 96. In *Methods of Information in Medicine*.

Soriyan HA, Akinde AD, Farewo FO, Adekunle MA, Oristoberu AO, & Korpela M. 1996. M and Kernel as appropriate technology in health care in Africa. *M Computing, 4(1)*, 15–19.

Section 6
Telemedicine and Telehealth

Chapter 28
Telemedicine: Taking a Leadership Role
   John C. Scott and Neal I. Neuberger   *435*

Chapter 29
Telemedicine at the Veterans Health Administration
   Roger H. Shannon and Daniel L. Maloney   *447*

Chapter 30
Telehealth for the Consumer
   Peter Groen   *466*

Chapter 31
Digital Imaging Within and Among Medical Facilities
   Ruth E. Dayhoff and Eliot L. Siegel   *473*

28
Telemedicine: Taking a Leadership Role

JOHN C. SCOTT AND NEAL I. NEUBERGER

Introduction

In 1995, the Center for Public Service Communications (CPSC) was asked by the Veterans Health Administration (VHA) to review private sector trends in telemedicine and to assess telemedicine trends and projects within the VHA. The study gathered information regarding current telemedicine-related activities and plans for future program expansion. Site visits were made to key centers chosen for the range of telemedicine technologies and applications being undertaken or contemplated.

Based in Arlington, Virginia, the CPSC provides telecommunications consulting services in the fields of telemedicine, disaster management, and humanitarian assistance to government and the private sector. As a company, CPSC has established a reputation as a lead organization, nationally and internationally, helping to interface activities of federal government and international agencies with those of nongovernmental organizations, including academic health centers, hospitals, other health providers, humanitarian agencies, and commercial enterprises. Center for Public Service Communications principals have considerable background and experience working in the fields of telecommunications and healthcare delivery in a variety of policy arenas and service areas.

This chapter draws upon the CPSC knowledge base to address two major areas: (1) national trends affecting telemedicine and (2) innovative telemedicine projects in the VHA. The discussion of national trends focuses on managed care, shared decision making, population-based health, alternative sites for care delivery, state involvement, medical education, human factors, and community hospitals. The section on the VHA identifies characteristics that are unique to the VHA, such as its managed care-like setting and primary-care emphasis, and recommends areas in which the VHA can play a leadership role in telemedicine, including interagency projects, telemedicine residencies, leveraging statewide networks, and evaluating new technologies.

As a large multi-facility system and the nation's largest telemedicine laboratory, the VHA can serve as a model to smaller enterprise systems as they work to realize the benefits of telemedicine.

National Trends Affecting Telemedicine

The rapid evolution of telemedicine occurs in an environment marked by change. Support for innovation is slowly transitioning from federally funded projects to private sector funds for local initiatives. Traditional fee-for-service health care is being replaced by managed care, new strategic alliances, and joint venture arrangements. Restructured healthcare organizations are emphasizing patient empowerment and nontraditional sites for care delivery. These changes in turn affect how healthcare organizations plan and implement telecommunications networks.

Increasingly, telecommunications is seen as means to enhance individual control in an otherwise unwieldy and at times impersonal healthcare system.

While major systems and managed-care organizations have been slow to adopt new telecommunications technologies, professional organizations are beginning to pay attention. At a policy level, the American College of Radiology, American Hospital Association, American Medical Association, American Osteopathic Healthcare Association, Federation of State Medical Boards of the United States, Health Information and Management Systems Society, and Voluntary Hospitals of America, Inc. have all launched attempts to identify issues and opportunities in the field. Another indication that these major organizations are getting serious about their involvement in telemedicine is that a coalition led by the American Hospital Association, the Association of Academic Health Centers, the Association of American Medical Colleges, and the National Rural Health Association has joined to help implement provisions to the recent Telecommunications Act of 1996 granting discounted rates for rural healthcare networks.

Further, the Federation of State Medical Boards of the United States recently drafted a "Model Act to Regulate the Practice of Telemedicine or Medicine by Other Means Across State Lines." They have also joined with the Agency for Health Care Policy and Research to develop further recommendations for how to deal with issues of state licensure.

Managed Care

Hospital systems and managed-care organizations are just now beginning to realize that telemedicine can help extend lower cost health services to an expanded population base and capture new market shares. With increased movement throughout the country toward managed-care models, it is expected that a whole new set of professionals and additional practice settings will require new information management and communications tools. Business as usual is no longer acceptable, and radical changes underway are evidence of the insistence of major employers, insurers, and the patients they cover. This is especially true for specialized (and traditionally expensive) services such as radiology, pathology, psychiatry, and urgent care. Often these services are not available in underserved rural and urban areas of the United States at any price, necessitating costly patient or clinician transfer to achieve coverage. To the extent that telemedicine can help increase access and reduce costs while preserving quality, the three major goals of systemwide healthcare reform may be addressed at least in part by using these new management and service delivery tools.

Results of an invitational consensus conference on telemedicine and managed care recently convened by the CPSC are further evidence of rapidly escalating interest. The meeting of major leaders in the emerging managed-care industry and experts in telecommunications discussed issues and opportunities facing the diffusion of telemedicine into managed care, strategies alliance, and joint venture-like practices. Key among the recommendations were:

- "Population-based" services are critical to the success of managed-care-like organizations, and information technologies should be used to help facilitate the changing view of healthcare delivery from an acute care to a prevention focus. The consumer should be the focus of effort with regard to telecommunications in managed care.
- Telecommunications must be used to enhance rather than further erode the public's trust in the healthcare sector.
- A uniformly accepted economic model for evaluating the effectiveness of telemedicine is needed.
- Science-based information is needed indicating which conditions and medical treatments best lend themselves to using telecommunications tools.
- Government will continue to play a stimulative, start-up, research and development role in helping to overcome barriers to telemedicine. The marketplace will then have to take over development and diffusion of technologies and applications, after which point government should be expected to step back into the process, acting a regulator in the public's interest.
- Complicated issues of equity with regard to the provision of health services must be addressed, so that new telecommunications are used appropriately to provide service to people in need in consideration of regional, ethnic, and other demographic and situational variables.
- Public and private sector groups must work together in addressing efficiency issues in terms of maximizing access to healthcare knowledge. This includes a balancing of priorities including a review of technologies and

applications with regard to marginal effectiveness, impact on price, quality, consumer preferences, and other considerations.

Just as leading telecommunications companies are beginning to strategically align themselves with rivals in a new telecommunications era, so too are most major healthcare concerns in an ongoing fight to preserve market share within a declining revenue base. Organizations are increasing their responsibility for the provision of vertically integrated services, increasing their capacity to provide a continuum of care that ranges from preventative to acute to long-term services. Massive industry restructuring including consolidations and mergers like Columbia/HCA, SunHealth and AmHS/Premier, and Aetna and US Healthcare will likely usher in telemedicine programs of unprecedented scope and scale.

Shared Decision Making

Many major healthcare organizations are adopting a patient-centered approach, which emphasizes empowerment and shared decision making. The Foundation for Informed Medical Decision Making, headed by Dr. John E. Wennberg of Dartmouth, gathers and disseminates information about the nature and potential outcomes of alternative treatments, so patients and physicians can share in making informed medical decisions. In the provider setting, computer-based programs allow patients to move through the programs of most interest, providing information about their conditions along with unbiased descriptions of the potential benefits of alternative surgical and nonsurgical interventions. Videotapes, including interviews with patients of similar experiences, allow consumers to seek answers to their questions in their own homes.

Population-Based Health

The success of population-based health programs depends in large part on the collection, analysis, use, and communication of health-related information. In April 1995, the Public Health Service sponsored a first ever conference at the National Library of Medicine on using of National Information Infrastructure (NII) technologies to address information problems of population-based health. Among the barriers identified by conferees were:

- A lack of nationally uniform policies to protect privacy while permitting critical analytic uses of health data;
- A lack of nationally uniform, multipurpose data standards that meet the needs of the diverse groups who record and use health information;
- Organizational and financing issues that make it difficult to integrate information systems or bring potential partners together.

Alternative Sites for Care Delivery

Today many of the traditional caregiving settings are being questioned as to cost and appropriateness. Physician extenders and new diagnostic and therapeutic interventions make it easier to provide care in alternative sites, including ambulatory settings, nursing homes, patient homes, and small accessible clinics in outlying areas.

According to Ace Allen, M.D., an oncologist with the Telemedicine Project at University of Kansas and editor of *Telemedicine Today*,

Telemedicine would seem to be an ideal way of supplying the observational and cognitive skills of a nurse, while eliminating the 'wind-shield time' that eats up so much of a visiting nurse's day. (1995, p. 26)

This is no small consideration, given the nearly half billion home health visits in 1993 estimated by the National Association of Home Care (Allen, 1995).

Tele-home health projects are underway through a number of vendors and sites, including American Telecare, Inc. (Minnesota), md/tv Inc. (Florida), Tevital, Inc. (Pennsylvania), and H.E.L.P. Innovations (Kansas). According to Dr. Jay Sanders, President of the American Telemedicine Association (in comments on Capitol Hill at a meeting of the Senate House Ad Hoc Steering Committee on Telemedicine and Health Informatics), modifying existing cable infrastructures or using ISDN will in time allow interactive medical care to be delivered in a patient's home. What Sanders calls a home medical channel will support diagnostic and monitoring modalities and applications, along with various scopes and peripherals.

Already, online service providers, dial-up servers, and direct Internet access services provide interactive information to patients in their homes. A PC-based system at the University of Wisconsin called Comprehensive Health Enhancement Support System (CHESS) offers health information and problem-solving tools, including modules on breast cancer, AIDS, sexual assault, alcoholism, and stress management. The Harvard Community Health Plan provides 24-hour access to an interactive database that guides Burlington HMO members through questions about their conditions and advises them whether further medical attention is required. Other at-home online services include American Online's Better Health and Medical Forum and CompuServe's Health and Fitness Forum.

Other nonclinical settings are being actively tested for telehealth services. The University of Texas Medical Branch at Galveston and at the East Carolina University Correctional Telemedicine program offer alternatives to costly and disruptive transfer of prisoners for healthcare service. Even mobile telemedical services are being contemplated for such services as magnetic resonance imaging (MRI), children's preventative health, and other services. These would function as potential remote sites for tertiary-care centers that could provide instant specialty consultation to diagnostic or preventative healthcare vans out in the field.

State Involvement

Today the states, already involved in telemedicine, stand to take on newly expanded roles in the delivery of government financed health services. The CPSC, Intergovernmental Health Policy Project (IHPP) of George Washington University, and the Telemedicine Information Exchange (TIE) at Oregon Health Sciences Center have tracked telemedicine activities in Georgia, Iowa, Kansas, Louisiana, North and South Carolina, New Mexico, Oklahoma, and Texas. Both CPSC and TIE remain ongoing sources of current information with regard to telemedicine activities.

Some states have enacted restrictive or enabling licensure or other regulatory provisions; others have been actively involved in public–private partnerships that have set up statewide telemedicine and related networks. Using an informal case study approach, the IHPP recently issued a two-volume report identifying activities in the 30 states they surveyed. In addition, the National Governors Association and Western Governors Association (WGA) have each expressed interest in the issue. At its 1994 winter meeting, the Western Governors established a Telemedicine Policy Review Group consisting of senior state officials, telemedicine experts, and other interested parties. A WGA *Telemedicine Action Report* was issued, which sets forth recommendations in the areas of infrastructure planning and development, telecommunications regulation, reimbursement for telemedicine services, licensure and credentialing of telemedicine providers, medical malpractice liability, and confidentiality of electronic patient records.

Medical Education

Many of the early and ongoing telemedicine projects have been based at or operated in conjunction with academic health centers (AHCs). A partial listing includes programs at AHCs and/or medical schools in Arizona, Arkansas, Colorado, Georgia, Indiana, Kansas, Louisiana, Michigan, Nebraska, New Hampshire, New York, Oregon, Pennsylvania, South Carolina, Virginia, and Washington. Today AHCs face uncertainties in a changing healthcare marketplace. As government continues to ratchet down payments under Medicare and Medicaid (including set asides for graduate medical education and residency slots), AHCs may find it difficult to sustain their current level of activities.

Telecommunications infrastructure is expensive, and much of the work to date has depended upon federal funds. On the one hand, AHCs may respond to budgetary constraints by cutting back on or eliminating the telemedicine development work they have begun. On the other hand, they may continue their telemedicine investment as a means of obtaining referrals from outlying areas and expanding their market share. The decision hinges upon their willingness to shift their technology focus from pure or basic scientific or health services research to the economics of the situation they face. The decision will determine whether telecommunications will serve their mission.

Human Factors

A major issue facing telemedicine is the degree to which healthcare providers and administrators adopt new telecommunications technologies into their everyday practice. As federal program staff know, many of the grant projects they administer are significantly underutilized relative to the technology investment made.

Human factors are critical to the successful incorporation of telecommunications in health care. Development efforts have addressed standards, policies, and other issues confronting the field, yet few have looked at the subtle questions of leadership, organization, individual and institutional comfort level, training, and related matters. One such study was conducted by CPSC last January for the Federal Office of Rural Health Policy (ORHP), is available in draft form as The Human Dimension of Telemedicine: Reports on the ORHP Workshop on Barriers to Practitioner Acceptance.

A prime concern for healthcare organizations in both the public and private sectors, clinical acceptance is currently being addressed by the Institute of Medicine in its evaluation of telemedicine. Such acceptance will more likely occur with:

- User-friendly technology design and implementation;
- Better needs assessments;
- Identification of lead change agents in institutions;
- Societal readiness;
- Better teacher-training creating a more technology literate professional staff;
- Better understanding of overall changes taking place in the healthcare system and in the individual delivery environment.

Some who have closely studied the situation feel that human factors reengineering is in order.

Community Hospitals

Some few studies have attempted to quantify the extent of market penetration of telemedicine in US community hospitals. The Oregon-based Telemedicine Information Exchange and the Center for Public Service Communications maintain anecdotally acquired information in listings of telemedicine projects, but little formal effort has been made by the federal government or private industry to gauge the dissemination of various telemedicine technologies and their utilization. Using a survey he conducted of 50 telemedicine programs in North America, Dr. Ace Allen (1995) reported a total of 2110 interactive patient consults via videoconferencing with physicians in 1994, roughly twice as many as in 1993 according to his estimates. In rank order, consultations were most common for psychiatry, cardiology, dermatology, medical specialties, pediatrics, and surgical specialties, especially orthopedics. While instructive as a broad measure, these numbers are most likely substantially underestimated.

Conclusion to Section on National Trends

Healthcare reforms within the public and private sector have resulted in concerted efforts to increase access to quality and result in more cost-effective service. New information technologies, including telemedicine, are viewed as tools that can contribute to the successful implementation of HMOs, strategic alliances, and new joint-venture arrangements. They are also viewed as mechanisms capable of supporting increased patient participation and shared decision making.

At the same time, there are many barriers, real and perceived, to the ubiquitous, interoperable, and scaleable implementation of telemedicine throughout the United States. Among these, physician clinicians in the private sector have identified the inability to get direct insurance reimbursement for other than face-to-face consultations, and potential liability problems with regard to interstate practice of medicine. In contrast, VHA staff are salaried and licensed throughout the VHA medical system.

Telemedicine in the VHA

System Characteristics

The VHA has pioneered innovative uses of communications technologies in health service delivery, with a wide range of technologies and applications spread across the United States. Certain characteristics qualify the VHA to take a leadership role in telemedicine. These are, in brief, the following:

- Largest laboratory with 173 hospitals and over 400 out patient centers;
- Open system allowing applications without cross-state practitioner licensure barriers;
- Successful implementation of "model" telemedicine initiatives designed to resolve cost, quality, and access issues, for example, teleradiology, cardiac pacemaker surveillance, etc.;

- Managed-care-like setting;
- Primary-care emphasis;
- Patient involvement;
- Network approach that seeks to maximize potential by collaboration with a variety of local health institutions, including academic health centers and Department of Defense medical installations.

Telemedicine Defined

In recognition of the increasing role telecommunications and information technologies will play in providing VHA services, the Associate Chief Medical Director for Operations, with the support of the Under Secretary for Health, appointed the Task Force on Telemedicine in March 1995. The Task Force adopted its definition of telemedicine:

Telemedicine involves the use of telecommunications technology as a medium for the provision of medical services to sites that are at a distance from the provider. The concept encompasses everything from the use of standard telephone service through high-speed, wide bandwidth transmission of digitized signals in conjunction with computers, fiber optics, satellites, and other sophisticated peripheral equipment and software.

This definition is sufficiently broad to encompass the breadth of programs currently being utilized within VHA, including cardiac pacemaker monitoring and evaluation, nuclear medicine interpretation, teleradiology, patient data exchange, nurse teletriage systems, patient dial-up access to scheduling and pharmacy information, and continuing medical education, to name only a few.

VHA as Telemedicine Lab

Today, the VHA is the nation's largest operational telemedicine laboratory. As in other areas of health services management and clinical practice, the VHA has much to offer with regard to telemedicine in a managed-care-like environment.

If a collaboration between and among the new regional-based Veterans Integrated Service Networks (VISNs) is encouraged, as expected, VAMC telemedicine programs could serve as laboratories for telemedicine within and outside the VHA, in managed-care and other settings. Further, and with fewer barriers than in the private sector, the VHA could establish model evaluation programs and share their results with other healthcare organizations. Veterans Health Administration telemedicine could also be of help as efforts continue to define and refine the vision for a national information infrastructure. The combination of the scope and diversity of VHA telemedicine efforts and the competency and commitment of VHA staff should result in technology, software, and applications that are valuable and instructive to national telecommunications infrastructure development efforts.

Indeed, VHA could easily create from among its individual sites the largest truly national telemedicine network demonstration. Such a demonstration of a large multipoint system would help to address barriers facing the field and to identify opportunities for VHA and private sector telemedicine.

Historically, the VHA has encouraged the development of innovative applications of information technologies at the local, or VAMC level. With the transition to the VISN structure, it is expected that each VISN and each VISN director will use a uniform set of standardized metrics to ensure measurable progress toward agreed upon goals. These metrics will be used across the agency as the basis for program evaluation, resource allocation, and management decision making.

The VISN structure, therefore, represents a challenge for VAMC telemedicine programs as they seek to interface closely with emerging statewide networks, state offices of rural health, and other state and local organizational entities focused on telemedicine and related efforts. To the extent possible, VAMCs should be aware of (and be creative in developing electronic and programmatic linkages with) the non-VHA telemedicine efforts that are rapidly developing within many of the 50 states.

To support telemedicine, the VHA's Chief Information Officer (CIO) should encourage information sharing and invite Information Resource Management (IRM) directors and lead clinicians to discuss institutional approaches to telemedicine and exchange experiences. Venues might include VISN-wide telemedicine meetings, realtime and virtual, using networking tools.

Leadership Areas

As described above, over the course of recent months, the VHA has been transitioning from a facility-based system to a new system of VISNs. Applications of telecommunications and information technologies currently pervaded the VHA system as a whole and are healthy and growing within individual medical centers. Increasingly, as the VISNs gain vitality, decisions on the types of applications and the degree to which telemedicine will be "practiced," and by whom, will be significantly influenced by the VISN directors.

Based on CPSC's tour of VAMC telemedicine programs, the following are a few recommendations made in the report.

Interagency Projects

Veterans Health Administration should continue to play a leadership role among federal agencies involved in intra- or extramurally funded telemedicine. Leveraging of federal resources between, for example, the Department of Defense and Veteran's Administration facilities and programs in a given region would make great sense. As an example of this, the VHA is working collaboratively with the Department of Defense to establish joint videoconference networking for the Tripler Army Hospital's Akamai project in the Pacific.

While some technology transfer and sharing has occurred (especially in the area of digital imaging), additional cooperation and joint programming might be encouraged at several levels.

IRM Telemedicine Residency

Many VHA centers are affiliated with academic health centers and connected, in varying ways, by telecommunications links. In keeping with the principle of autonomy that characterizes decisionmaking at VAMCs, the relationship varied depending on requirements (and needs) of either the VHA or academic health center, and was dependent upon the budgetary or extrabudgetary funding available. In most, if not all, of the programs observed, one of the significant mutually beneficial elements of the relationship was clinical residencies. Medical students get good clinical experience by staffing VHA hospitals, while the VHA gets the advantage of well-educated medical students.

It might provide useful for VHA to use the clinical residency model to telemedicine and to develop an internship program linking the management information systems offices of the VHAs with the computer sciences programs at the academic health centers for a formal technical program in telemedicine and health informatics. In such cases, the VHA would give clinical guidance and program structure in exchange for high-quality engineering support. This may be particularly timely because as the many organizations and agencies—especially, for example, the Department of Defense—have been in the process of down- (or "right-") sizing, computer sciences students who once saw these groups as possible employers are looking for other opportunities.

Challenge of Leveraging Statewide Networks

Recent trends in the development of telemedicine indicate a move from federally sponsored, grant-supported programs to operational systems funded by the states themselves (or to private sector-funded programs on a smaller scale). However, the political and service-area boundaries of VHA medical centers within VISNs do not necessarily correspond to the geographic boundaries of states. The case of West Virginia is illustrative, with VHA medical centers located in three different VISNs, each then associating with other VAMCs outside the state.

Evaluating New Technologies

The VHA could take a leadership role in evaluating new technologies, modalities, and applications for the remote delivery of health services.

Personal digital assistants, new wireless technologies, multimedia applications over the Internet, desktop videoconferencing, and others offer great potential to the field. Working collaboratively with private industry, VHA could help to bring new products and services into the practice of clinical care for the benefit of veterans and others.

Conclusion

The VHA has played and will continue to play a key role in telemedicine innovation. A nationwide multi-facility system, the VHA is uniquely qualified to serve as model for other sectors, governmental and private, as they strive to respond to trends toward managed care. The newly created VISNs can continue the evolution started in the VHA and provide leadership in key areas.

Reference

Allen A. 1995. Home health care via telemedicine. *Telemedicine Today, 3(2)*, 26.

29
Telemedicine at the Veterans Health Administration

ROGER H. SHANNON AND DANIEL L. MALONEY

Introduction

Telemedicine in the Veterans Health Administration (VHA) is a broad concept with extensive possibilities. Telemedicine is neither an isolated program, nor a list of projects that will ameliorate this condition or that. In concept and technology, telemedicine actually promises a new time-independent geography in which to operate a large, integrated healthcare organization. It is a fundamental development that must be wisely woven into the fabric of the entire enterprise.

This chapter focuses on telemedicine at the enterprise level and examines the interaction of the enterprise with activities elsewhere in the national and international environment. The chapter thus looks inward and outward. Healthcare nationwide is changing dramatically, and the VHA itself is transforming its organization and methods of providing health services to its clients. In this changing context, telemedicine technology is a focal point of interest because it offers capabilities to:

- Control costs while improving access, and
- Sustain quality improvement.

Telemedicine also makes it possible for the veteran to experience satisfaction with the care provided and to take pride in the organization supplying it.

Background

Telemedicine is not new; it enjoyed a period of popularity in the 1970s. Grants were awarded for numerous projects, and demonstration projects were often fairly successful. However, the technology was expensive, and the United States was entering a period of public reaction to the rising cost of medicine. The Great Society was launched and healthcare corporations were undergoing substantial growth. With the passage of PL93-641 in the mid-1970s, the Health Care Financing Administration (HCFA) was created and quickly brought to bear a powerful curb on cost. Telemedicine did not have benefits that were sufficiently proven to justify the cost 25 years ago, and so it lost popularity. Only recently has widespread interest returned, fueled by the dramatic explosion of the Internet with its World Wide Web (WWW) and a new belief that telemedicine can cost-effectively use new communications technology and new business concepts to transform the way health care is delivered.

Dynamic Setting

Three universal societal trends are radically reshaping all health delivery systems today. First, the explosive advance of communications technology has democratized and enhanced interpersonal exchange. Second, healthcare delivery is dramatically shifting from small to large business. Finally, large business, including medical enterprises, is being reengineered to flatten hierarchical organization, increase efficiency, and restructure production teams to meet client-driven objectives.

In 1994, VHA commenced reengineering to decentralize authority and decision making in a manner that will cultivate an increasingly virtual organization. At the time of this writing, a major delegation of authority from the Central Office to the newly installed directors of 22 geographically defined Veterans Integrated Service Networks (VISNs) is being implemented. Headquarters has been downsized and flattened. Its attention has focused on issues of significance to the enterprise as a whole, policy and standards, and accountability. Network directors now are under performance contracts and responsible for accomplishing their missions within guidelines established to assure coherence of the VHA as an integrated enterprise.

As might be expected in a time of new local power and flexibility, local initiatives are flourishing. Telemedicine plans are prominent among these. They are to be encouraged; but, at the same time, it is essential to determine what must be coordinated or standardized to assure that the enterprise is served by a coherent, flexible network of communications and critical common data. There must be an enterprise concept of telemedicine with which the local projects are compatible.

Evolution of Telemedicine in the VHA

Telemedicine does not mean the same thing to all people. If viewed narrowly it equates to videoconferencing, often with the thought that it will be used to reach relatively isolated providers and patients. That, indeed, is one important function of telemedicine, but the evolving concept embraces all technologies and processes that diminish the obstacles of distance and time.

The heart of the telemedicine model is the consulting room, which varies little from project to project, or from federal to private sector. It is a typical examining room with a place for care provider and patient. The usual tools are there—stethoscope, otoscope, ophthalmoscope. A close-up and wide-angle video and a document scanner are normal. An electrocardiogram (ECG) transmitter is common, and there are often a fax machine, a personal computer and a printer, all online. And, of course, there are appropriate video monitors. Applications, potential and actual, extend beyond this nucleus.

The basic tools and techniques of remote care and consultation, as generally recognized, in and outside the VHA, have been well reviewed by Perednia and Allen (1995).

Service programs reported as being used in VHA include the following:

- Cardiac pacemaker monitoring and evaluation;
- Electrocardiogram interpretation;
- Nuclear medicine interpretation;
- Teleradiology;
- Patient data exchange;
- Networked e-mail;
- Video teleconferencing;
- Missing patient registry;
- Continuing medical education;
- Telephone care by healthcare providers;
- Nurse teletriage systems;
- Multimedia patient data;
- Patient dial-up access to scheduling and pharmacy information;
- Remote system access;
- Hospital inquiry to the national Veterans Benefits Administration eligibility database;
- Online inquiry to identify locations providing prior VHA treatment;
- National data reports.

Imaging systems have a particularly interesting relationship to telemedicine. In some cases, teleradiology ventures have expanded to become general telemedicine systems. In others distant learning or clinical systems have been upgraded to include diagnostic imaging components. Development of imaging systems in the VHA has followed several pathways to reach the present sophisticated, state of multimedia computing. These include Picture Archiving and Communication Systems (PACS), teleradiology products, and an imaging component of DHCP, which interfaces with the medical record. PACS visuals make a significant contribution to the power of telemedicine technology. They encompass radiology images, photographs, and text documents, as well as other forms of images and graphics. Digital images of endoscopy, pathology, dermatology, and other specialties are also part of electronically constructed working environments.

At its heart, telemedicine is a communications concept. Communications infrastructure is discussed in detail elsewhere in this volume, but a few concepts of importance should be emphasized in this overview of telemedicine in the VHA. No artificial restriction should be placed on what technology can be included in support of telemedicine. Discussions of virtual organizations and telemedicine tend to emphasize the use of high technology. Still, phones, fax, and video are bedrock technology. Suites of high technology often include lowly components, and such is the case for telemedicine.

The VHA infrastructure is evolving to include visual presentations and images. Developmental work focuses on technical and content integration. The latter has the power to enhance the view of the patient as well as to facilitate the work of the care provider and manager. Clearly, the VHA system must be broader than that traditionally included as part of the internally developed Decentralized Hospital Computer Program (DHCP). It must and already does include modules of commercial offering. A notable example is the development at the Baltimore VAMC of a gateway between a commercial PACS and DHCP's integrated imaging and record system. The Clinical Information Resource Network (CIRN) is also noteworthy for integration and support of the VISN approach.

Local-area networks (LANs), wide-area networks (WANs), ground lines, microwave, and satellites are all options. Service should come to the provider's place of need, by direct attachment or wireless transmission. A special infrastructure in the form of the Internet and WWW has begun to assume monumental importance throughout the medical world including the VHA. The term "intranet" is being used to apply Internet conventions and technology to internal networks.

The possible use of the WWW as a medium for an electronic medical record draws upon attributes of the WWW that are particularly suited to objectives of telemedicine. Hence, telemedicine projects make excellent testbeds for WWW computer-based patient record development. Useful attributes of this technology include the following: platform independence, natural collaboration, security, resource availability, and self-organization. Platform independence allows flexibility and ease of network communication that facilitates the virtual organization. Collaboration with others, in and outside the VHA, in developing and maintaining databases and decision support tools is simplified by the WWW environment. Consequently, many more easily accessible resources are available to any WWW-based system. Organization of complex information environments becomes a natural function of system maintenance through the proper placement and removal of links to the changing areas of interest among the medical sites on the web. However, there are a number of projects using more traditional technological approaches that point to the need for the enterprise regulation of local projects regardless of their size.

What's Been Going On—Scenarios in the Making

Hybrid Open Systems Technology Telemedicine Projects

The Hybrid Open Systems Technology (HOST) program has 14 projects, nearly half of the total number, which contain an element of telemedicine.

- Baltimore, MD, is evaluating a gateway designed to link a commercial PACS with DHCP, including its imaging system. This has major significance in VISN 5's initiative to consolidate facilities and provide remote services.
- Livermore, CA, accesses programs and materials in Palo Alto for patient education regarding eye problems.
- Philadelphia's extensive integration project includes teleradiology.
- Fayetteville, AR, is developing telepathology.
- Iron Mountain, MI, is doing the same with robotic control of the remote microscope and the capability of teleconferencing.
- Muskogee, OK, is working on computerized telephone triage.
- Northampton, MA, will be installing teleradiology for rural support and will be compatible with a larger Boston center.
- Richmond, VA, is developing a four-hospital teleradiology consulting network.
- Sepulveda, CA, is evaluating a teleradiology link with West Los Angeles.
- Syracuse, NY, is linking their digital angiographic unit to DHCP with provision for transmission to remote referral sites for viewing.
- Washington, DC, is collaborating with the Department of Defense (DoD) to share remote digital and video dental images among four widely separated sites.
- Washington, DC, is also using the WWW to access Agency for Health Care Policy and Research (AHCPR) Clinical Practice Guidelines.
- Washington, DC, linked a central ECG server to DHCP for use by several facilities.
- West Haven, CT, is the focus of a regional healthcare alliance to network digital Echocardiography.

Local Initiatives

At present here are many local telemedicine initiatives in telemedicine within the VHA. Ambitious and wide ranging, they illustrate service features and technical attributes that are important to have consistent across the enterprise. Six scenarios, drawn from actual VHA projects, are briefly presented below to further illustrate issues that make it imperative to consider telemedicine at an enterprise level. A full list of projects can be accessed from the VA Homepage at *http://www.va.gov*. This document is updated periodically and provides a rolling appendix to this chapter.

Scenario 1

An urban VA medical center (VAMC) has been a center for excellence in the development of telemedicine. It has evolved a highly successful multimedia information system over the period of a decade. Picture Archiving and Communications Systems were carefully studied. Detailed specifications were developed and a commercial PACS was acquired. Concurrently, an imaging system was developed as part of DHCP. Long and difficult negotiations produced agreement upon and development of the gateway between the two systems. The agreement set the precedent of using the DICOM standard throughout the VHA. The hospital achieved filmless imaging with exchange between a dedicated PACS and the hospital information system. As part of the managerial consolidation of that medical center with two smaller centers during VHA reorganization, the PACS system was extended to afford remote coverage for the satellite operations. With establishment of the VISN governance, telemedicine capabilities were expanded and the VAMC became the center for integrated telemedicine development and testing.

Issues of special interest

With the DICOM gateway in place, this center is not only consolidating the management of several medical centers but is also using high technology to consolidate and optimize the resources of the VISN. It has evolved its infrastructure so that a front-end expenditure has been leveraged to extend the system at low marginal cost. Some interface challenges will be posed, but they will be based on international standards and exist within the VHA, thereby requiring only negotiation with the vendors in conjunction with larger purchases.

Scenario 2

Widely separated VAMCs in two adjacent VISNs have a rural clientele that comes from an area served by a private sector healthcare organization. The private organization is experienced in the placement of rural primary-care practitioners and the use of telemedicine techniques to support them. The state is supportive of collaborative solutions to health care.

Issues of special interest

The use of contract services to create a virtual VHA rural network of major significance is an appealing solution to rural outreach. The VHA can use outside services but maintain an integrated record and oversee patient care. Collaboration between two VAMCs and other agencies will be fostered.

Two VISNs will have to cooperate in following similar conventions. As opposed to scenario 1, the preexistence of the private sector network will necessitate negotiation of development and cost sharing of the technical interfaces involved. An enterprise policy supporting appropriate standards will simplify the process.

Scenario 3

A large and a small VAMC are linked by commercial telemedicine equipment to support remote patient interviewing and consulting. There are close collaborative ties with DoD, which has a major medical center in the area and a center that develops and tests advanced telemedicine technology. A large state medical school supports an extensive telemedicine service throughout the state and is heavily involved in telemedicine research and development. The three institutions collaborate on telemedicine and other medical ventures with strong encouragement by the state.

Issues of special interest

The VAMC provides remote support for a smaller facility in many specialized areas. Its consultation rooms were set up at the same time with a single integrator's equipment, which is essentially a turnkey solution. Collaboration with the other agencies in the future will require somewhat more sophisticated planning, but the environment is rich in expertise. A standards-based approach on the part of all players will be important to local success and as an example to others for whom they will serve as a model. Research possibilities are clearly nurtured in this type of setting.

Scenario 4

A single VISN involving three states has several major VA medical centers and a number of outpatient facilities in a mixed rural-urban setting. A heavy concentration of military personnel with advanced military healthcare facilities is in the area. In addition, distances between these entities vary greatly because some sites are remote. Two major medical school affiliates are involved. At least two of the VAMCs have traditionally operated independently of one another but will now fall under the same territorial governance.

Issues of special interest

This VISN faces complex organizational and collaborative issues as it reengineers its operations. Some facilities are being consolidated; some are being linked. There is an opportunity to optimize resources by use of the technology and development of some outsourcing to form a virtual organization. Sharing of infrastructure will be a prominent feature of the collaboration offering an economy of scale and the possibility of cost-effective satellite transmission because of the geographic spread. With several major medical schools and medical centers, this is fertile ground for peer-to-peer specialty consultation and collaborative research studies.

Scenario 5

A consortium that includes one VAMC in a VISN, a university, a not-for-profit organization, DoD, a single state, and private industry wishes to build a network of health-related services that will be accessible to any client, regardless of the client's degree of isolation. The consortium already has access to several developed projects that are candidates for component applications. It wishes to expand later and to develop a model that can be replicated.

Issues of special interest

A driving force in this project is strong interest in setting up a mechanism to achieve and measure enhancement of outcomes in terms of cost, quality, access, and satisfaction for clients and for providers. Consequently, the prospects for productive health services research are excellent. Collaboration is so far exemplary. A strategy to develop an infrastructure that can be leveraged for the more tenuous aspects of the projects is in place and is being structured to optimize area resources. The project is ambitious enough to realize many economies of scale. There is a major outreach to rural areas.

Scenario 6

There are a number of nested projects in this VISN that include telemedicine development by the VHA, states, universities, private and public sector hospitals, medical organizations, telephone companies, and other federal agencies. There is a strong military presence. Veterans Administration medical centers in two states that previously operated independently and launched separate telemedicine initiatives now wish to consolidate to unify the VISN. Had there been a national policy at the time the projects were conceived, the integration might have been greatly simplified.

Issues of special interest

Of interest here (in addition to previous comments on similar situations) is a marked difference between the sophistication of infrastructure in the states involved. The collaboration of the telephone companies and the VAMCs appears to be well accepted by all parties and will constitute an important interface example for the enterprise as it develops its position on telemedicine. Integration efforts will identify important areas to be served by standards. This complex also has potential for a high profile in the fourth mission of the VHA, which is to back up the military in time of crisis.

The questions raised by the scenarios emphasize the need for VHA-wide coherence of telemedicine management because they cannot be wholly dealt with within one VISN or be settled differently from one VISN to another. Only if there are enterprise-wide information and overarching management will it be possible to know outcomes of telemedicine techniques.

Challenges

Since the healthcare legislation of the 1970s, policy has targeted cost, quality and access, roughly in that order. Depending on the situation, any one of the three cornerstones of policy may be the primary concern. Certainly, if one falls below some threshold of acceptability, regaining balance becomes an urgent priority. A fourth cornerstone is client satisfaction. Today, business reengineering principles—particularly the notion that business is client driven—is being applied to health care. At last, patient–client satisfaction is beginning to receive the attention it deserves.

Over the past three decades, as medicine has become big business, business practices have emerged that often require difficult value judgments at the point where business, morals, social welfare, and individual interests intersect. It is not simple to balance business imperatives with traditional medical ethics, but it can be done. Cost must be minimized; quality must be maximized; access must be maximized; patient satisfaction must be maximized. Success will be measured by the ability of the enterprise to balance these four imperatives. Optimization, therefore, hinges upon complex value judgments—judgments that will be only as valid as the data on which they are based.

Evaluation and Accountability

Accountability is based on valid evaluation of methods, tools, performance, and outcomes. These must be criteria driven. Assessment requires measurable indicators of how closely reality matches the target model. The criteria and indicators must be known by each level that is to be evaluated, and solid data must be collected to be fed back to complete the accountability loop.

A good framework for evaluation can be used not only to evaluate performance but also to discern new methods and applications. For example, the system presented by Fryback and Thornbury (1991) raises the questions that must be asked when a large enterprise, like the VHA, bears significant responsibility for the delivery of health services to a population while delegating major authority for action to its field employees.

Thornbury and Fryback's efficacy framework is hierarchical; each criterion is subsumed by the next. Its six basic levels are as follow:

- Does the object or method of interest deliver the intended technical outcome?
- When used, does the technology enable accurate results?
- Does it affect provider thinking?
- Does it affect provider action or therapeutics?
- Does it affect patient outcome?
- Does it meet the requirements of the enterprise/society?

The first five of the six criteria are of obvious importance to the providers in the field. The sixth is the focus of the enterprise. Most importantly, all the rest are subsumed under the sixth criterion, and the enterprise—as the accountable entity—must be responsible for coherence and integrity of all of the criteria throughout the enterprise as a whole.

Problems in Evaluating Systems: The Issue of Context

Within this evaluative framework, some questions can be difficult to answer, particularly the two concerning patient outcome and societal significance. Quality, cost-effectiveness, and patient satisfaction, three of the four cornerstones of health care, are buried in the two most difficult areas to measure. All criteria involve value judgments that vary with society's mood. Evaluations can be especially problematic when undertaken before a concept, its technology, and its methods are fully diffused, that is, before the context (environment) is mature. Novel systems are particularly difficult to assess quantitatively and accurately. Expert judgment and simulation can be helpful, but evidence is soft when major investment and adversarial opinions must be taken into account. This is the challenging environment in which telemedicine will evolve.

Competing Themes

In the VHA, several dominant organizing themes must be consolidated in the process of balancing values while integrating and optimizing the healthcare delivery system.

Strategic Health Groups

In a radical departure from traditional practice patterns, the VHA is replacing the classic stovepipe structure of healthcare delivery based on medical specialties with healthcare delivery groups based on services and personnel that often combine in everyday healthcare delivery. These new sections, termed strategic health groups (SHGs), emphasize the integration of healthcare delivery for a given problem, rather than the commonality of practitioner' learning and skills. Expected SHGs within the Office of Patient Care Services are primary care, acute care, geriatric/extended care, rehabilitation, mental health, nursing, diagnostic services, and support services. Other groups will undoubtedly be added.

Strategic health groups are divided into product lines. Each product line consists of a related group of programs or services that meet the needs of a specific patient population. Product lines include all of the VHA Special Programs that address needs that are of especial importance to veterans. The special programs are blind rehabilitation, geriatrics and long-term care, homelessness, Persian Gulf veterans programs, post-traumatic stress disorder, preservation amputation care and treatment (PACT), prosthetics and orthotics, readjustment counseling, seriously mentally ill, spinal cord dysfunction, substance abuse, and women veterans programs. However, other product lines will also be developed. Such lines might be Organ Transplant, Cardiothoracic Surgery, etc. Integral to this reorganization is a deliberate shift to primary, noninstitutional care and preventive medicine. Details of this reorganization will have evolved by press time, but its basic objectives will endure. So too, will its implications for the patient record. As can be seen from the summary of product lines, SHGs will tend to weaken the integrity of the patient record unless care is taken to integrate a patient's data at the outset.

Patient Record Integrity

Key to the validity of an electronic medical record, is the requirement that data concerning any given patient/client be integrated with respect to different aspects of the patient and be carried over the patient's lifetime. Under this holistic concept, the individual becomes the organizing principle of the integrated patient record. This is in contrast to single provider-driven systems lacking linkages to other modules or other providers, which have historically neglected to maintain the integrity of the patient's record.

Individuality Within the Enterprise

A system as large as the VHA can bring creative diversity to bear on the problems at hand, as development of the DHCP exemplifies. Nevertheless, the VHA cannot afford to be a collection of fiefdoms. It must optimize its health delivery system through integration while maximizing individual authority at the points of service to clients.

Coherence of the Enterprise

As the VHA is reengineered to reap the benefits of decentralized authority in rendering health care, it becomes imperative that the framework in which the caregivers operate be coherent throughout the enterprise. Interoperable data and communication systems are requisite if the system is to operate across VISN lines or if VISNs are to be amenable to redistricting. It is just as imperative to have a seamless system of accountability and to have data-supported evidence of uniform quality of service and management throughout the enterprise.

Implementing Systems Management

We in VHA are one system that we have historically reduced to parts and treated as disjunct issues. We, like many organizations, have solved problems with resources rather than rigorous analysis and planning. Now, our resources are becoming finite. Our system is closing—losing its slack.

The need for convergence of operations at the enterprise level has been emphasized throughout this chapter. In concrete terms, a few obvious reasons to insist on headquarters mastery of the organization are the need for:

- Accountability to the people through Congress;
- Consistent efficacy;
- Virtual organization;
- Cross-VISN requirements;
- Economies of scale;
- Interagency collaboration;
- Fourth VHA mission to back up the DoD;
- Redistricting as patterns change.

The VHA can meet these and like needs only by understanding and successfully applying system concepts.

"System" Concepts

Several systems management principles are presented here for those who wish to review them: A system is a set of elements and relationships that are fully related. This is easy to say but not always easy to appreciate. Weinberg (1975, p. 162) offers the definition as, "a collection of parts, no one of which can be changed." A little more abstract but instructive example is the set of equations (von Bertalanffy, 1968, p. 56) shown in Figure 29.1.

The notion of open and closed systems is also helpful. The former do not exchange across their boundaries; the latter take from and give to their

$$\frac{\delta Q_1}{\delta t} = f_1(Q_1, Q_2, \ldots Q_n)$$

$$\frac{\delta Q_2}{\delta t} = f_2(Q_1, Q_2, \ldots Q_n)$$

$$\frac{\delta Q_n}{\delta t} = f_n(Q_1, Q_2, \ldots Q_n)$$

FIGURE 29.1. Loosely, Qs here represent the state of system elements; δs represent relations. Every element is dependent on functions of all the others and itself. If any change occurs, there is a response that ripples throughout the system. This is the essence of how systems work. (No one thing can be changed.)

environments freely. In the case of a closed system, no information—nothing—crosses the external boundaries. This is like a chemical reaction in a sealed vessel. An open system is like an animal, taking in food and discarding waste as needed. The environment is an unlimited resource and depository.

As resources become limited, an open system begins to close. Its freedom diminishes. It must plan its boundary exchanges. It will necessarily have to pull itself together and optimize its functions to survive.

Management Considerations

If we choose to use technology to optimize our system, we can no longer afford the luxury of reductionism in our management. At the least, our reference architecture—the model we use to understand our structure and function—must clearly express its unity.

Decentralization and local empowerment foster disassociation of a unified organization unless a conscious first priority effort is made to design and implement a sparse but robust systematic weld between the enterprise headquarters and the field. What are some of the necessary elements of this system scaffold?

First, there must be a clear picture of what the enterprise is about, what must be its outcome, what will be the attributes held in common by its operational replications (VISNs) in the field. How will compliance with systemic enterprise needs be assured?

Discussions regarding the organization of central services suggest that there will be a number of enterprisewide information utilities, a set of policies and standards and support services including start-up assistance.

Modeling and Simulation are yet to be ignited. Conceptual modeling has begun, however; and newly appointed groups of system architects and technical integrators promise to be excellent resources as the enterprise models its telemedicine initiative.

Centrally relevant considerations for modeling and simulation include the ability to study function and cost and to quantitate the impact on every cost center. Synthesizing experience from different projects can give a baseline for evaluating telemedicine within predicted rather than existing environments and for evaluating alternative approaches before implementation. The function can help develop change and deployment strategies, plan for maintenance and upgrades, and design and test aspects of interoperability. Finally, modeling and simulation can facilitate selection of new methods and technology as well as associated buy–build decisions.

Strategic Systems Approach

In applying systems management, it is a first principle that an enterprise is always a single system, even though an optimal enterprise system—one with all the outcomes balanced—probably does not exist. Outcomes vary, quality varies, cost and client satisfaction vary.

The chief executive officer (CEO) is the person accountable for the performance of the enterprise, and for the successful optimization of the organization as it achieves the targeted outcomes. The CEO can delegate authority to those on the action line, as is the current practice in organizational reengineering, but responsibility for the organization as a whole remains at the top. Decentralization has great potential and great risks. Performance and outcome must be held to the enterprise framework. Decentralization of authority cannot be a decentralization of power. Power to dispatch his responsibility must remain with the CEO.

To retain that power, the CEO will need good information about critical factors of the system and expect those responsible for aspects of the enterprise outcome to be accountable in turn. If outcomes are not those that have been defined and expected, the CEO will adjust the enterprise system so that the outcomes are corrected. Information and accountability that reflect a coherent understanding of the enterprise system are a management essential. The other side of the coin of roll-up data is the need for operatives in the field to know exactly what is necessary to support the enterprise's system requirements.

Virtual Organization

Assuming all the foregoing are in order, then the CEO can avail himself of the benefits of virtual organization, particularly with the help of technology

A virtual organization is a community of employees, specialists, contractors, suppliers, and others who are associated by agreements in a cohesive team. Team members are committed to achieving common objectives even though they may work for a variety of legally separate organizations or for the same organization at sites that normally preclude collaboration. Teams are the model organizational form; they accomplish their purposes through rich communication and information access with the extensive help of high technology. This cross organizational team structure flattens the management hierarchy, empowers local personnel, and puts decision points close to the action. Flexibility is enhanced, and virtual teaming dramatically increases interaction between VHA and the world outside. The central requirement of the virtual organization is informed use of high technology for communication of sufficient scope and quality to eliminate the constraints of time and space. The kinds of technology and techniques that enable the collaboration characteristic of virtual business in its most advanced form are basically the same for telemedicine. The precepts of the virtual organization and the qualities and goals of telemedicine are remarkably consistent, if not identical.

Reorienting Health Service in Time and Space

The VHA-wide redirection of primary responsibility for operations from medical centers to regional network directors opens a major opportunity to reallocate resources, both physical and human. Rather than a way to reach a remote and presumably underserved population, telemedicine becomes a means of reevaluating the way an entity relates to the population it serves. Telemedicine cannot only repattern the way clients relate to the establishment, but also repattern provider relationships and equipment distribution. Seven instances of medical center consolidation have been undertaken in the first year of reorganization. Several of these centers have used various forms of teleradiology (one of the facets of telemedicine) to enhance services. One pattern of revision concentrates all of the radiologists in the largest center while another leaves a radiologist in one of the smaller institutions. In the former case, telemedicine extends the services of a traditionally structured professional group. In the latter case, technology consolidates geographically separated professionals into a virtual group.

There are many possibilities for developing virtual professional groups with technology currently available. Though we know little about the social dynamics of such reorganizations, we are beginning to discover positive social aspects of LANs, telecommuting, the Internet and WWW, to say nothing of telephones and fax machines. The endless variations in patterns, degrees of separation, and levels of autonomy await discovery.

The point is, that although we have extended our range of interaction with the communications technology of the past half century, we are only now on the verge of freedom from the temporal and physical barriers to social intercourse. This freedom will empower us to take dramatic steps in redefining the medical workplace.

Policy, Standards, Feedback

Policies, standards, and operational guidelines form the common language. Metrics of performance and outcomes must be developed and appropriate data collection mechanisms put in place. The data must be accessible centrally and applied against the criteria of accountability. This is the substance of the accountability loop previously described. Adjustment of operations using the knowledge from the feedback of valid data reflecting a well-designed system is the essence of success in decentralizing and organization and exploiting the potential of virtual operations.

Infrastructure

To develop the ubiquitous infrastructure for a liberated healthcare system with telemedicine components, funds will have to be creatively reprogrammed. Return on the investment will have to come from full utilization of the technology. However, optimizing resource distribution in operations already being performed provides a way to leverage funds now being spent on inefficiencies of travel and communication.

Consider travel, for example. Planning within the VHA is able to incorporate savings that the private sector tends to discount; specifically, the VHA often covers transportation costs of patients. Both groups cover transportation costs of healthcare providers yet neither group currently provides optimal contact between provider and client. However, the solution is more complicated than merely extending communications.

Provision or Direction?

In empowering field organizations to act in a decentralized organization, the choice of which enterprise requirements will be met by the field under direction and which will be supplied by central purchases and services is immensely challenging. Reliable uniformity, available expertise, and economies of scale must be considered when the options are weighed.

Current Status

Internal Status

Top management in the VHA is committed to seeing that telemedicine is coordinated so that there is coherence throughout the enterprise. A facilitator has been involved in projects for a year at the time of this writing with the expectation that a high-level telemedicine official will be appointed soon. Several other clinicians and technical professionals have monitored significant aspects of developments. A task force was formed last year to study telemedicine. It reviewed the literature and inventoried projects planned and in progress. Its recommendations included the need for evaluation (Grigsby, Kaehny, Sandberg, Schlenker, & Shaughnessy, 1995) and for centralized guidance for many aspects of telemedicine. Information resources now under priority development will be pillars upon which the broad concepts of telemedicine will depend. Among these are the Clinical Patient Record System (CPRS), Work Station Pilot with graphical user interface (GUI) interface, the CIRN, the Ambulatory Care Information System (ACIS), Decision Support System (DSS), and Primary Care Management Module (PCMM). Clearly, telemedicine is an example of technology transforming the conditions under which health care is provided.

Presently the Office of Information Resources Management takes an inventory of projects every 6 months and updates the project database, but no set of metrics has been developed to provide a means to acquire roll-up data and meaningful monitoring. With the recent arrival of the first VHA Chief Information Officer (CIO), study, formulation, and implementation of these and additional concepts are feasible.

Interagency Collaboration

A natural exercise to increase knowledge of VHA developers, encourage de facto standardization by consensus, and learn about the problems and possibilities implicit in virtual operations is to collaborate with other federal agencies, professional organizations, and VHA academic affiliates. The possibility of extending some of these concepts to the private sector is also being explored.

Some of the areas where the VHA, either formally or informally, maintains contact are listed briefly below.

White House Health Care Reform

The VHA participated in the White House process to draft healthcare reform legislation early in its administration. Short-term information resource focus was on administrative systems and savings. Long-term focus was on big payoff from clinical information systems, decision support, and improved quality of care. Telemedicine was a specific part of these plans.

Health Subgroup of National Information Infrastructure (NII)

This subgroup, known as the Health Information Application Work Group (HIAWG), is a federal government interagency group that evolved through the merger of several independent initiatives. It addresses problems of coordinating initiatives across agencies and fosters interagency communication about information resource issues. The group authored a white paper for the Vice President on NII interests in standards, telemedicine, concurrent education.

Telemedicine Subgroup of HIAWG

This is a further evolutionary step for the previous committee by which it was reconstituted under the Department of Health and Human Services (HHS). The committee developed and published a strategic road map for telemedicine. It is developing a WWW tool to collect and present federally funded telemedicine projects. It is collecting telemedicine evaluation criteria.

High Performance Computing and Communications

This is an interagency program to coordinate advanced High Performance Computing and Communications (HPCC) initiatives to fund research and purchases.

Healthcare Data Card

The G-7 is an international group of the seven largest economic powers, who have agreed to cooperate in developing global projects. It was launched in February 1995 with health care as the eighth of eleven themes. A VHA representative serves on the subgroup dealing with healthcare data cards that are expected to be important telemedicine devices.

Management Advisory Board on the Health Passport Project for the Western Governors Association

The VHA has served in an advisory capacity to the Western Governors Association (WGA) on the feasibility of using card technology to improve delivery of state and federal supported services to citizens. Telemedicine is one objective.

Other Liaisons

A number of other liaison activities occur through personal activities of VHA personnel who are active in one or another of the VHA telemedicine programs. These involve committee participation, grant reviews, and consultative exchanges. Examples of organizations touched in this manner include: National Science Foundation (NSF), National Library of Medicine (NLM), Institute of Medicine (IOM), General Services Administration (GSA), (DoD), Center for Devices and Radiological Health (CDRH), National Telecommunications and Information Administration (NTIA), and North Carolina Healthcare Information and Communications Alliance (NCHICA).

Conclusion

The VHA is accountable as an enterprise to the people through Congress and must be coherent, consistent, and flexible throughout. To be successful as a flat organization with extensive decentralized authority, it must have a system of policy, standards, guidelines, and measures of accountability that are unambiguously related to the enterprise model and easily implemented. The enterprise itself is held accountable for its outcomes, and must be in control of its telemedicine.

Telemedicine is the key to virtual organization of healthcare delivery. By using high technology solutions to the problems posed by time and space, telemedicine creates a new environment for health care. New tools can be funded by dollars currently spent on travel, limit lost time for expensive personnel, curtail hospitalization of those who must be closely followed for chronic diseases (Bindman et al., 1995), support efforts to inform patients, and generally decrease the inefficiencies of static, self-contained organizations. Infrastructure and tools can provide access to more cost beneficial, quality health care through virtual organizations that will be eminently more satisfactory to the Veteran client—sick and well.

References

Bindman A, Grumbach K, Osmond D, Komaromy M, Vranizan K, Lurie N, Billings J, & Stewart A. 1995. Preventable hospitalizations and access to health care. *JAMA*, *274*(July 26), 305–311.

Fryback D, & Thornbury J. 1991. The efficacy of diagnostic imaging. *Medical Decision Making*, *11*(Apr–Jun), 88–94.

Grigsby J, Kaehny M, Sandberg E, Schlenker R, & Shaughnessy P. 1995. Effects and effectiveness of telemedicine. *Health Care Financing Review*, *17*(Fall), 115–131.

Perednia D, & Allen A. 1995. Telemedicine technology and clinical applications. *JAMA*, *273*(February 8), 483–488.

von Bertalanffy L. 1968. *General system theory: Foundations, development, applications.* New York: George Braziller.

Weinberg G. 1975. *An introduction to general systems thinking.* New York: John Wiley & Sons.

30
Telehealth for the Consumer

PETER GROEN

Telehealth in the Department of Veterans Affairs

Telehealth systems will become a reality for veterans and other consumers in the coming decade. Systems in place now or "on the drawing board" take advantage of a wide range of digitized voice, data, and video applications. With them, our veterans can obtain information, talk with physicians, check their pacemakers, request service, inquire about benefits, and perform other tasks from their homes or offices using telecommunications technologies. These new systems allow us to meet the changing expectations of our customers.

The Customer

In any industry, whether it be manufacturing, tourism or health care, you have to know who your customer is and focus on his needs and requirements. For the Department of Veterans Affairs (VA), the identity of our primary customer is deeply ingrained in our corporate culture. Every employee knows we are here "to care for him who shall have borne the battle and for his widow and his orphan...." In other words, the American veteran is our primary customer. It is not the many business and other internal and external parties we work with daily. In *Vision for Change* (1995), Dr. Kenneth Kizer, VA's Undersecretary for Health, clearly differentiates between these parties whom he identifies as stakeholders versus customers.

Once we have identified our customers, we have to understand them and get to know them. What are they like? What are their major characteristics? What do they want? Need? As we enter the year 2000, we can categorize our customers in the VA the following way. There are the World War II and Korean War veterans who are all over 60 years of age, generally retired from business life, and growing fewer in numbers with each passing year. Then there are the veterans from the Vietnam War era who are generally between 40 to 55 years in age and still active in today's work force. Finally, there are the post-Vietnam War veterans who fought in Grenada and more recently in Operation Desert Storm. These veterans are in their 20s and 30s. They were born and raised in this Information Age.

The aging veterans of World War II and the Korean War grew up in the age before computers. Computers were not a tool they had extensive experience with daily in a hands-on mode. The telephone, the postal system, and face-to-face meetings were the primary methods used to communicate with others. Access to the Internet and use of e-mail were not something associated with this group's lifestyle and normal business practices. That does not mean, however, that they are not very much aware of what computer technology is capable of and how it has affected their lives and the world in general. Certainly, as they have aged and come into more frequent contact with healthcare institutions, they have seen for themselves how computers and information technologies are used to assist in the healthcare process.

The veterans of the Vietnam War era have seen the Information Age ushered in and have played an active role over the past 20 to 30 years developing and introducing computer systems into almost every industry and every facet of our daily lives. These veterans have seen tremendous changes over the years. As they age and begin to interact increasingly with healthcare institutions, their expectations of service may differ somewhat from that of the World War II veterans. They expect their healthcare institutions to have computerized patient records, automated clinic scheduling, computerized laboratories, and other sophisticated tools to ensure efficient and effective service to meet their needs.

The younger Desert Storm veterans have grown up playing with computers since birth (e.g., Atari, Commodore, and Nintendo). Children of the Information Age, they are active on the Internet and have moved beyond simple e-mail to groupware products, video mail, video conferencing, telecommuting, and now virtual reality systems. By the time these veterans start aging significantly and begin to interact extensively with healthcare systems, they will expect instantaneous access to their healthcare providers. They will expect their providers to be online and available 7 days a week, 24 hours a day, and accessible via voice, data, or video communication connections. These veterans will expect many of their needs to be met by artificial intelligence systems that diagnose and prescribe care. They will not necessarily expect to need a physician unless specialized hands-on care or expertise is required.

VA Online

History

The 1950s saw the first introduction of computer technology into the Department of Veterans Affairs (VA). During the 1960s VA used computer technology to automate many of its internal operations (e.g., payroll, accounting, inventory). In the 1970s the focus was on the automation of the Veterans Benefits Administration (VBA), bringing up the Target system and computerizing Compensation & Pension, Education, Insurance, and other benefit modules. In the 1980s the focus was on the nationwide automation of the Veterans Health Administration (VHA) and its medical centers and the implementation of the VA Decentralized Hospital Computer Program (DHCP). The 1990s could be characterized as the decade VA began to design systems specifically for the use of veterans and began to put VA online for the 21st century.

Vision

More than a concept, VA ONLINE is a vision of VA doing business in the 21st century that is rapidly becoming a reality. With VA ONLINE, the veteran comes first. Online services are available to the veteran 24 hours a day, 7 days a week via voice, data, or video communications gateways. The basic premise for the VA ONLINE concept is a commitment to develop and implement information systems specifically designed for the use of our customers, the veterans, their dependents, and the general public. While VA will continue to maintain and enhance existing internal operational information systems (e.g., payroll, laboratory, and pharmacy), it now has a wealth of information stored in its computer systems that customers should be authorized to access on their own. The technology to provide this access is commercially available today. Further, veterans, their dependents, and the public in general have become attuned to using the new technologies that are now available.

The VA ONLINE concept involves a coordinated, collaborative effort by the VBA, VHA, National Cemetery System (NCS), and other organizational components of the VA to acquire, develop, and implement systems specifically for veterans. Included under the umbrella of the VA ONLINE initiative are communication gateways for voice, data, video, and multimedia; each of these categories includes a number of applications.

Communications Gateways

Voice Communications

Plain old telephone systems (POTS) and private branch exchange (PBX) switches continue to be the primary electronic communications tools used by most people and organizations. Almost everyone has access to a telephone that can be used to dial up and connect to the VA. Every VA medical center (VAMC) is equipped with voice mail, interactive voice response (IVR) modules, paging, and other features to enhance communications.

Interactive voice response systems, such as those already installed at over 80 VAMCs and VBA offices across the country, allow veterans to renew prescriptions, schedule clinic appointments, and inquire about benefit programs. With these systems, the patient can interact directly with a hospital's computer and central database using voice communications technology, normally accessed by a toll-free number.

Toll-free 800 service is a customer service used by VAMCs across the country. It allows veterans or the general public to dial and connect to a VAMC at no cost. The caller is then routed through the center's PBX switch to a specific clinician or office as needed. It is a fairly straightforward system that most people are familiar with in this country.

Interactive toll-free 800 service involves more than just making a free phone call to the VA to talk to someone. It involves using a combination of interactive computer technologies to complement simple telephone service. When a veteran dials the toll-free 800 number, the system is answered by an IVR system that informs him of a number of services that are available at the touch of a button. These include connecting to the hospital's computer system to make or cancel an appointment, renew a prescription, or inquire about some portion of his medical record. Other services available at the touch of a button include having selected information mailed or faxed (via a faxback server), leaving voice mail messages with an individual, connecting to a particular clinician or office, or various other options.

Data Communications

The VA ONLINE Bulletin Board Service (BBS) is a computer application already developed and implemented by the VBA, VHA, IRM, NCS, and other key offices. It is available to veterans and the general public and contains a wide variety of information such as the *Veterans Benefits Manual*, late-breaking news of interest to veterans, new legislation related to veterans programs, online inquiries about veterans benefits, special forum areas to exchange information, a NCS burial locator module, information on veterans healthcare programs, hospital volunteer programs, and many other items. With the addition of a more secure front end, access control features, and encryption technology, it will be possible for veterans to one day enter their personal identification number (PIN), view their master veteran record, and conduct much of their business with the VA online.

The VA ONLINE Internet WWW server is also operational and may someday replace the VA BBS. The VA's home page on the WWW was developed, implemented, and is supported by the VHA, VBA, NCS, and other key offices in the VA. It contains a wealth of information related to veterans programs and the VA. In the coming years, with appropriate security measures, veterans will be able to use the home page as their entry into the VA. There will be less need to travel physically down our nation's highways to the nearest VA regional office or medical center. The veterans will be able to travel to the medical center via the digital communications highways, and will be able to make a clinical appointment, renew a prescription, or communicate with a clinician about their healthcare problem using a home computer.

Computer software, such as the self administered tests for patients in the DHCP Mental Health Package (MHP) and the education benefits inquiry module to be pilot tested by the VBA, have been designed specifically for use by the veteran. Numerous other applications will begin to be developed in the coming years related to physical and vocational rehabilitation, interactive health education modules for patients, self-enrollment in programs, and applications we cannot yet imagine.

Kiosks continue to move forward, slowly finding their niche in the information distribution and retrieval arena. In the future, kiosks will be located in most federal buildings, post offices, malls, and other public places. Their capabilities will be enhanced and they will be connected to the network of government systems that contain general information as well as personal information. The kiosks will become one more tool citizens can use to inquire about government benefits, request specific services, or view their personal files.

Video Communications

Interactive patient TV, much like the interactive TV modules in modern hotels, could be implemented to allow patients to see what their next meal is, when the doctor is coming, and what medical exams have been scheduled for them. They could use the remote control unit of their TV set to select education or entertainment programs from the hospital's virtual library, order items from the medical center's store, or even to connect back to desktop video conferencing units at the homes of their high-tech families and friends for face-to-face conversations.

Interactive cable TV projects are cropping up all over the country, like the system the Sunflower Cablevision and the University of Kansas Medical Center have teamed to develop and implement in Lawrence, Kansas. It is an interactive, full-motion, two-way video link between home-based patients in that town and healthcare professionals back at the medical center. The VA could greatly expand its reach and meet the needs of veterans by partnering with local cable TV companies. Veterans could communicate or interact with the VA from the comfort of their homes using their TV sets.

Televideo conference room systems are at the high end of televideo technology. These are high-quality, expensive video communications systems normally used when group meetings or consultations are required. This technology could be used by veterans and their clinicians at remote VA outpatient clinics or community clinics when they need to consult with specialists back at a VA medical center hundreds of miles away. That is just one of the many examples where these systems might be used.

Desktop televideo conferencing systems, at the lower end of the televideo conferencing spectrum, will have a tremendous long-term impact on the way we do business in the future. the number of personal computers with this capability will grow from a few hundred thousand to more than a hundred million in the coming decade. Using this technology, patients will be able to communicate face-to-face with clinicians or other VA employees using ISDN or regular phone lines. In home healthcare situations, such as hospice care, a personal computer equipped with televideo capabilities may be loaned to the family so they can interact with clinicians at the primary healthcare facility regularly without requiring expensive, time-consuming visits to the home by a nurse or physician.

A veterans cable TV channel could become a reality soon. Partnering with state veterans affairs organizations, veterans service organizations (e.g., American Legion, Veterans of Foreign Wars, etc.), Department of Defense (DoD), and other organizations, enhances the possibility of setting up a cable TV channel dedicated to veterans. The success of other TV channels with a specific focus (e.g., travel channel, history channel, etc.) supports this idea. The first tentative steps in this direction have already been taken.

Multimedia Communications Gateways

Interactive multimedia 800 services embody many of the aspects of communications services already described. It allows the customer to dial a toll-free 800 number using a telephone, personal computer, or video communication device. The network is capable of detecting what communications medium is being used and routing the customer to the most appropriate services offered. These could be a PBX voice communications switch, bulletin board, Internet Web server, interactive voice response system, videoconferencing system—even a human being if needed or desired.

The Future of Telehealth

We have only just begun to identify the technologies that can be brought to bear upon the services we provide to our customers. Wireless technologies continue to be enhanced. Satellite services are growing, and the Internet Web is expanding. Virtual reality technology is entering the marketplace. These technologies and others will come together to complement the existing communications tools and networks already in place.

The future of telehealth hints at tremendous change, change we can only begin to envision. Will we need the number of clinicians that we have today? Will the focus of health care shift back to the home? What can we expect? And when?

One thing we do know, VA will continue to be right there, riding the wave into the future.

Reference

Department of Veterans Affairs. 1995. *Vision for change: A plan to restructure the Veterans Health Administration.* Washington, DC: Author.

31
Digital Imaging Within and Among Medical Facilities

RUTH E. DAYHOFF AND ELIOT L. SIEGEL

Introduction

Image storage and communication are very important in the practice of medicine. In the past, storage and communication normally have taken place at a single institution. However, as medical institutions reengineer their practices to increase cost-effectiveness, image communication is playing a greater role in medical care. Digital imaging, telemedicine, medical networks, and the electronic patient record are emerging as important technologies for the multimedia electronic patient record.

Physicians use many different kinds of medical data when treating patients. These include text, drawings, images such as x-rays and endoscopic pictures, and graphical data such as electrocardiograms. Images constitute an important part of the medical record of the patient, but they are not normally stored in the patient's chart because of their awkward size or nonpaper media. Rather, they are stored throughout hospitals by the various medical departments that collected them. Typically, text is used to describe and interpret images in the chart, but nonstandard terminology and the inherent difficulty in describing a complex image with words prevent full data communication.

With today's technology, it is possible to digitize and store medical images in networked computer systems. As shown in Figure 31.1, the images can be viewed on workstations located throughout a hospital, along with patient record text (Dayhoff, 1993; Dayhoff & Maloney, 1993). The Department of Veterans Affairs (VA) has implemented such a system, called the Decentralized Hospital Computer Program (DHCP) Imaging System, which has been running at the VA Medical Center (VAMC) in Washington DC since 1990. The DHCP Imaging System is interfaced to a commercial radiology Picture Archiving and Communications Systems (PACS) at the Baltimore VA Medical Center, which has been in continuous operation since 1993 (Kuzmak & Dayhoff, 1994).

The Washington, DC VAMC has been operating the DHCP Imaging System for over 5 years, capturing over 100,000 images in cardiology, gas-

474 Telemedicine and Telehealth

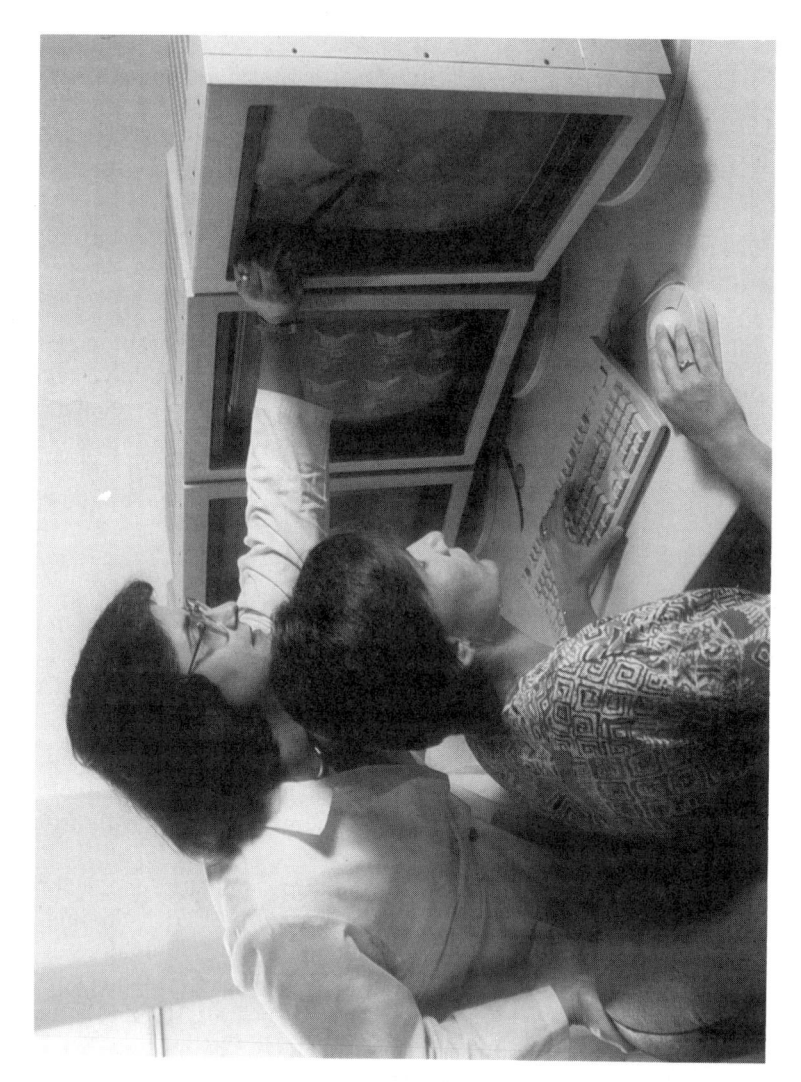

FIGURE 31.1. Radiologists review images at a four-monitor computer workstation.

troenterology, pulmonary endoscopy, hematology, surgery, podiatry, dentistry, and selected scanned radiology films. Implementation of direct Digital Image and Communications for Medicine (DICOM) digital image acquisition of Computed Tomography (CT) and Computed Radiography (CR) images is underway. Images are routinely used by clinicians during conferences, rounds, medical procedures, and follow-up care.

At the Baltimore VAMC, the DHCP Imaging System is utilized for capture and storage of surgery, pathology, gastroenterology, bronchoscopy, dermatology. A commercial radiology PACS system acquires radiology images and supports filmless operation. All radiology images are transferred from the commercial radiology PACS system to the DHCP imaging system. Both clinical and radiology PACS workstations are located throughout the hospital for display of the over 800,000 images that are available online.

The ability to transmit images among institutions can meet a variety of existing staffing and consultation needs, as well as reduce the cost of outside review. It allows subspecialty physicians to provide consultation to sites lacking specialists. Image transmission offers a promising extension to quality control procedures. Sites where one or two physicians must cover night and weekend call schedules can use teleconsultation to reach on-call physicians at their residences, allowing staff to provide rapid response at lower cost. Teleconsultation can also serve as a tool to facilitate "peer review" or the provision of a second opinion for imaging studies as is required in most hospitals.

All VA facilities are connected by a wide-area packet-switched network consisting of 23,000 miles of fully digital optical fiber network with four backbone nodes and 22 tributary nodes 1. By the end of 1996, all VA medical facilities will have been connected by frame relay running at 1.544 megabits/second. In addition, T1 lines have been installed between centers that share a large amount of data.

The VA is now organizing patient-care subnetworks across the country and is beginning to consolidate medical services within the networks. Thus, a medical department located at one facility will provide patient-care services for all locations within its network. Three medical centers in the Baltimore area have consolidated recently. This reengineered patient-care process is dependent upon superb communications between locations, with transmission of text, images, and other multimedia data.

User Requirements for Digital Imaging Systems

To obtain optimal benefits from new technologies such as digital imaging, telemedicine and the electronic patient record, some changes are required to the processes involved in providing medical care. Additional staff time may be necessary during the adjustment period to achieve savings later.

Users require "easy to use systems" that do not interrupt their workflow or patient relationship, maintain completeness of patient information, deliver high-quality data, and protect privacy and the security of patient information.

Ease of Use

Data entry has always been the most difficult and time-consuming aspect of electronic medical records. The Institute of Medicine's study of the computer-based patient record (CPR) identified the development of a technology that is sufficiently powerful and appropriate to the needs and preferences of healthcare professionals so that they can—and will—enter medical and other healthcare data directly into the computer, as the single greatest challenge in implementing the computer-based medical record (Dick & Steen, 1991, p. 142). Data entry includes input of textual and multimedia data.

The capture of images can be a relatively easy procedure (Dayhoff, 1993). The DHCP imaging system workstations are used during a medical procedure such as an endoscopic, surgical, or physical examination, to digitize images from a video source. The user identifies the patient and enters any pertinent descriptive information related to the images. This simple procedure takes about 30 seconds and the result is more informative than providing a textual description of the image. Typically, the clinician selects images that are significant to the patient's diagnosis or treatment course.

Images may be captured directly from a medical device such as a CT or magnetic resonance imaging (MRI) scanner, a PACS system, or an electrocardiography system. If a device has a standard interface, image capture may be done automatically, requiring little or no human intervention. With an automatic interface, generally all images are captured, including those that may be less significant to the patient's care.

Complete Patient Information

Treating clinicians will use images, reports, and other data from the patient record to make decisions about patient care. Treating clinicians need a variety of patient information provided by many specialties in many formats. Using paper charts, clinicians experience difficulties in assembling complete patient information. Online medical records have helped this problem by providing simultaneous access by multiple users to a record that cannot be misplaced. As images and other multimedia data are added to the online medical record, its completeness and depth of information far surpasses that of any paper chart.

Data entry and retrieval are dependent upon the correct identification of the patient. When data are provided from multiple systems, possibly at multiple locations, this can be difficult. With interfaced systems, one patient registry must be considered the master database and information should be provided to the other system. In many telemedicine cases, both systems maintain the master for their site. Often a patient will not yet be known to the patient registry at the consulting site. It must first be determined that the patient is not already registered, then the patient may be registered as a new patient. This allows the linking of patient information at each site and between sites. The DHCP imaging system performs new patient registration at the remote location after completion of an automated search for potential matching patients and manual verification that no matches exist.

When performing a teleconsultation, the clinician does not have the same benefits of patient contact when compared to an onsite clinician. Consultation from a distant location requires at least the same complete integrated patient record available to onsite physicians. Both images and associated text must be available and their association must be maintained. Commercial teleconsulting systems typically have been unable to provide this. The VA's client–server architecture allows communication with hospital information systems at multiple locations. Patients can be registered from remote sites, medical record information can be sent, and reports can be returned. The VA's software infrastructure includes integration structures at each medical center to support this capability.

Quality of Data

Viewing of clinical images on workstations is done for two major purposes. Images are examined to make medical diagnoses, for example by a radiologist, pathologist, or dermatologist. Images are also referred to during medical decision making, such as when a treating clinician examines all aspects of the patient's condition to select a treatment regimen. The image quality required for these two types of processes may be different. The radiologist interpreting a diagnostic imaging study needs to detect or exclude all abnormalities. The treating clinician may need to view a particular lesion described in the radiology report to evaluate the degree of abnormality, determine the impacts of the lesion on the patient, and select a treatment strategy.

Some specialties have defined a minimum resolution and required number of digitized bits for diagnosis. The requirements that radiologists have for computer workstations are somewhat different from those of other users. Radiologists have medical requirements for spatial resolution and brightness for the image display and an operational requirement for retrieval and display speed that are directly related to performance.

The American College of Radiology (ACR) Standard for Teleradiology (1994, pp. 168–173) stipulates that it is the responsibility of the radiologist to "provide images of sufficient quality to perform the indicated task" and to "satisfy the needs of the clinical circumstance." The ACR standards stipulate that the "display system" should be 2 K × 2 K × 8 bits or better and that it should support tools to perform window/level (contrast/brightness), magnification, image invert, image rotate, and linear measurement. The monitor is required to have a brightness of "at least fifty foot lamberts." This suggested standard for teleradiology can be met with a PACS workstation using a monitor that can display a 2 K × 2 K image. Alternatively, the requirement that the "display system be 2 K by 2 K × 8 bits or better" could be fulfilled using a video buffer that could hold an 8-MB image and a lower resolution monitor such as a 1.6 K by 1.2 K pixel display. This would require that the physician viewing the image use the workstation's magnification tools to review the full image dataset.

In addition to the medical requirements for a very high level of image quality, a radiology workstation must offer similar or faster image throughput and handling than a conventional film viewbox or alternator. A four-monitor workstation is required for review of most conventional radiographs and cross-sectional studies to permit rapid review of current and reference images from previous examinations. A workstation with fewer monitors requires many more keystrokes and more time and consequently constrains productivity of the radiologists. Images as large as 8 MB (a typical chest radiograph) should be available in less than 3 seconds and it should be easy for the user to arrange these images for comparison purposes. The database structure should be hierarchical and all imaging studies should be accessible on all workstations.

These "special" requirements for image quality and throughput by the radiology department currently can only be met with relatively costly workstations that would be prohibitively expensive as generic imaging workstations throughout a medical center. The solution to this constraint is the use of multiple types of workstations. The VA has mixed subspecialty workstations, such as the commercial PACS at the Baltimore VA, with the DHCP imaging workstations. These standard off-the-shelf workstations can be modified to include multiple 2000 pixel monitors rather than the usual single less expensive 1200 pixel monitor.

The quality of data required for diagnosis and treatment varies by specialty, and many specialties have not performed studies to determine the image quality required. Additional variables are important in other specialties. For example, in pathology, important variables include color quantization, spatial resolution, brightness, contrast, gamma, focus, convergence, and color balance (Black-Schaffer & Flotte, 1995). The issue may be further obscured because the necessary image quality will vary with the types of abnormalities being distinguished. Telemedicine makes requirements more complicated as there may be a need for video teleconferencing for patient interaction and identification of significant lesions, and still image transmission for diagnosis at a higher resolution. At the present time, there are no defined requirements for "reference" quality images used by treating clinicians.

Digitized medical images can require particularly large computer files. A number of systems use image compression, with or without irreversible loss of data, to reduce image storage and transmission requirements. Some studies have been done to measure the ability to make diagnoses using various modes of image compression.

Privacy and Security

Finally, privacy and security of patient data are critical. The process of image transmission can increase the risk of privacy invasion or corruption of data. In addition, the data may be stored at two locations rather than one following the consultation because both clinicians may continue to participate in the patient's care. This dual storage also increases the privacy risk.

There are legal requirements for long-term storage of images. Images stored digitally must be able to be read throughout the required time period. Changes in technology may require copying of images from one long-term storage device to a newer one when technology upgrades are made.

Importance of Integration

There are a number of areas where integration is important to the success of imaging systems. The clinical user wants to be able to access all data from any workstation at their facility. This means that online medical record information from the hospital information system (HIS) or HIS components must be available on the workstation. A straightforward input procedure must allow data acquisition from a variety of sources, including independent digital imaging systems such as PACS. Image data must be accessible by patient and by procedure. Users must be able to access related information of all kinds easily without numerous manual operations. Useful data include clinical information, laboratory results, radiology and pathology reports, results of previous similar studies. For examples, see Table 31.1.

Hospital Information System

The HIS used by the VA has allowed the integration of patient record and multimedia data for display on user workstations. The VA uses a hospital information system developed and maintained by its own staff called the Decentralized Hospital Computer Program (DHCP). The DHCP is written in the American National Standard (ANS) language known as "M" (formerly Massachusetts General Hospital Utility Multi-Programming System [MUMPS]) and runs on a variety of hardware platforms. The DHCP is in the public domain and is used in virtually all 172 VAMCs, as well as in a derivative form by the Department of Defense, the Indian Health Service,

TABLE 31.1. Integration of data required by radiologists.

- New and previous radiology study images
- Imaging reports
- Reason current study was ordered
- Bun and Creatinine, other laboratory values
- Old pathology reports
- Patient clinical information
- Allergy information

other government agencies, and private hospitals worldwide. This internal system development approach provides compatibility among systems in all of the VA facilities and allows the VA to enhance its HIS to meet constantly changing user and administrative requirements.

The VA's DHCP system is different from multivendor systems in that it is an integrated system based on a central set of software development tools. All of DHCP's many software modules, such as the laboratory, pharmacy, radiology information system (RIS), surgery, medicine, health summary, clinical record, and local and regional registry modules, are based on the VA toolkit and adhere to its rules. Clinicians can use the health summary module to define custom datasets for display on the workstation or for printing. These datasets may draw elements from any of the patient data modules and may be restricted with time limits.

The VA has adopted a new client–server workstation architecture to allow its clinical workstations to communicate with its hospital information system. Transmission Control Protocol/Internet Protocol (TCP/IP) messages are used for communications between VA-developed "broker" software located on the HIS and the workstations. All messages are processed through the broker software that performs the requested operations on the HIS and returns the results to the workstation via a TCP/IP message. Security log-on and server connection are also handled by the broker software. With the proper security privileges, a workstation user may connect to any HIS server on the wide-area network. Workstations currently run Microsoft Windows. Software running on the workstations is written in Borland's Delphi, except for integrated off-the-shelf products. The use of the VA toolkit and the broker software has allowed the integration of multimedia data.

Image Acquisition Systems

Images may be captured using workstation hardware, or they may be transmitted from independent systems. Imaging workstation capture within the VA is done in a number of locations including the cardiology department, the gastroenterology endoscopy lab, the bronchoscopy examination room, the surgical pathology reading room, the hematology research laboratory, the dermatology and dental clinics, and the operating room. Images are generally collected by consulting services to meet their own needs as well as the needs of the referring physicians. Radiology and nuclear medicine images are best provided across ACR-NEMA standard interfaces (1994).

The VA has interfaced commercial radiology PACS with its hospital information system using the ACR-NEMA standard. This interface is currently being upgraded to DICOM version 3 (Kuzmak & Dayhoff, 1996).

The ACR-NEMA standard, bidirectional interface between the Baltimore VA Medical Center's commercial PACS and the hospital information system (DHCP) has been in place for approximately 3 years. This interface passes both images and text data, such as patient demographics, orders, and reports, between the systems. The interface has resulted in several major benefits to the hospital and radiology department and is considered to be a critical component in the ability to operate the radiology department and hospital in a filmless environment.

The major benefit of this interface is the ability to maintain a high level of consistency between the HIS/RIS and commercial PACS databases. Manual entry of patient information into the commercial PACS inevitably results in the creation of databases that differ from the "master" HIS/RIS database. The lack of an interface increases the likelihood of a single patient's images being stored in multiple locations with different spellings of the name or different versions of the patient identification. Additionally, the interface prevents double entry of patient demographics and ordering information, which results in savings in time and personnel resources.

Another important advantage of the integrated interface between the PACS and the HIS/RIS at the Baltimore VAMC is that physician orders placed on the HIS/RIS serve as a "trigger" for transfer of relevant historic images from the long-term storage (optical jukebox) to short-term storage making them available more quickly for review. The resulting elimination of waiting times for images to be fetched from the optical jukebox significantly enhances radiologist and clinician productivity. These images are usually transferred to short-term storage well before the patient actually undergoes the imaging examination. Because all orders are placed on the HIS/RIS and transferred to the PACS database and workstation, radiologists are able to view requests for new imaging studies as well as old images and reports. This has resulted in paperless, in addition to filmless, operation for the radiologists.

Imaging reports are entered into the HIS/RIS system and then transferred to the PACS via the interface. As reports are edited and verified on the HIS/RIS system, these changes are also transmitted to update the commercial PACS database. This mechanism prevents double entry of reports and makes all radiology reports available to all users of the PACS and the HIS/RIS.

In addition to the integrated text interface, there is an interface between the PACS image database and the DHCP imaging database. This permits the HIS/RIS to retrieve each image generated on the PACS for display using the DHCP imaging workstation. These workstations are used at Baltimore for integrated display of all imaging modalities including radiology and nuclear medicine.

The use of the text and image interfaces to the commercial PACS has consequently resulted in substantial improvements in data accuracy and integrity, efficient use of personnel, and economic savings. The experience at the Baltimore VAMC suggests that a robust bidirectional interface from a commercial PACS to the hospital information system and complete automation of the data transfer is essential for successful operation of the system (Dayhoff, 1995).

Display Workstation

The display workstation must provide integrated access to the many types of data of the medical record (text, color and gray scale still images, motion video sequences, audio, and graphics) and to live video teleconferencing for telemedicine. The user must be able to access all this information with a single log on and a single patient lookup. Finally, a variety of data views should be provided to suit the purposes of the user.

A medical center may use more than one type of display workstation to meet different users' requirements. However, users expect to be able to access all data from each workstation. The VA's multimedia display workstation meets the requirements for integrated display of medical record text and multimedia data. In addition, high resolution (at least 2000 by 1600 pixel) radiology workstations are necessary for radiologic diagnosis.

The DHCP imaging workstation provides integrated data to the user. The users sign onto the workstation with their HIS security log-on codes. The patient is identified once, in the same manner as on the HIS. To meet the various needs of clinical users, two major access modes are provided: access by patient and by procedure. A user must be able to access data by patient, viewing all of the patient's image and text data; this "visual chart" capability, shown in Figure 31.2, is generally used by treating clinicians. A user must be able to view data related to a clinical procedure, including images and report. This mode is generally used by specialists, such as radiologists, pathologists or cardiologists, when producing procedure reports. The DHCP imaging workstation allows both modes of access.

The workstation must meet clinicians' needs for true color displays of 16 million colors per pixel to handle pathology, dermatology, and endoscopy images; for display of 12-bit gray scale images; for display of other multimedia data such as motion video, audio, and electrocardiogram data; and for HIS data display. Digital image capture may be performed using the workstations.

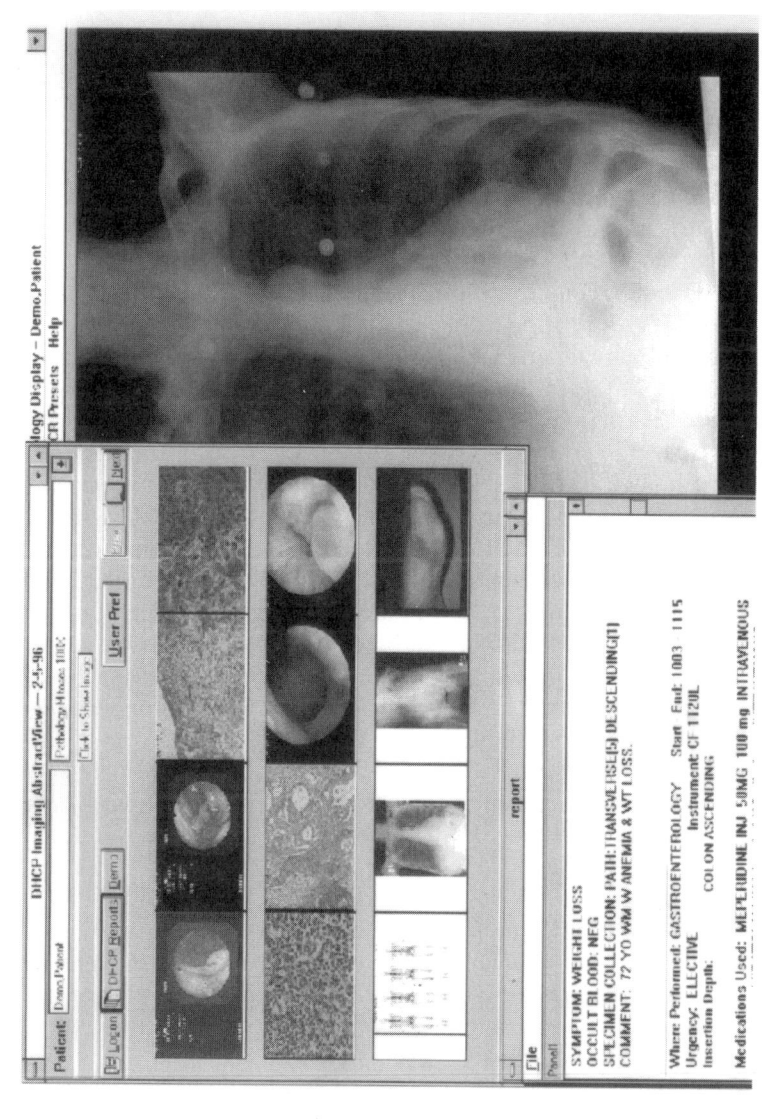

31. Digital Imaging Within and Among Medical Facilities   483

FIGURE 31.2. DHCP imaging workstation screen displays a patient's "visual chart," including tiled thumbnail image summary, textual procedure report, and full view of x-ray image.

Digital Imaging Benefits Within a Medical Facility

The Baltimore VAMC converted from conventional film-based operation in the radiology and nuclear medicine departments to the use of a hospital-wide PACS to achieve filmless operation. All of the images from the various modalities in the radiology department including computed radiography, fluoroscopy, Ultrasound, CT, MRI, digital angiography, and nuclear medicine are sent via a DICOM or ACR-NEMA gateway into a large central server that uses a Redundant Array of Inexpensive Disk (RAID) architecture. Images are also immediately archived onto a 1-terabyte optical jukebox. Additionally, images from the cardiac catheterization laboratory can also be sent to the PACS. Forty-three PACS workstations are located throughout the medical center including the operating rooms, the intensive care units, the emergency room, patient clinics, physician "team" rooms, the auditorium and the medical media department. The system has a bidirectional interface to the HIS/RIS (DHCP), which houses the "master" patient database and patients' electronic medical records.

The DHCP Imaging System has a similar architecture and infrastructure, with fast magnetic Windows NT servers for rapid image access and a slower optical disk jukebox for long-term image storage. An additional 50 DHCP imaging workstations are located throughout the medical center in a similar distribution to the PACS workstations. These DHCP imaging workstations can bring up all of the radiology, nuclear medicine, and cardiology images from the PACS and can also display pathology, endoscopy, dermatology, and bronchoscopy images in addition to intraoperative photographs, and digitized medical documents. The image review using DHCP workstations for radiology is typically performed after the study has been interpreted and reported by the radiologists.

A hospital-wide prospective study of the operations of the PACS was performed, which examined several parameters related to the radiology department and digital image utilization by the medical center (Siegel, 1993, 1995; Siegel & Pickar, 1994). These data were acquired through a combination of an analysis of the DHCP database, direct observation, and clinician and radiologist surveys. Data collected before and after implementation of filmless operation were compared to determine the impact of the PACS on the radiology department and hospital.

The implementation of the PACS was found to result in a number of operational improvements in both the radiology department and the hospital in general, as shown in Table 31.2 (Siegel & Brown, 1994; Pomerantz, Siegel, Protopapas, & Reiner, 1995; Siegel, 1995; Siegel, Denner, Pomerantz, Reiner, & Protopapas, 1995; Siegel, Pomerantz, Reiner, & Protopapas, 1995). The average time from when a study was performed to when it was read dropped from approximately 1 day to less than 1 hour. The "lost" image rate, defined as the number of studies not interpreted within 48

31. Digital Imaging Within and Among Medical Facilities

TABLE 31.2. Radiology imaging system benefits.

Radiology imaging system benefits at the
Baltimore Veterans Affairs medical center

- Allows realtime interpretation of studies
- Process automated from order to report
- Decrease in lost films
- Film retake rate decreased
- Radiology utilization increased
- Film room workload decreased dramatically
- Allows more specialization by radiologists
- Some decrease in radiation dosage to patients

hours of being performed, dropped an order of magnitude from about 8% to approximately 0.6%. The "film" retake rate dropped from approximately 5% to below 1%. Radiologist productivity increased by more than 20% and technologist productivity also increased. For example, CT technologist productivity increased by approximately 35% after the transition to filmless operation. Although the average length of stay and the average daily census dropped by more than 10% after introduction of the PACS, it is difficult to determine to what extent this was due to the PACS or to the many other external forces acting to decrease these parameters. Clinician surveys indicated a high level of satisfaction with the filmless operation with 93% preferring the PACS to film and only 3% preferring film to digital operation. Most clinicians indicated that the PACS resulted in substantial savings in the use of their time. Preliminary economic estimates of the impact of the PACS have suggested that the savings in film and personnel and increased productivity are largely offset by the depreciation of the costs of the equipment and the annual maintenance costs. However, the savings in physician time spent attempting to find and review images are considerable and probably amount to the equivalent of two to three physicians at a medium-sized facility such as the Baltimore VAMC.

A study done on the DHCP Imaging System at the Washington VAMC found that the system "was changing clinical work in ways that physicians considered beneficial" (Kaplan, 1995). See Table 31.3. Clinical users felt that the system improved patient care because images always were available. It conveyed clinical information better than written reports and improved communication among specialists. It decreased the time required to reach consensus at conferences because everyone saw the same images.

Telemedicine: Digital Imaging Among Medical Facilities

Digital imaging systems within hospitals create a firm foundation for telemedicine systems. With incremental investment, digital images and other data may be communicated between sites. Telemedicine may be

TABLE 31.3. Imaging system benefits.

Imaging system benefits found at the Washington, DC Veterans Affairs medical center

- Complete patient data allows better care
- Simultaneous availability of images
- Increased communication among clinicians
- Fewer repeat procedures
- Important role in conferences
- Continuing medical education on the job
- Patients can see their condition and participate in treatment decisions
- Automatic peer review

based on point-to-point communication where two sites communicate with each other. In a system like the VA medical care system, it is important that medical centers can communicate with all other sites. In addition, sites that work more closely together may run point-to-point communication lines for special purposes. Compliance with standards is particularly important as more components are added to a complex system.

Importance of Evolving Standards

Standards are especially important for integration of systems and for teleimaging between locations. Standards will allow systems to communicate with any other standard system, not just those purchased at the same time or from the same vendor. The DICOM standard is an important standard for medical imaging. It is being used not only for radiology imaging, but also for cardiology, gastroendoscopy, and pathology. This standard describes all communication aspects of digital image production, storage, retrieval, transmission, and display in a medical environment, and how these functions are related to various information systems (Kuzmak & Dayhoff, 1993).

Several other technology developments are helping in system integration. For example, published application programmer interfaces (API) allow one software developer to incorporate software written by others. The ability to mount servers across wide-area networks allows integration of data storage. Internet communications allow broad access to data and service sources based on standards.

Telemedicine as an Extension of Digital Imaging Systems

The use of telemedicine within and among hospitals in the VA system holds the promise of a number of advantages in quality of patient care and in increasing efficiency of patient care resulting in cost savings. One of the best examples of telemedicine within and between medical centers is the PACS and teleradiology network at the Baltimore VAMC.

Three VA hospitals in Maryland recently combined to form a single institution under a single director with clinical services that serve all locations. This "integration" resulted in the combination of the three radiology and nuclear medicine departments into a single "service." Teleradiology is being used to support this new organization.

This new Maryland teleradiology system is fully integrated with the shared DHCP hospital information system. A DICOM version 3 interface is used to send DICOM messages over a T1 line from the Baltimore VAMC HIS to the Perry Point VAMC, about 40 miles away. For each radiology order, the Baltimore DHCP system sends three separate messages, the receiving interface places the information in a local database and provides worklist information to a Computed Radiography x-ray acquisition device. Further efforts underway are to implement the new DICOM Modality Worklist interface directly to acquisition devices. Images will be sent in DICOM format from the Fort Howard and Perry Point Medical Centers to Baltimore using a high speed "T1" telecommunications line. The images are simultaneously stored at the sending site on local servers to minimize subsequent traffic on the T1 lines. The images are interpreted using the commercial PACS and then sent to the DHCP imaging system at Baltimore. Here they flow into the integrated patient medical record in DHCP to be integrated with other images and patient data. The full electronic patient record is then available at all sites.

The additional cost for PACS at these medical centers will be approximately 5% of the price of the PACS at the central site, Baltimore. This "hub and spoke" configuration is expected to result in substantial additional savings across the wide-area network. Other "spokes" are planned for the future in addition to modifications to upgrade the capacity of the server and long-term storage at Baltimore.

The transition to filmless operation has successfully demonstrated that "reinvention" of the process of image acquisition, storage, and display using sophisticated computer systems can result in significant operational improvements. The economic savings are minimal in the radiology department, but become substantial when one evaluates hospital-wide or network-wide operations. A wide-area network such as the one used by the three Baltimore medical facilities could serve as a model for a distributed imaging database and system. Benefits that are expected from such a system include a greater degree of availability of subspecialty expertise that is more uniformly available across the many VA facilities. A wide-area imaging network would effectively bring the "experts" to the patients, rather than requiring the much less efficient transportation of patients to multiple sites. The ability to more easily provide back up and weekend and nighttime coverage will also increase both the quality of patient care and will result in significant savings in personnel and contracts for non-VA radiology coverage.

Conclusion

Systems integration is a key factor in system efficiency, usability, and user acceptance. Integration may require system architectures that include interfaces to stand-alone systems. Information available in one system can be used to assist the users of another system. From the user's perspective, all data are available from a single source, the workstation. Because of integration, data that are put into any part of the system serves all users. This results in a system that contains the necessary "critical mass" of information. Automatic mechanisms to extract relevant information for any particular situation add value to the system. Users are more likely to find what they need from the system, therefore they look first to the system, and find it cost-effective to place information in the system. Reaching this critical mass is essential to achieve the maximum benefits from an integrated patient record system.

The VA has built its information systems in progressive steps. It extended its hospital information system to produce an imaging system, and built on its imaging systems to create a teleradiology system. External systems were integrated as necessary. This progressive approach avoids redundancy and lowers costs. It also allows process reengineering to take place in steps.

System implementation and reengineering often proceed in concert. Reengineering often requires changes to manual processes to achieve the greatest benefits from automation. For example, the manual processes at a single medical center must be modified to utilize onsite imaging systems. It is straightforward to extend the system to support teleradiology and allow similar reengineering of an entire network of medical centers.

Many institutions today are undergoing reengineering to improve cost-effectiveness. In many cases, consolidation of patient-care services at multiple locations is being utilized. The effects of distance and time are diminished through the use of technology. Teleimaging systems are well-positioned to support this consolidation and distant collaboration of clinicians. For example, teleradiology can provide better round-the-clock coverage, better subspecialty coverage, the ability to better distribute workload, and increased efficiency and productivity of radiologists.

It is important to look beyond the immediate benefits when assessing cost-effectiveness of systems. The transition to filmless operation at the Baltimore VA Medical Center has successfully demonstrated that "reinvention" of the process of image acquisition, storage, and display using sophisticated computer systems can result in significant operational improvements. The economic savings are minimal in the radiology department, but become substantial when one evaluates hospital-wide or network-wide operations.

References

American College of Radiology Digest of Council Actions. 1994. *Section Iiw. Standard for Teleradiology*, pp. 168–173.

Black-Schaffer S, & Flotte TJ. 1995. Current issues in telepathology. *Telemedicine Journal, 1(2)*, 95–106.

Dayhoff RE. 1993. The electronic medical record: Data capture and display methods for images, electrocardiograms, scanned documents, and text. *Proceedings of the Image Management and Communications Conference.*

Dayhoff RE. 1995. Integration of medical imaging into a multi-institutional hospital information system structure. MEDINFO 95 Proceedings, pp. 407–410.

Dayhoff RE, & Maloney DL. 1993. Exchange of VA medical data using national and local networks. *Annals of the New York Academy of Sciences, 670*, 62–63.

Dick RS, & Steen EB. 1991. *The computer-based patient record*. Washington, DC: National Academy Press.

Kaplan B. 1995. Information technology and three studies of clinical work. *ACM SIGBIO Newsletter, 15(2)*, 2–5.

Kuzmak PM, & Dayhoff RE. 1993. Implementing the digital image and communications for medicine (DICOM) protocol in M. *M Computing, 3(3)*, 33–40.

Kuzmak PM, & Dayhoff RE. 1994. A bidirectional ACR-NEMA interface between the VA's DHCP integrated imaging system and the Siemens-Loral PACS. *Proceeding of the Medical Imaging Conference of the International Society for Optical Engineering (SPIE).*

Kuzmak PM, & Dayhoff RE. 1996. An architecture for MUMPS-based DICOM interfaces between the Department of Veterans Affairs HIS/RIS and commercial vendors. *Proceedings of the Medical Imaging Conference of the International Society for Optical Engineering (SPIE).*

National Electrical Manufacturers Association (NEMA). 1994. *Digital Imaging Communication in Medicine (DICOM). NEMA Standards Publication PS 3.* Washington DC: NEMA.

Pomerantz S, Siegel EL, Protopapas Z, & Reiner BI. 1995. Experience with PACS in the operating room in a filmless hospital. *Proceedings of the 1995 Image Management and Communications Conference, IEEE Computer Society Press.*

Siegel EL. 1993. Plunging into PACS. *Diagnostic Imaging, 15(2)*, 69–71.

Siegel EL. 1995. Impact of filmless radiology on the Baltimore VA Medical Center. *Proceedings of the 1995 Military Telemedicine Symposium.*

Siegel EL, & Brown A. 1994. Preliminary impacts of PACS technology on radiology department operations. *Proceedings of the Annual Symposium on Computer Applications in Medical Care*, pp. 917–921.

Siegel EL, Denner J, Pomerantz SM, Reiner BI, & Protopapas Z. 1995. PACS and medical media in a filmless hospital. *Proceedings of the 1995 Image Management and Communications Conference.*

Siegel EL, & Pickar E. 1994. The transition to the filmless imaging department: Early experience at the Baltimore VA hospital. *S/CAR 94 Society of Computer Applications to Assist Radiology*, p. 5.

Siegel EL, Pomerantz SM, Reiner BI, & Protopapas Z. 1995. PACS in a digital hospital: Experience with filmless operation at the Baltimore VA Medical Center. *Proceedings of the 1995 Image Management and Communications Conference.*

Index

A

Achievable Benefits Not Achieved (ABNA), 23, 25
Administrative activities
  anesthesiology system, 301–302
  and clinical workstations, training of users, 157–158
  security, 59, 60–61, 66–67, 104–105, 127–132, 479
AIDS Information Center, 316
Ambulatory care
  future policy, 154
  history of, 148–150
  implementation of policy, 150–153
    change, resistance to, 152
    components of project, 150–151
    data, capture and integrity of, 152–153
    obstacles, 152
    workgroup, formation of, 150
Ambulatory Care Program, 231–233, 239
  databases
    DHCP, 233
    Health Summary (HS), 233, 234–235
    Outpatient Clinic File (OPC), 233–234
    Patient Treatment File (PTF), 233
  data collection, 236
    optical scanning, 236–238
    other methods, 238–239
  data packages, existing
    demographics, 234
    laboratory tests, 234
    pharmacy, 234
    radiology, 234
  data packages, new or developing
    patient history, 236
    Patient-Care Encounter (PCE), 235–236
    Problem List, 235
    Progress Note, 235
    Uniform Ambulatory Medical Care Minimum Data Set (MDS), 236
American National Standard (ANS) computer language, 51
Anesthesiology system, 293, 304–305
  administrative activities, 301
  and Joint Commission on Accreditation of Healthcare Organizations, 302–303
  quality management, 301–302
  scheduling, 301
  Electronic Anesthesia Recordkeeping system (EAR), 299–305
    data exchange, 305–306
    expansion in operating rooms, 304–305
  intraoperative, 296–300
  postoperative, 300
    data collection, 301
    patient tracking, 300–301
    recovery room, 300
  preanesthesia, development of automated systems, 293–294
    data retrieval, 295–296

medical records, all-electronic, 295
and text integration utility (TIU), 296
ANS computer language. *See* American National Standard (ANS) computer language
Application protocols, 90
Application Requirement Groups (ARGs), 140–141
Automated Data Processing Application Coordinators (ADPACs), 71, 105, 143
Automated Information Systems security program, structure of, 118–119
elements of, 119–120
federal laws and regulations, 120–121

B

Baltimore VAMC, evaluation of radiology, 400–401
Background jobs. *See* TaskMan
Bronx VAMC HOST anesthesiology project, 297–299
Bulletins, transmission of, 65–66
Business process redesign, 142

C

Cardiology, applications in, 275–276
database, 285
benefits, clinical and research, 286–288
expert panel, input from, 285–286
enhancements, 288
images, addition of, 288–289, 290
teleconsultations, 289, 291
pacemaker, surveillance of, 276–277
malfunction, impending, 281–284
telemedicine protoypes, 280–281
telephone-based, 277–280
CARE decision support system (CDSS), 203–226
action statements, 209–211
components of, 204–205

debugging tools, 217–218
language compiler, 212–217
modularity, evolution of, 220–221
protocol-authoring environment, 211–212
compilation process, 213–217, 218
Operator Precedence Table, 213–214
source code, 213
State Transition Table, 213–214
protocols, impact of, 222–225
on clinical trials, 224–225
on inpatient care, 223–224
on outcome research, 224
on preventive measures, 222–223
procotols, invocation of, 218–220
repository of clinical data, 205–206
syntax, overview of, 206–209
CARE-GUIDE, 342–344
Chicago, University of, automated preanesthesia system, 294
Classic improvement cycle, 21
ClassMan, 105
Client-server systems, 93–94
Clinical Applications Requirement Group (CARG), 15
Clinical decision support, 183–184, 185, 201
history of in DHCP, 184–191
modules, examples of, 184–187
notifications, examples of, 187–191
standards of care, 192
goals, 195–197
and expert systems, 1980199
and Medical Logic Modules (MLMs), 195–198
and metadata dictionary, 197–198
and new notifications, 199–201
order checks, 195
systems, evolution of, 192–194
knowledge or ruled-based, 192–193
and HELP System, 193–194
and Iliad, 193–194
Knowledge Data Systems (KDS), 194
Clinical information, dissemination of, 109–110

Index 493

Clinical needs, 16–17
Clinical Reminders, 235
Clinical workstations
  development of
    benefits derived, 156–157
    purpose, 155–156, 162–163
    technology used, 156
  implementation of
    methodology, 158
    participants, 157–158
    resources, acquisition and maintenance of, 158
    resources, deployment of, 158–159
  methodology
    evaluation of, 159–161
    tactical steps, 159
  problem solving, 161–162
  technology
    education, 161
    teleconferencing, 161
    telemedicine, 161
Clozapine, 184–185
Collaboration, fostering of, 91–93
Communication protocols, 90
Computer-based Patient Record Institute (CPRI), 41
Computer Fraud and Abuse Act of 1986, 120
Computerized Patient Record System (CPRS), 137–139
  design of, 139–142
    clinician acceptance, 141
    user needs, 140–141
    usability, testing of, 141–142
  organizational issues, 142–145
    ADP coordinators, 143
    business process redesign, 142
    clinical coordinator, 142
    committees, formation of, 145
    coordinator council, 144
    Information Resource Management office, 145
    management policies, 142
    personnel, redefined roles of, 142–143
  technical issues, 145–147
    data content, 145–146
    event processing queries, 146–147
    message formats, 146–147
    vocabulary, standardization of, 145–146
Computer Matching and Privacy Protection Act of 1988, 120
Computer Security Act of 1987, 120
Computing resources, principle of local control of facility-based, 43–45
Conceptual integrity, in system design, 386–387
Connectivity, 59–60
Controlled representation of data, 164–165
  challenges to functionality, 165
  standards, creation of sufficient, 168–169
  healthcare data interchange, 172
  implication of, 180–181
  information, meaning of, 165–166
  knowledge processing, 166
    applications of, 170–172
    barriers to, 166–170
    business process layer, representation of, 176–178
    controlled clinical vocabularies, 172–173
    medical language, unified system, 174
    online information, 174–176
  VA experience with, 178–180
  and Clinical Lexicon, 179
COTS technologies, 110, 399–400
CPRS. *See* Computerized Patient Record System
Current Procedural Terminology (CPT), 145–146

D

Data, segregation of local and national, 59
Databases
  ambulatory care program, 233–235
  cardiology, applications in, 275–276, 285–288
Database, shared across facilities, 81–86
  architecture of, 82–83
  data administration, 82

integration agreements (IAs),
library of, 83–84
and database administrator (DBA),
role of, 81–82
and consolidated systems, 84
and distributed systems, 84–86
and object-oriented designs, 86
Data collection, ambulatory care
program, 236–239
Decentralized Hospital Computer
Program (DHCP), 33, 55–56,
385–386, 406–408
achievements of, 40–41
adaptability, 389–392
attributes of, 57–60
conceptual integrity
in system design, 386–387
evolution of, 41–43, 387–388
involvement in, 60
evolution of VA, 392–393
facility, scalability to size of, 71–74
reusable components, 72–73
system performance, 73
infrastructure, 60–71
database, growth of, 73–74
disruptions, response to, 76–78
and non-DHCP systems, 49, 67
strategic redundancy, 74–76
tensile strength, 78
modular applications, design of, 45,
66
principles of, 43–55
public domain, 389
updating, 69–70
Deming, W. Edwards, 20
Department of Defense (DoD),
collaboration with VA, 91–92
Developer environment, 58–59
Decision support. *See* Clinical
decision support
DHCP. *See* Decentralized Hospital
Computer Program (DHCP)
Digital imaging, 473–475, 488–489
benefits of, 484–485
integration, 479
Hospital Information System
(HIS), 479–480
image acquisition systems, 480–
482

workstation, display, 482
and telemedicine, 485–488
user requirements, 475–476
data, 477–479
patient information, 476–377
security, 479
use, ease of, 476
Disk space, 101
Display workstation, 482
Disruptions, response to, 76–78
Distributed systems, 84–86
Downtime. *See* Decentralized
Hospital Computer Program,
infrastructure, strategic
redundancy
Duke University, intraoperative
records, 296

E
Edsall, David, 297
Electronic Anesthesia Recordkeeping
system (EAR), 299–305
Electronic Communications Privacy
Act of 1986, 120
Electronic Error and Enhancement
Reporting (E3R), 106–107
Electronic mail. *See* FORUM;
MailMan
Entity lists, 206
Establishment of VISN, values guiding,
4
Expert system. *See* Clinical decision
support
Extensible Editor, 212, 213

F
Federal laws and regulations
concerning security, 120–
121
Finland, installation in, 428, 431
Florida, University of, automated
preanesthesia system, 294–
295
Formal symbol systems. *See*
Controlled representation
FORUM (e-mail system), 100, 106,
110–111
Freedom of Information Act, 120
Functional and Independence

Measurement System (FIMS), 332
Functionality, management of expansion of, 102–103
  agency-driven demands, 104
  mission-driven demands, 103–104
  user-driven demands, 103

G
Generic Code Sheet (GCS), 68
Germany, installation in, 428, 431

H
Health Care Information System (HCIS), 164
  symbolic control, implications of, 180–181
Health Employer Data and Information Set (HEDIS) measures, 153
Health Summary (HS), 46, 233, 234–235, 268–269
Help desks, 111
HELP System, 193–194
Hippocrates, 23–24
Hospital Information System (HIS), 479–480
HOST. *See* Hybrid Open Systems Technology
Hybrid Open Systems Technology (HOST), 110
  clearinghouse, 399
    COTS technologies, 399–400
    successes, 400–401
  establishment of, 397
  historical perspective, 395–396
  implementation of, 397–399
    field projects, 397
    HOST Laboratory, 399
    technologies, identification of promising, 397–398
  telemedicine, 401–402, 451
  and VISNs, 403–404

I
Ile-Ife Project, 431
Iliad, 193–194
Indiana VAMC, and legacy system, 203

See CARE decision support system (CDSS)
Indian Health Service (IHS), 33, 405
  information systems, initiatives in, 405–406
  See Patient Care Component; Resource and Patient Management System
Information Resources Advisory Council (IRAC), 14–16, 108, 140–141
  Clinical Applications Requirement Group (CARG), 15
  Integration and Technology Applications Requirement Group, 15
  Management Applications Requirement Group, 15
Integrated database with a data dictionary, principle of, 52–53, 59
Integrated Data Communications Utility (IDCU), 124–125
Integrated information system. *See* Library network (VALNET)
Integration agreements (IAs), library of, 83–84
Integration and Technology Applications Requirement Group, 15
International installations, 431–432
  principles of, 426
  relevance of, 431
  utilization of
    database purchasers, 427
    quantitative survey, 428–431
    satisfaction survey, 427–428
International teleinformatics networking program, development of, 31–34
  program plans, 32–33
  users, 33
  workstation, 33. *See also* Clinical workstations
Interwest Quality of Care, Inc., 31, 32
Iron Mountain, MI, VAMC, remote telepathology and consultation imaging system, 402

J

Johnson, Martin, 42
Joint Commission on the
    Accreditation of Healthcare
    Organizations (JCAHO), 256,
    302–303, 308–310, 336

K

Kaiser Permanente, 3
Kernel Installation and Distribution
    System (KIDS), 70–71, 121–
    122, 139
Kizer, Kenneth, 232, 234, 235, 466
Knowledge Data Systems (KDS), 194
Knowledge processing, 166–178

L

LDS Hospital, Salt Lake City, UT,
    29, 31
  and HELP System, 193–194
Lexicon Utility, 364–375
  contents, 364–366
  unresolved narratives, 366–375
  user options, 366
Library network (VALNET), 308, 328
  changes in
    Chief Information Officer (VA),
      308, 309
    information, improved access to,
      308–310
    information management
      functions, 309–310
  network structure, 310–312
    customers, 312–314
    outreach activities, 314–315
    shared information services, 315–
      316
  network services, 316–317
    audiovisual software delivery
      program, 325
    cataloging of materials, 317–318
    interlibrary loans, 324–325
    journal collections, 320–324
    online public access catalogs,
      319–320
    satellite TV network, 326–327
    "union" listing, 318–319
Local software selection, principle of,
    45–46

applications, selection of at each
    facility, 45–46, 66
Health Summary, 46
Lockheed-Martin Western
    Development Laboratory, 31,
    32

M

M (programming language), 71–72,
    87–88, 89, 93, 104, 158–159
MailMan, 62–63
Maloney, Dan, 285
Management Applications
    Requirement Group, 15
Management needs, 17–18
  costs, determination of, 17
  management information, sharing
    of, 18
  services, comparison of at different
    facilities, 18
McDonald, Clement, 204
Medical records, electronic, 16–17,
    249–251, 295, 350–353
  See Computerized Patient Record
    System (CPRS)
Modules
  nursing, 254–271
  surgery, 241–245
Multi-facility system
  evolution of, 39–43
  goals, validity of, 423
  implementation effects, 423
  integration project, 416–417, 419–
    422
  multisite effects, 423
  technology effects, 423
  See also Decentralized Hospital
    Computer Program
Muskogee, OK, VAMC,
    computerized telephone triage,
    402

N

National On-Line Information
    System (NOIS), 107
Nimmo, Robert, 42
Nursing information systems, 253–
    254, 273
  administration software, 264

Index   497

communication, 266
education, 267
fiscal information, 265–266
manhour module, 265
patients, classification of, 265
personnel, 264
quality improvement, 267–268
recruitment, 266
research, 266–267
workloads, 265
clinical package
  intake and output information, 261, 263–263
  pathway software, 256–259
  patient assessment, 263
  patient-care plan, 255–256, 257, 258
  treatment record, 263–264
  vital signs, measurement of, 259–261, 262
modules, 254–255
  adverse reaction tracking, 269
  Computerized Patient Record System (CPRS), 269
  consult/request tracking, 270
  dietetics, 271
  Health Summary (HS) module, 268–269
  laboratory, 271
  mental health, 270
  and North American Nursing Diagnosis Association (NANDA) diagnostic taxonomy, 254
  and Nursing Minimum Data Set (NMDS), 254
  pharmacy, 269–270
  problem list, 270
  progress notes, 270
  social work, 271
  surgery, 270–271
point-of-care approach, 272–273
training, 271–272

O

Ohio State University, intraoperative records, 296
Omnibus Health Care Act of 1973, 231

O'Neill, Joseph T., 42
Online services, 467–469, 174–176, 319–320
  data communications, 469–470
  multimedia gateways, 471
  video communications, 470–471
  voice, 468
Outpatient Clinic (OPC) File, 233–234

P

Pacemaker, surveillance of, 276–77
  malfunction, impending, 281–284
  telemedicine protoypes, 280–281
  telephone-based, 277–280
Pantheon Health Equity Corporation, 31, 32
Patch Module, 107
Patient Care Component (PCC), 410–411, 415–416
  data integration
    operational, 413–415
    technical, 411–413
  multi-facility integration project, 416–417, 419–422
    data integration alternative approaches, 417
    implementation framework, 418
    system architecture, 418–419
Patient-Care Encounter (PCE), 235–236
Patient data, exchange of, 67–68. *See also* Computerized Patient Record System
Patient-oriented care, 16–17
Patient Treatment File (PTF), 233
Philadelphia VAMC HOST
  anesthesiology project, 300
  commercial integrator, 401
Picture Archives and Communication Support (PACS) radiology system, 110
Point-of-care approach, 272–273
Privacy. *See* Security
Privacy Act of 1974, 120
Problem List software, 235, 349–350, 379
  key features, 350
  electronic medical record, role in, 350–353

498   Index

features of
  List Manager, 353–359
  management options, 359–364
  lexical data models, more
    advanced, 375–378
  Lexicon Utility, 364–375
    contents, 364–366
    unresolved narratives, 366–375
    user options, 366
Progress Note, 235
Prototyping, 82

Q

Quality assessment, 330–332, 346–347
  creed, 330–331
  Decentralized Hospital Computer
    Programs
    local programs, 338–340
    national programs, 335–338
  hybid programs
    local and external system, 342–344
    national and external system, 340–342
  quality
    definition of, 332–333
    evaluation of, 333–335
  stand-alone programs
    local, 345–346
    national, 344–345
Quality Improvement (QI), 21–23
  "outcomes-based," system, 26
  and teleinformatics, 31–34
  See also Quality assessment;
    Teleinformatics
Quality Improvement Checklist
  (QUIC), 335–338
Quality management (QM)
  generic categories, 20, 22
    innovative systems, 20
    traditional systems, 20
  history of, 19–20
  teleinformatics, 20–21. See also
    Teleinformatics

R

Rapid prototyping, 100
Read system, 173
Reference files, access to, 62

Regenstrief Institute for Health Care, 204
Remote Access Data Entry Network
  (RADEN), 405
Remote Procedure Call (RPC)
  Broker, 68–69, 83–84
Resource and Patient Management
  System (RPMS), 405–406
  benefits, 424
  costs, 424
  DHCP software, use of, 408
  drawbacks, 424–425
  implementation of
    cooperative VA-IHS activities, 410
    hardware and software
      strategies, 408–409
    status of, 409–420
Robustness. See Decentralized
  Hospital Computer Program,
  infrastructure, disruptions,
  response to

S

Saginaw, MI, VAMC, commercial
  DHCP-dental interface, 401
Salt Lake City VA Information
  Resources Management Field
  Office, 349
Salt Lake City Ventilator Review, 346
San Diego Transfusion Unit, 345–346
San Diego VAMC HOST
  anesthesiology project, 299–300
Science Applications International
  Corportion (SAIC), 40–41
  and Composite Health Care
    System (CHCS), 40–41
Security, 59, 60–61, 66–67, 95, 104–105, 132, 479
  administrative controls, 127–131
    background investigations, 128
    computer security policies, 127
    contingency plans, 130
    contract specifications, 129
    incidents, handling of, 130–131
    positions, descriptions of, 128
    program compliance, assessment
      of, 131

risk, management of, 127–128
suitability determinations, 128
training and awareness
  programs, 129
user account management, 129
Automated Information Systems
  security program, structure of,
  118–119
  elements of, 119–120
  federal laws and regulations,
    120–121
  challenges to, 116–118
  physical controls, 131–132
  technical controls, 121–127
    access, control of, 121–122
    audit trails, 122
    backups and testing, 124
    encryption, 123–1
    file access, 123
    Integrated Data
      Communications Utility
      (IDCU), 124–125
    network, security of, 125–126
    operating system utilities, 12
    package, integrity of, 122
    program/application security, 121
    signature codes, electronic, 122–123
    VA/Internet Gateways, 125
    viruses, protection against, 126
    vulnerability, testing of, 127
  technical controls
    audits, 117
    electronic signatures, 117
    menu and option management,
      117
    password management, 117
Servers, creation of, 65
Software flexibility, principle of, 47–50
  consortium of participating sites,
    49, 67
  customization, ease of, 47–48
  national code, modification of, 48–49
Standards, promotion of, 94–96
  numbering systems, establishment
    of, 95
  and security, 95

standards development groups, 94
and terminology, 95–96
Standards-based development,
  principle of, 50, 51, 58
  advantages of, 89, 91
  application protocols, 90
  collaboration, fostering of, 91–93
  communication protocols, 90
  importance of, 87
  information systems, role in
    supporting, 93–94
  origins of, 87–89
  See also Decentralized Hospital
    Computer Program
Strategic Information Systems Plan
  (SISP), 108–109, 249–250
Strategic planning, importance of, 113
Strategic redundancy, 74–76
Support teams, creation of, 111–113
Surgical systems, 240
  reports, 245–247
  risk assessment, 247–249
  surgery module, 241–245
    information data fields, 241–244
    main menu, 240, 241
Systems, 98–115
  development of
    design principles, 99
    implementation of, 100
    infrastructure, 99–100
    support, 191
  evolution of
    applications, addition of, 101,
      102
    customer feedback, 106–107
    customer support, 105
    functionality creep, management
      of, 101, 103–104
    security program, development
      of, 104–105
    standards, development of, 104
    training needs, 105
  growth of, 107–113
    clinical information,
      dissemination of, 109–110
    commercial applications, 110
    users, communication with, 110–111
    support services, menu of, 112

support teams, creation of, 111–113
principles of, 113–115
Strategic Information Systems Plan (SISP), 108–109, 249–250

T
TaskMan, 64–65
Teleconsultations, 289, 291
Telehealth, 466, 471–472
  customer, definition of, 466–467
  online, VA services, 467–469
    data communications, 469–470
    multimedia gateways, 471
    video communications, 470–471
    voice, 468
Teleinformatics
  diagnostic outcomes, improvement of, 26–29
  and computerized expert systems, 26
  seriousness of errors, 26–28
  and outcomes management, 21–26
    classic improvement cycle, 21
    diagnostic outcomes, 24, 25
    and quality improvement strategy, 21–22
    patient health status, 23–24
    quality management (QM), 19–21
    and local area networks, 20
    and wide area networks, 20
  therapeutic outcomes, improvement of, 29–31
  See International teleinformatics networking program
Telemedicine, 435–436, 443, 462–464
  and digital imaging, 486–488
  and HOST, 401–402, 451
  and management considerations, 459
    strategic systems approach, 459–462
  national trends in
    care delivery, alternative sites for, 439–440
    community hospitals, 441–442
    decision making, shared, 438
    human factors, 441
  managed care, 436–438
  medical education, 440–441
  population-based health, 438
  state involvement, 440
  and pacemakers, surveillance of, 276–281
  and the VHA, 447–450
    competing themes, 456–457
    evaluation of, 455–456
    HOST projects, 451
    leadership role, 444–446
    local initiatives, 451–455
    system characteristics, 442–443
    systems management, implementation of, 458–459
    as a telemedicine lab, 443–444
Terminal-based information systems, 93
Text integration utility (TIU), 296

U
Unified Medical Language System (UMLS), 96, 174–176, 185
Uniform Ambulatory Medical Care Minimum Data Set (MDS), 236
User-directed software development, principle of, 53–55
  and Special Interest Users Groups (SIUGs), 53–55
User environment
  features of, 64
  menus, 64
  standardization of, 58
User's Toolbox, 66

V
VA Clinical Lexicon, 179
VA FileMan, 58, 61–62, 82, 123, 276
VALNET. *See* Library network
Vendor independence, principle of, 50–52, 57–58
  and ANS computer language, 51
Veterans Health Administration (VHA), 3–13, 39–40
  and software, 98–99
Veterans Integrated Service Network (VISN), background of, 3–13

business plan, development of, 10–11
and clinical workstations, 161–163
information technology, use of, 10
management assistance council, 7
Resource Planning and Management (RPM), 12
system learning, 11–12
"Virtual healthcare organization," 5
Viruses, protection against, 126

VISN. *See* Veterans Integrated Service Network

W

Washington Information Resources Management Field Office (IRMFO), Silver Spring, MD, and VA/Internet Gateway, 125
Western Governors Association, 464

Contributors

CURTIS L. ANDERSON
Computer Specialist, VA IRM Field Office, Salt Lake City, UT, USA (anderson.curtis@forum.va.gov)

RUSTY W. ANDRUS
Associate Director, Software Production, VA IRM Field Office, Salt Lake City, UT, USA (andrus.rusty@forum.va.gov)

MARION J. BALL, EdD
Professor, University of Maryland School of Medicine, and Vice President, First Consulting Group, Baltimore, MD, USA (mball@fcgnet.com)

GALEN L. BARBOUR, MD
Director, Planning, Education and Performance Improvement Office, VA Medical Center, Washington, DC, USA

GAIL BELLES, AAS
Acting Director/Deputy Director, Medical Information Security Service, VA Medical Center, Martinsburg, WV, USA (belles.g@forum.va.gov)

WENDY N. CARTER, MLS
Director of Library Programs, Veterans Health Administration, VA Central Office, Washington, DC, USA (carter.wendy@forum.va.gov)

A. CLAYTON CURTIS, MD, PhD
Chief Information Systems Architect, VA Medical Center, Boston, MA, USA (curtis.clayton@forum.va.gov)

RUTH E. DAYHOFF, MD
Director, Advanced Technology, VA IRM Field Office, Silver Spring, MD, USA (dayhoff.ruth@forum.va.gov)

Contributors

JOHN G. DEMAKIS, MD
Director, Midwest Center for Health Services and Policy Research, Hines Veterans Health Administration, Hines, IL, USA, and Professor, Clinical Medicine, Loyola University of Chicago, Stritch School of Medicine, Maywood, IL, USA (demakis.john_@hines.va.gov)

JUDITH V. DOUGLAS, MA, MHS
Director, First Consulting Group, Baltimore, MD, USA (jdouglas @fcgnet.com)

SUSAN H. FENTON, MBA, RRA
Health Information Manager, Health Administration Service, Veterans Health Administration, Rochester, MN, USA (fenton.susan@forum.va.gov)

ROSS D. FLETCHER, MD
Chief of Cardiology, VA Medical Center, Washington, DC, USA (fletcher.ross.@forum.va.gov)

HOLLY M. FORCIER
Computer Specialist, Anesthesiology Service, VA Medical Center, San Diego, CA, USA (Holly4Seer@aol.com)

THOMAS L. GARTHWAITE, MD
Deputy Under Secretary for Health (10A), Veterans Health Administration, Department of Veterans Affairs, Washington, DC, USA

JOAN GILLERAN-STROM, RN, BSN
Coordinator of Automated Information Systems, Nursing Service, VA Medical Center, Hines IL, USA (gilleran-strom.joan-m@forum.va.gov)

PETER GROEN
Director, Telecommunications Support Service, Silver Spring, MD, USA (groen.peter@forum.va.gov)

KENRIC W. HAMMOND, MD
Staff Psychiatrist, VA Puget Sound Health Care System (American Lake Division), Tacoma, WA, USA (khammond@u.washington.edu)

CHRISTIANE J. JONES, MLS
Chief Information Officer, VA Medical Center, Biloxi, MS, USA (jones. chris@forum.va.gov)

SHUKRI F. KHURI, MD
Professor of Surgery, Harvard Medical School, and Chief of Surgery, VA Medical Center, West Roxbury, MA, USA (khuri@brockton)

KENNETH W. KIZER, MD, MPH
Under Secretary for Health, Veterans Health Administration, Suite 800, Department of Veterans Affairs, Washington, DC, USA

ROBERT M. KOLODNER, MD
Associate Chief Information Officer, Veterans Health Administration, Department of Veterans Affairs, Washington, DC, USA (kolrob@mail.va.gov)

MARGARET ROSS KRAFT, MS, RN
Chief SCI/Rehabilitation Nursing, VA Edward Hines Hospital, Hines, IL, USA (kraft.m@hines.va.gov)

BARBARA LANG, RN, MSN
Project Manager, VA IRM Field Office, Hines, IL, USA (lang.barbara @forum.va.gov)

MICHAEL J. LINCOLN, MD
Associate Professor of Medicine and Adjunct Associate Professor of Medical Informatics, University of Utah School of Medicine and Salt Lake City VA Medical Center, Salt Lake City, UT, USA (mlincoln@medinfo.med. utah.edu)

DANIEL L. MALONEY
Director, Technology Services, VA IRM Field Office, Silver Spring, MD, USA (maloney.dan@forum.va.gov)

MATTHEW MANILOW
Computer Specialist, Bronx VA Medical Center, Bronx, NY, USA (manilow.matthew_nmn@bronx.va.gov)

DOUGLAS K. MARTIN, MD
Health Services Research and Development, VA Medical Center, Indianapolis, IN, USA

CHRISTOPHER MCMANUS
Systems Manager, Eastern Pacemaker Surveillance Center, VA Medical Center, Washington, DC, USA

MARY E. MEAD, RN
Nursing Informatics Coordinator, Syracuse VA Medical Center, Syracuse, NY 13210, USA

SHARON CARMEN CHÁVEZ MOBLEY
Director, Campus Management, Information Systems Center, Grand Prairie, TX, USA

TOM MUNNECKE
Assistant Vice President, Science Applications International Corporation, San Diego, CA, USA (tom_munnecke@epqm.saic.com)

NEAL I. NEUBERGER
Senior Partner, Center for Public Service Communications, Arlington, VA, USA (neal_ian@access.digex.net)

CATHERINE N. PFEIL, PhD
Associate Director, Software Production, Veterans Health Administration, San Francisco, CA, USA (pfeil@forum.va.gov)

VIRGINIA S. PRICE
Program Analyst, VA IRM Field Office, Tuscaloosa, AL, USA (price.ginger@forum.va.gov)

ROBERT H. ROSWELL, MD
Network Director, Florida-Puerto Rico Veterans Integrated Service Network, Bay Pines, FL, USA (roswell.robert@forum.va.gov)

FRANKLIN L. SCAMMAN, MD
Professor and Chief, Anesthesiology Service, Iowa City VA Medical Center, Iowa City, USA (scamman.f@iowa-city.va.gov)

CAMERON SCHLEHUBER
Database Administrator, VA IRM Office, Salt Lake City, UT, USA (schlehuber.cameron@forum.va.gov)

JOHN C. SCOTT, MS
President, Center for Public Service Communications, Arlington, VA, USA (jcscott@access.digex.net)

ROGER H. SHANNON, MD
National Director, Veterans Health Administration, Radiology, Clinical Associate Professor, Radiology, Duke University, Durham, NC, USA (shannon.roger@forum.va.gov)

ELIOT L. SIEGEL, MD
Chief, Imaging Service, VAMHCS, VA Medical Center, Baltimore, MD, USA (esiegel@umabnet.ab.umd.edu)

ROY H. SWATZELL, JR., MS
Director of Operations, VA IRM Field Office, Birmingham, AL, USA

BOBBIE D. VANCE, PhD
Chief, Nursing Service, VA Medical Center, Atlanta, GA, USA (vance.bobbie-m@forum.va.gov)

STEVEN A. WAGNER, MBA
Strategic Architect for IRM Policy and Planning Office, Office of the Chief Information Officer, VA Medical Center, Manchester, NH, USA
(wagner.steven@forum.va.gov)

JOHN W. WILLIAMSON, MD
Director, Regional Medical Education Center, Salt Lake City, UT, USA
(williamson.j@rmec-slc.va.gov)

Contributors

Curtis L. Anderson

Rusty W. Andrus

Marion J. Ball

Galen L. Barbour

Gail Belles

Wendy N. Carter

Contributors 509

A. Clayton Curtis

Ruth E. Dayhoff

John G. Demakis

Judith V. Douglas

Susan H. Fenton

Ross D. Fletcher

Contributors

Holly M. Forcier

Thomas L. Garthwaite

Joan Gilleran-Strom

Peter Groen

Kenric W. Hammond

Christiane J. Jones

Shukri F. Khuri

Kenneth W. Kizer

Robert M. Kolodner

Margaret Ross Kraft

Barbara Lang

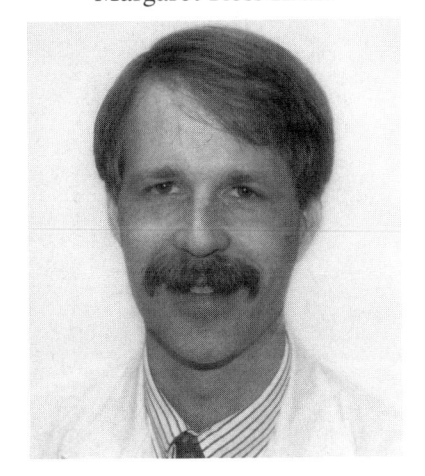

Michael J. Lincoln

512   Contributors

Daniel L. Maloney

Matthew Manilow

Douglas K. Martin

Christopher McManus

Mary E. Mead

Sharon Carmen Chávez Mobley

Contributors 513

Tom Munnecke

Neal I. Neuberger

Catherine N. Pfeil

Virginia S. Price

Robert H. Roswell

Franklin L. Scamman

Cameron Schlehuber

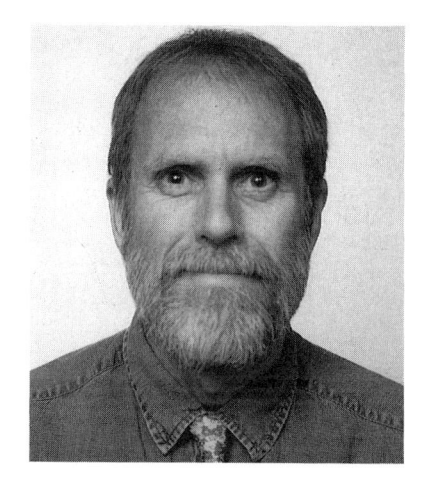

John C. Scott

Roger H. Shannon

Eliot L. Siegel

Roy H. Swatzell, Jr.

Bobbie D. Vance

Steven A. Wagner

John W. Williamson